Catalogue of Indian Insects

THE EDITOR

Thomas Bainbrigge Fletcher (1878-1950) was an English entomologist. He worked in the Royal Navy; he became an expert on the "microlepidoptera" and was appointed as the second Imperial Entomologist in India, succeeding Harold Maxwell-Lefroy at the Imperial Agricultural Research Institute at Pusa. He was a meticulous naturalist and very careful on matters of systematics and taxonomic nomenclature.

He produced a list of publications on India Entomology and Catalogue of Indian Insects. He also wrote general works on entomology including Some South Indian Insects, and a Veterinary Entomology for India and hints on collecting and preserving insects. He was Fellow of the Linnean Society, the Royal Entomological Society, Zoological Society of London and a President of the Cotteswold naturalists field club.

Catalogue of Indian Insects

Volume 3

–Editor–

Thomas Bainbrigge Fletcher

2016

Daya Publishing House®

A Division of

Astral International Pvt. Ltd.

New Delhi – 110 002

Catalogue of Indian Insects (5 Volumes Set)

Publisher's note:

The book "***Catalogue of Indian Insects***" by Thomas Bainbrigge Fletcher was published in 1921. But presently complete set not available anywhere. We have tried to reproduced the complete valuable set from different sources in a form as close to the original possible. This historical reproduction fulfill this long felt needs of students, teachers and researchers engaged in the field of entomology, zoology and other related discipline.

Cataloging in Publication Data—DK
Courtesy: D.K. Agencies (P) Ltd. <docinfo@dkagencies.com>

Catalogue of Indian insects / editor, Thomas Bainbrigge Fletcher.

volumes cm
First published in 1921.
ISBN: 978-93-86071-08-8 (Vol. 3)
ISBN: 978-93-5130-922-2 (Set) (International Edition)

1. Insects—Catalogs and collections—India. I. Fletcher, T. B. (Thomas Bainbrigge), 1878-1950, editor.

QL468.2.C38 2016 DDC 595.70954 23

Published by : **Daya Publishing House®**
A Division of
Astral International Pvt. Ltd.
– ISO 9001:2008 Certified Company –
4760-61/23, Ansari Road, Darya Ganj
New Delhi-110 002
Ph. 011-43549197, 23278134
E-mail: info@astralint.com
Website: www.astralint.com

Printed at : **Replika Press Pvt. Ltd.**

डा. एस. अय्यप्पन
सचिव एवं महानिदेशक
Dr. S. AYYAPPAN
SECRETARY & DIRECTOR GENERAL

भारत सरकार
कृषि अनुसंधान और शिक्षा विभाग एवं
भारतीय कृषि अनुसंधान परिषद
कृषि एवं किसान कल्याण मंत्रालय, कृषि भवन, नई दिल्ली 110 001

GOVERNMENT OF INDIA
DEPARTMENT OF AGRICULTURAL RESEARCH & EDUCATION
AND
INDIAN COUNCIL OF AGRICULTURAL RESEARCH
MINISTRY OF AGRICULTURE AND FARMERS WELFARE
KRISHI BHAVAN, NEW DELHI 110 001
Tel.: 23382629; 23386711 Fax: 91-11-23384773
E-mail: dg.icar@nic.in

Foreword

With rapidly increasing human population that hardly matched by the rate of increase in agriculture production, the per capita availability of food and other products is bound to fall short. Therefore, increasing production per unit area is the most plausible alternative to sustain self sufficiency in food, fodder and fibre. Insects have got both positive and negative impacts in agriculture. Insects become pests when they interfere with our welfare, aesthetics or profits. A reduction or loss in yield and quality of crops is partly due to result of insect feeding activities and as vector/host to pathogen transmissions. On the other hand these small creatures serve us a lot as pollinating agents of many important agricultural crops, as biological controlling agents of pests and weeds as well as by producing honey, silk and many other materials essential to us.

This book 'Catalogue of Indian Insects' is divided into 5 volumes and covers 29 parts. This historical compilation is a dedicated effort to present the different aspects of the insects in a very concise and relevant manner. I believe that the students and scientist working in the area of entomology, agriculture, fisheries, zoology, botany and veterinary science will find this book as a source of useful and authentic information.

I appreciate the efforts of Daya Publishing House, A Division of Astral International (P) Ltd. New Delhi.

(Dr. S. Ayyappan)
Secretary (DARE) & Director General (ICAR)
Krishi Bhawan, New Delhi – 110001

Preface

There is a immense number of insect species on the planet. The class Insecta is a large and diverse as well as important ecologically. When compared to other groups, the number of insect species is much larger and represents more than three quarters of all known animal species. Most insects are either beneficial or harmless. Yet there are large numbers which are economically important as pests, pollinators, natural enemies and other beneficial insects like honey bees, silkworms, etc. Insects are fascinating to investigate, their classification is complex. It important to learn the main differences among insects so you can distinguish one group from another. Knowledge of insect classification is of primary importance in carrying out proper control procedures and quality pest management. Proper insect classification and identification is extremely important. If a beneficial or harmless insect is improperly identified, the resultant unnecessary and undesirable pesticide application usually disrupts natural control agents.

Insects deliver a host of ecological services fundamental to the survival of humankind. Insects play an important role in plant reproduction. Thousands of pollinator species have been identified and almost all of these are insects. Over 90 percent of the flowering plant species depend on pollinators. Insects play an equally vital role in waste biodegradation. They clean up dead plant matter, breaking down organic matter until it is fit to be consumed by fungi and bacteria. Insects also play a significant role in decomposing animal carcasses and manure. Insects have a large array of natural enemies, predators and parasitoids, keeping them under economic threshold levels. Due to large number of insects species it is necessary that these catalogued so as to enable locating the species into correct classification. One shall also know what are the species knows from a specific group. Especially in case of Indian Insects where large number of species are known. These are important Catalogue of Indian Insects as a compilation serves this purpose.

"Catalogue of Indian Insects" is divided into 5 volumes and covers 29 parts. Volume 1 deals with Part I to Part IX. It covers Acrydidae, Culicidae, Bombyliidae, Trypetidae, Nitidulidae, Stapylinidae, Lasiocampidae, Amatidae and Zygaenidae. Volume 2 gives a detailed account of Part X to Part XVII. This volume covers Stephanidae, Brenthidae, Tabanidae,

Cicindelidae, Palpicornia, Cecidomyidae, Cosmopterygidae and Yponomeutidae. Volume 3 is devoted to Part XVIII i.e. Carabidae. Volume 4 covers in detail of Part XIX to Part XXV. This volume deals with Gyrinoidea, Alucitidae, Lycidae, Phalonidae and Chlidanotidae, Chalcidoidea, Evanidae and Thysanoptera. Volume 5 of the book describes the Part XXVI to Part XXIX. This volume deals with Serphoidea, Isoptera, Anthribidae and Asilidae.

This historical reproduction fulfill the long felt needs of students, teachers, researchers engaged in the field of entomology, agriculture, ecology, fisheries, forestry, paleontology, zoology, botany, plant science, medical and veterinary science, towards knowing insects.

We are extremely grateful to Dr. S. Ayyapppan, Secretary and Director General, Indian Council of Agricultural Reserarch, New Delhi for providing a foreword. The cooperation and support of Dr. V.V. Ramamurthy, Principal Scientist (Retd.), Division of Entomology, IARI, New Delhi is duly acknowledged.

All suggestions for future improvement for this book will be highly solicited.

Publisher

Contents

VOLUME 3

VOLUME 4

VOLUME 5

PART XVIII-CARABIDAE.

Abacetus* Dejean.

Abacetus Dej. Sp. iii. 1828. 195; Lac. 315; Chaud. SEZ. 1859. 126; id., Essai monographique sur le genre Abacetus, BM. 1869. i. 355; Tchitch. HER. xxxii. 1898. 519, 531, and 538; id. HER. xxxv. 1902, 506; Andr. MB. 44.

Distrigus Dej. Sp. iii. 1828. 191; Lac. 316.

Astigis Ramb. Fn. ent. Andalousie 1838. 95.

Distrigodes Motch. BM. 1864. ii. 353 (part).

Caricus Motch. BM. 1866. i. 394; Andr. TES. 1928. 14.

O*	**aenigma** Chaud. Mon. 358; Fairm. AF, 1888. 336.	Hong Kong. Cochin-China.
X*	*marginicollis* † Bates (not Chaud.) AF. 1889. 276; Andr. TES. 1921. 178.	Laos., Cambodia. Tonkin
X*	**aeratus** Tchitch. HER. xxxiv. 1900. 281.	Tonkin.
G*	**amplicollis** Bates AG. xxvii. 1889. 106; id. AG. xxxii. 1892. 360.	Chota-Nagpur; Barwa. U.P.: Almora, Upper Gumti Valley, Timli (Siwalik Hills), Dehra Dun. Burma: Katha, Teinzo, Bhamo.
Q	*impictus* Tchitch. HER. xxxv. 1901. 57; Andr. AM. (9) xvii. 1926. 256.	
Q	**annamenis** Tchitch. HER. xxxvi. 1903. 53.	Annam.
G*	**antepunctatus** Bates AG. xxxii. 1892. 362.	Burma: Katha, Teinzo.
O*	**antiquus** (*Argutor*) Dej. Sp. iii. 1828. 246; Chaud. Mon. 391; Andr. RIM. xxii. 1921. 344.	Tranquebar. Pondicherry. Barkuda. I. (Lake Chilka). Colombo., Burma.

O*	**atratus** (*Distrigus*) Dej. Sp. iii. 1828. 194; Chaud. Mon. 358.	Pusa. Berhampur.
J*	*costatus (Distrigus)* Nietn. AM. (3). ii. 1858. 176; Andr. AM. (9). xix. 1927. 103.	Murshidabad. Balugaon., Bandra Kasara., Madras "Coromandel". Madura. Ceylon : Colombo. Mamaduwa, Murunkam, Vavuniya.
G*	**batesi** × Andr. AM. (9). xvii. 1926. 256.	Karen Hills.
	maculipes † Bates (not Chaud.) AG. xxxii. 1892. 361; Andr. TES. 1921. 148.	
L	**bipunctatus** *(Distrigodes)* Motch. BM. 1864. ii. 355; Chaud. Mon. 386; Andr. TES. 1928. 21.	Colombo.
G*	**birmanus** *(Loxandrus)* Bates AG. xxvii. 1889. 106; *(Abacetus)* id. AG. xxxii. 1892. 358.	Muzafferpur. Bhamo.
G*	**bisignatus** Bates AG. xxvii. 1889. 105 ; id., AG. xxxii, 1892. 360.	Bhamo, Shwegu.
J*	**ceylanicus** *(Drimostoma)* Nietn. AM. (3). ii. 1858. 178; Tchitch. HER. xxxiv, 1900. 263; Andr. AM. (9) xix. 1927. 102.	Ceylon: Colombo, Puttalam, Negombo, Galle, Kotte.
B*	*lioderes* Bates AM. (5) xvii. 1886. 144.	
X*	**chalcopterus** Tchitch. HER. xxxiv. 1900. 283.	Tonkin.
O*	**convexiusculus** Chand. Mon. 385; Tchitch. HER. xxxiv. 1900. 286.	Celebes. Salwaty. Sumatra.
O*	**cordicollis** Chaud. Mon. 357.	
B*	*atratus*† Bates (not Dej.) AM. (5). xvii. 1886. 143; Andr. TES. 1921. 148.	Central Provinces Nagpur, Kanpa. Kodaikanal. Tranquebar. Bangalore. Ceylon: Galle, Colombo, Hambegamuwa, Hambantota, Galapitagawa, Mamaduwa, Vavuniya.

O*	**cyathoderus** Chaud. Mon. 373; Bates AG. xxxii. 1892. 359.	Dacca. Palon.
G*	**cyclodes** Bates AB. 1892. 232.	Sardah. Chota Nagpur.
J*	**dejeani** *(Distrigus)* Nietn. AM. (3). ii. 1858. 178; Chaud. Mon. 390; Andr. AM. (9). xix. 1927. 103.	Kierpur (Purnea). Central Provinces: Nagpur, Munser, Raipur, Kanpa. Mhow. Bandra. Madras:, Trichinopoly. Bangalore Manaparai., Madura. Ceylon: Colombo. Anuradhapura, Puttalam, Weligama, Trincomali, Nalanda, Kotte, Murunkam, Hanaratgoda.
B*	*relinquens (Argutor)* Walk. AM. (3). ii. 1858. 204; Andr. TES. 1919. 189.	
B*	*antiquus*† Bates (not Dej.) AM. (5). xvii 1886. 144; Andr. TES. 1921. 148.	
O*	**dilutipes** Chaud. Mon. 383; Bates AF. 1889. 276; Vuillet Ins. ii. 1912. 19.	Siam., Annam. Cambodia.
L	**dorsalis** *(Astygis)* Motch. BM. 1865. ii. 229; Chaud. Mon. 397; Andr. TES. 1928. 21.	Nagpur., Belgaum. Madura., Tranquebar. Barkuda I. (Lake Chilka). Toungoo. Tenasserim.
L	**femoralis** *(Distrigodes)* Motch. BM. 1864. ii. 354; Chaud. Mon. 386; Bates AG. xxxii. 1892. 363; Andr. TES. 1928. 21.	Tranquebar., Bhamo. Rangoon., Martaban. Tenasserim.
L	**flavipes** *(Coelostoma)* Motch. BM. 1865. ii. 288; Chaud. Mon. 392; Andr. TES. 1928. 21.	"East Indies".
G	**foveifrons** Bates AG. xxxii. 1892. 360.	Houngdaran Valley (Tenasserim).
O*	**guttula** Chaud. Mon. 374; Bates CRB. 1891. 334; id. AG. 1892. xxxii. 361. Tchitch. RRE. 1901. 88; Andr. MB. 45 ; id. EMM. 1926. 69.	Throughout Northern India and as far South as Central Provinces. Kandy, Bhamo, Katha, Kawkareik.

B* *quadriguttatus* † Bates (not Chaud.) AM. (5). xvii. 1886. 143 ; Andr. MB. 45 (note); id. Spol. Zeyl. 1924. 137.	Persia.
O* **haplosternus** Chaud. BM. 1878. ii. 25.	Katrain., Almora. Ranikhet., Haldwani. Nagpur. Hoshangabad. Siam., Sumatra.
O* **hirmocoelus** Chaud. Mon. 372; Bates AG. xxxii. 1892. 360; Andr. TES. 1921. 148.	Dum-dum., Nagpur. Balaghat., Poona. Rambha (Ganjam). Burma; Rangoon, Senmigion.
G **illuminans** Bates AG. xxxii. 1892. 359.	Kawkareik.
O* **impressicollis** *(Distrigus)* Dej. Sp. iii. 1828. 193; Dej. et Boisd. Ic. et HN. ii. 401. t. 125. f. 1; Chaud. Mon. 359.	Decca Pusa.
V *dostojewskii* Tchitch. HER. xxxii. 1898. 18; id. HER. xxxiv. 1900. 273.	
Q **indrapoerae** Tchitch. HER. xxxvi. 1903, 56.	Sumatra. Java.
G* **insolatus** Bates AG. xxxii. 1892. 359.	Khapa (Balaghat). Rangoon., Palon. Malewoon (Tenasserim).
B* **leucotelus** Bates TES. 1873. 283; Tchitch. Ab. xxix. 1897. 75.	Annam. China., Japan.
X* **longulus** Tchitch. HER. xxxiv. 1900. 289.	Tonkin.
I* **lucifugus** Andr. RIM. xxvi. 1924. 116.	Assam: Siju Cave.
O* **maculipes** Chaud. Mon. 384.	Martaban. Tenasserim.
O* **marginicollis** Chaud. Mon. 359; Bates AG. xxxii. 1892. 359.	Assam: Kamrup Farm, Sibsagar. Sylhet. Burma: Rangoon, Tharrawaddy, Hopin, Teinzo, Pyinmana, Pakokku or Seikpyu, Tavoy.
X* **ornatus** Tchitch. HER. xxxiv. 1900. 279.	Tonkin.

O*	**pallipes** Chaud. Mon. 286; Bates AF. 1889. 276.	Malabar.
B*	*nictneri* † Bates (not Chaud.) AM. (5). xvii. 1886. 145; Andr. TES. 1921. 148; id. Spol. Zeyl. 1924. 136.	Ceylon: Colombo, Negombo, Dehiwela, Kotte., Martaban. Cochin-China. Sumatra.
X*	**picescens** Tchitch. HER. xxxiv. 1900. 290.	Cochin-China.
L	**picipes** *(Distrigus)* Motch. BM. 1865. ii. 228; Chaud. Mon. 392; Andr. TES. 1928. 21.	"East Indies".
X*	**pictus** Tchitch. HER. xxxiv. 1900. 281.	Tonkin.
B*	**placidulus** *(Agonum)* Walk. AM. (3). ii. 1858. 203; Andr. TES. 1919. 188.	Calcutta. Khargpur. Berhampur Coast. Kierpur (Purnea).
B*	*infixus (Selenophorus)* Walk. AM. (3), ii. 1858. 204; Andr. TES. 1919. 190.	Chandipur. Madras: Rambha,
J*	*aeneus (Distrigus)* II Nietn. (not Dej.) AM. (3). ii. 1858. 177; Andr. TES. 1921. 148; id. AM. (9). xix. 1927. 102.	Satpara, Barkuda I. (Lake Chilka), Madras Pondicherry, Mahé,
M*	*hindustanus (Pogonus)* Motch. BM. 1864. ii. 192; Chaud. AB. xiv. 1871. 38; Andr. TES. 1928. 8.	Tranquebar. Travancore:
	nietneri × Chaud. Mon. 392.	Karunagapalli.
B*	*carinifrons* Bates AM. (5). xvii. 1886. 144; Andr. TES. 1919. 189.	Ceylon: Colombo, Galle, Puttalam, Weligama, Trincomali, Negombo, Anuradhapura, Kanthalay, Harrowapatam, Kotte, Murunkam, Elephant Pass.
O*	**politulus** Chaud. Mon. 369; Bates AG. xxxii. 1892. 361.	Burma: Rangoon, Senmigion, Victoria Point., Java.
O*	**politus** Chaud. Mon. 368; Bates AG. xxxii. 1892. 359.	Dacca Sahibganj, Saraghat., Rajmahal. Palon.
O*	**promptus** *(Distrigus)* Dej. Sp. iii. 1828. 195; Chaud. Mon. 370.	Nagpur., Madras. "Coromandel". Ceylon: Henaratgoda, Peradeniya.

O*	**quadricollis** Chaud. Mon. 382; Bates AF. 1889. 277; Tchitch. HER. xxxiv. 1900. 291.	Martaban. Tenasserim. Indo-China.
O*	**quadriguttatus** Chaud. Mon. 387; Bates AG. xxxii. 1892. 362.	Calcutta. Martaban.
X*	*lophoides* Bates AF. 1889. 277; Andr. TES. 1921. 178.	Kawkareik. Cambodia., Laos. Tonkin.
O*	**quadrinotatus** Chaud. Mon. 388.	Himalayas from Simla Hills to Bhutan.
O*	**reflexus** Chaud. Mon. 358; Andr. RIM. xxii. 1921. 344; id. MB. 45.	C. P.: Nagpur, Balaghat, Tamia (Chindwara), Khawasa (Seoni), Tumgaon (Raipur). Mhow., Belgaum. North Kanara. Madras: Rambha, Surada, Chipurupalle.
C*	**rubidicollis** *(Badister)* Wied. Zool. Mag. ii. 1. 1823. 58: Andr. TES. 1921. 172.	Dacca.
O*	*quadrimaculatus* Chaud. Mon. 380.	
J*	**rufopiceus** *(Distrigus)* Nietn. AM. (3). ii. 1858. 177; Chaud. Mon. 398; Andr. AM. (9). xix. 1927. 103.	Trichinopoly. Madura. Ceylon: Colombo, Kala-Oya River, Negombo, Matale, Kandy, Anuradhapura, Vavuniya.
B*	*henryi* Andr. Spol. Zeyl. 1923. 238.	
O*	**rufotestaceus** Chaud. Mon. 398.	Calcutta. Dacca.
L	**rufulus** *(Distrigodes)* Motch. BM. 1865. ii. 227; Chaud. Mon. 386; Andr. TES. 1928. 21.	Burma.
O*	**siamensis** Chaud. BM. 1878. ii. 26.	Siam.
Q*	**spurius** Tchitch. HER. xxxiii. 1900. 160; id. HER. xxxiv. 1900. 290 (note).	Java.

J*	**submetallicus** *(Distrigus)* Nietn. AM. (3). ii. 1858. 177; Andr. AM. (9). xix. 1927. 103; id. TES. 1928. 7.	Throughout India, Ceylon, and Burma. Nepal., Laos.
O*	*chalceolus* Chaud. Mon. 384; Bates CRB. 1891. 334; id. AG. xxxii. 1892. 361; Andr. TES. 1921. 179; id. MB. 45.	Cambodia. Tonkin.
G	**sulculatus** Bates AG. xxxii. 1892. 360.	Rangoon., Palon.
M*	**testaceipes** *(Caricus)* Motch. BM. 1866. i. 394; Andr. TES. 1928. 14.	Colombo.
O*	*anomalus* Chaud. Mon. 367; Bates AM. (5). xvii. 1886. 145.	
Q	**zarudnyi** Tchitch. RRE. 1901 88.	Seistan.

ABAX* Stephens.

Abax Steph. Ill. Brit. Ent. i. 67 and 124; Dej. Sp. iii. 1828. 378; Andr. TES. 1919. 91.

O	**tantillus** Fairm. AB. 1888. 12.	Yunnan.

ACINOPUS* Latreille.

Acinopus Latr. in Cuvier's Règne Animal (Ed. 2). iv. 1829 (April). 389; Dej. Sp. iv. 1829 (Dec.). 31; Lac. 287; Sem. HER. xxxii. 1899. 607; Tchitch. HER. xxxiv. 1900. 358; id. HER. xxxv, 1901. 230; Andr. MB. 30.

Haplacinopus Sem. HER. xxxii. 1899. 607.

Q	**laevigatus** Mén. Cat. rais. 1832. 128.	Kashmir; Sind Valley, Sonemarg, Srinagar. Bashahr: Poo., South Russia. Persia., Turkestan.
I*	*striolatus* † Bates (not Zubk.) SYM. 5; Andr. MB. 31.	
Q	*novorossicus* Sem. HER. xxxii. 1899. 604 and 609; Andr. MB. 31; Sem. RRE. xx. 1926. 42.	
A	var. **kashmirensis** Schaub. Col. Centralbl. 1927. 7.	Bashahr; Poo.
A	var. **inaequalis** Schaub. Col. Centralbl. 1927. 7.	Bashahr: Poo.

ACTENONCUS* Chaudoir.

Actenoncus Chaud. Essai monographique sur less Orthogoniens, AB. xiv. 1871. 126. — Siam. Java. Borneo.

L*	**ater** *(Lebia)* Cast. Et. Ent. 1834. 48.	
O*	*atratus* Chaud. Mon. 126; id. BM. 1878. ii. 7.	

ACUPALPUS Latreille.

Acupalpus Latr. in Cuviers' Règne Animal (Ed. 2) iv. 1829 (April). 391; Dej. Sp. iv. 1829 (Dec.) 435; Lac. 302; Motch. BM. 1864. ii. 205; Tchitch. HER. xxxiv. 1900. 365; id. HER. xxxv. 1901. 246.

Anthracus Motch. Käf. Russ. 1850. t. vii; id. BM. 1864. ii. 267.

Manicellus Motch. BM. 1864. ii. 207.

X* **annamensis** Bates AF. 1889. 272; id. AG. xxxii. 1892. 350.	Kulu., Almora. Ranikhet. C. P. Tumgaon (Raipur), Korai (Seoni) Madras: Kodaikanal. Ceylon: Weligama, Anuradhapura. Bhamo., Tenasserim. Java., Sumatra. Cochin-China. Annam., Tonkin. Formosa., Hongkong. Philippine Is.
N* **biplagiatus** *(Stenolophus)* Boh. ER. 12; Tchitch. HER. xxxv. 1901. 166.	China.
A* **brunnicolor** *(Thenarotes)* Sloane PLS. NSW. 1898. 466.	Borneo., Australia.
B* **derogatus** Walk. AM. (3). ii. 1858. 204; Bates AM. (5) xvii. 1886. 80; Andr. TES. 1919. 190.	Nuwara Eliya.
A* **horni** Andr. Spol. Zeyl. 1923. 237.	Sitapur (U.P.). Dehra Dun., Madras. Belgaum., Ceylon: Anuradhapura. Kyauktan. Mandalay. Tenasserim.
X* **ovatulus** Bates AF. 1889. 273; Andr. TES. 1921. 177.	Cochin-China. Annam.
fugax (Stenolophus) ‡ Lesne (not Dej.) Miss. Pav. 76; Andr. AM. (9). xiv. 1924. 590.	Siam.
G* **sinuellus** Bates AG. xxxii. 1892. 350.	Bhamo., Palon.

ADELOTOPUS* Hope.

Adelotopus Hope TES. 1834. 11; Wesw. Pseudomorpha et Adelotopus, genera duo anomaliae a familia Carabidarum synoptice tractata, RMZ. 1853. 403; Lac. 153.

B* **collaris** Waterh. TES. 1877. 2; Siam. Andr. TES. 1919. 197. — Siam.

Z **jacobsoni** Rits. NLM. xxxi. 1909. 255. — Java.

G **papuanus** Gestro. AG. 1893. 287. — New Guinea.

AEPHNIDIU* Macleay.

Aephnidius Macl. Ann. Jav. 1825. 23; Schm. Goeb. FCB 88; Lac. 308 ; Schaum BEZ. 1863. 76; Chaud. Etude monographique des Masoréides, des Tetragonodérides et du genre Nematotarsus, BM. 1876 ii. 15; Andr. MB. 60.

B* **adelioides** Macl. Ann. Jav. 1825. 23. t. 1. f. 7; Schm. Goeb. FCB. 88; Schaum BEZ. 1863. 78; Bates TES. 1873, 307; Chaud. Mon. 17; Bates AF. 1889. 282; id. AG. xxvii. 1889. 110; id. AG. xxxii. 1892. 404; Andr. TES. 1919. 159; Sloane PLS. NSW 1920. 322; Andr. TES. 1923. 51; id. AM. (9). xiii. 1924. 470; id. MB. 60. — South East Asia, including Japan. Australia.

J* *sericeus (Masoreus)* Zimm. GF. 121; Schaum BEZ. 1863. 78; Chaud. Mon. 17; Andr. AM. (9) xix. 1927. 99.

A *australis (Masoreus)* Sloane PLS. NSW. 1904. 535; id. DEZ. 1907. 472.

P* **fuscipennis** Schm. Goeb. FCB. 89; Schaum BEZ. 1863. 78; Chaud. Mon. 16; Bates AG. xxvii. 1889. 110; id. AG. xxxii. 1892. 404; Andr. TES. 1923. 51. — Rawalpindi. Dehra Dun. Dacca., Karimganj. Bhamo., Shwegu. Kawkareik. Tavoy., Penang. Siam., Cambodia.

J* **opaculus** *(Masoreus)* Zimm. GF. 120; Schaum BEZ. 1863. 78; Chaud. Mon. 19; Andr. AM. (9). xix. 1927. 99; id. AM. (9) xx. 1927. 271. — India., Ceylon. Burma., Andaman Is. Sumatra.

P* *simplex* Schm. Goeb. FCB. 89; Schaum BEZ. 1863. 78; Chaud. Mon. 22; Bates AG. xxvii. 1889. 110; id. AG. xxxii. 1892. 405; Andr. TES. 1923. 52; id. MB. 60. — Philippine Is.

J*	**pleuronectes** *(Masoreus)* Zimm. GF. 120; Schaum BEZ. 1863. 78; Chaud. Mon. 19; Andr. AM. (9). xix. 1927. 99.	"East Indies." Malabar. Coimbatore., Ceylon.
A*	**rubidus** Andr. AM. (9). x. 1922. 169.	Nagpur., Khandesh. Belgaum., Madura.
O*	**ruficornis** *(Mazoreus)* Chaud. BM. 1850 i. 452; Schaum BEZ. 1863. 77; Chaud. Mon. 18; Bedel Fn. med. 1904. 225. note (2).	"Western India" (Bedel). Mesopotamia. Arabia., Syria. Kurdistan.
G	**submaculatus** Bates AG. xxxii. 1892. 405.	Palon.

AEPSERA* Chaudoir.

Aepsera Chaud. BM. 1874. i. 28; Andr. AM. (9). xiv. 1924. 591.

Nanochlaenius Bates AG. xxxii. 1892. 321.

O*	**ferruginea** Chaud. BM. 1874. i. 30.	Rangoon.,
G*	*feae (Nanochlaenius)* Bates AG. xxxii. 1892. 322; Andr. AM. (9). xiv. 1924. 591.	Katha., Palon. Sumatra.

AGASTUS* Schmidt—Goebel.

Agastus Schm. Goeb. FCB. 30; Lac. 87.

P*	**lineatus** Schm. Goeb. FCB. 31; Gestro AG. 1875. 867; Bates AF. 1889. 280; id. AG. xxxii. 1892. 388; Andr. TES. 1923. 10.	Calcutta. Burma: Toungoo, Palon. Cochin-China. Cambodia. Philippine Is.
G*	**ustulatus** Gestro AG. 1875. 867.	Singapore. Borneo.

AGONOTRECHUS* Jeannel.

Agonotrechus Jeann. AM. (9). xii. 1923. 428; id., Mon. des Trechinae (iii), Ab. xxxv. 1928. 85.

A*	**andrewesi** Jeann. AM. (9). xii. 1923. 430. f. 17; id. Mon. (iii). 89. ff. 1324-7.	Darjeeling. Kurseong.
G*	**birmanicus** *(Trechus)* Bates AG. xxxii. 1892. 297; Jeann. AM. (9). xii. 1923. 429; id. Mon. (iii). 87. ff. 1319-23.	Karen Hills.

ALLOCOTA* Motchulsky.

Allocota Motch. Et. Ent. 1859. 29; Chaud. BM. 1877. i. 203.

G **aerata** Bates AG. xxxii. 1892. 425. — Bhamo.

Y* **cyanipennis** Heller PJS. xxiii. 1923. 305. — Philippine Is.

M **viridipennis** Motch. Et. Ent. 1859. 30; Chaud. BM. 1877. i. 205; Andr. TES. 1928. 21. — Singapore. Java., Borneo. Philippine Is.

AMARA Stephens.

A. *Amara* Steph. Ill. Brit. Ent. i. 67 and 126; Dej. Sp. iii. 1828. 456; Zimm. GF. 30; Lac. 332; Putz. SEZ. 1865. 332; id., Etude sur less Amara (Coll. Chaudoir), Mém Liége 1866. 174; id., Monographie des Amara de l'Europe et des pays voisins, Ab. xi. 1870. 6; Andr. TES. 1919. 91; id. MB. 42.

B. *Bradytus* Steph. Ill. Brit. Ent. i. 67 and 136; Zimm. GF. 37; Lac. 333, note (1); Putz. SEZ. 1865. 332; id. Mém. Liége 1866. 260; id. Mon. 86; Andr. MB. 42.

Cu. *Curtonotus* Steph. Ill. Brit. Ent. i. 67 and 138; Putz. SEZ. 1865. 333; id. Mém Liége 1866. 231; id. Mon. 80.

C. *Celia* Zimm. GF. 18; Lac. 333, note (1); Putz. SEZ. 1865. 333; id. SEZ. 1867. 172.

L. *Leiocnemis* Zimm. GF. 38; Lac. 333, note (1); Putz. Mém. Liége 1866. 197; id. Mon. 52.

Lm. *Leiromorpha* Gangl. Käf. Mitteleur. i. 1892. 313.

Cm. *Cumeres* Andr. MB. 98.

N. *Niphobles* Andr. EMM. 1926. 75.

N.A. **aenea** *(Carabus)* Deg. Mém. Ins. iv. 1774. 98; Andr. MB. 42.

U. *trivialis (Harpalus)* Gyll. Ins. Suec. i. (2). 1810. 140; Bates SYM. 13

— Palaearctic region. Kashmir: Sind Valley, Sonemarg, Gulmarg, between Gilgit and Nagy. Bajaura.

I* A. **ambigena** Bates PZS. 1878. 716; id. SYM. 14. — Kashmir: Pangong Valley between Tanktze and Chagra.

N.B. **apricaria** *(Carabus)* Payk. Mon. Carab. 1790. 125; Bates SYM. 17.	Palaearctic region Canada (Ganglbauer). Pamir.
O* *parallela (Leirus)* Chaud. BM. 1842 ii. 827; Bates SYM. 17.	Kashmir: Sind Valley Dras, Kargil, Leh, Sonemarg.
I*A. **bamidunyae** Bates PZS. 1878. 716; id. SYM. 14; id. Entom. 1891. Suppl. 9; Andr. MB. 43.	Pamir. Tibet: Gyangtse. Kashmir. Goorais Valley. Kulu: Koty. Spiti: Pulga.
B. **batesi** × Csiki Col. Cat. 104 Carab. 1929. 454.	Himalayas from Kashmir to Sikkim.
I* *compacta (Bradytus)* || Bates (not Lec.) PZS. 1878. 717; id. SYM. 17; Andr. MB. 42; id. EM. 1926. 69.	
B* Cm. **brucei** Andr. AM. (9). xi. 1923. 276; id. AM. (9). xix. 1927. 111.	Mount Everest Expeditions base camp.
B* *thalia* Andr. EMM. 1926. 69 and 76.	Kumaon: Milam (Gori Valley). Tibet : Supi River.
S* A. darjelingensis Putz. SEZ. 1877. 102; Andr. TES. 1922. 249; id. EMM. 1926. 69.	Punjab: Seraj. (Jalori Pass), Dharmasala. Kulu: Naggar and Parbatti Valley. Kumaon : Sunderdhunga Valley, Sukhatal, Gori Valley and Gorge, Bhurphu. Chulli Khud (Chakrata). Sikkim: Darjiling, Lachung. Khasi Hills.
?Q L. **elegantula** Tchitch. HER. xxxii. 1899. 659.	Darjeeling. Lachung.
I* L. **frivola** *(Leiocnemis)* Bates PZS. 1878. 717; id. SYM. 15.	Yarkand or Pamir.

I* L. **himalaica** *(Leiocnemis)* Bates PZS. 1878. 716; id. SYM. 14; Andr. MB. 43.	Kashmir: Dras, Kargil, Leh, Sonemarg. Bashahr: Poo. Simla Hills: Kotgarh, Matiana. Laboul: Sumdeo. Almora. Ranikhet.
B* Lm. **hypsela** Andr. AM. (9) xi. 1923. 277.	Mount Everest Expeditions base Camp.
O* L. **indica** *(Leiocnemis)* Putz. Mém. Liége. 1866. 216.	"North India."
A* Cm. **lamia** Andr. MB. 99. t. 3. f. 3.	Leh. Rukshu : Tsho-Morari.
A* A. **nila** Andr. MB. 96. t. 3. f. 2.	Spiti : Manikarn, Tchary-Djoni, Pulga, Tsho-ti. Kulu. Kandi., Koty. Kangra., Chamba. U.P.: Mundali, Sunderdhunga Valley, Pindar Valley, Konain, Chulli Khud, Sainj Khud.
B* A. **obscuripes** Bates TES. 1873. 294; Putz. CRB. 1875. 52; Bates TES. 1883. 242; id. CRB. 1891. 334.	Chota Nagpur: Konbir (Bates). Japan., Tonkin.
I* Cu. **pamirensis** *(Curtonotus)* Bates PZS. 1878. 717; id. SYM. 17, t. 1. f. 15.	Pamir: between Sirikol and Panga.
B* N. **Splendens** Andr. EMM. 1926. 69 and 75.	Kumaon: Milam and Burphu (Gori Valley), Girthi Valley, Yangti Valley, Shelshel.
I* L. **tartariae** *(Leiocnemis)* Bates PZS. 1878. 716; id. SYM. 15. t. 1. f. 14.	Between Yangilissar and Sirikol.
A* C. **tiruka** Andr. MB. 97.	Sonemarg. Leh.

AMAROSCHESIS Tchitcherin.

Amaroschesis Tchitch. Ab. xxix. 1897. 28 and 58; id. HER. xxxiv. 1900. 362; id. HER xxxv. 1901. 244 and note (67); id. HER. xxxvii. 1906. 262.

O* **chinensis** *(Zabrus)* Fairm. AF. 1886. 313; Tchitch. Ab. xxix. 1897. 29; id. HER. xxxvii. 1906. 268. — Yunnan.

?Q **delavayi** Tchitch. A. xxix. 1897. 29; id. HER xxxvii. 1906. 269. — Yunnan.

O **yunnana** *(Amathitis)* Fairm. Ab. 1887. 95; Tchitch. Ab xxix. 1897. 28; id. HER. xxxvii. 1906. 271. — Yunnan.

AMBLYSTOMUS* Erichson.

Amblystomus Erichs. Käf. Mark Brand. i. 1837. 59; Lac. 301; Woll. Col. Hesp. 1867. 13; Reitt. Rév. WEZ. 1883. 139 (palaearctic species); Tchitch. HER. xxxiv. 1900. 348; Sloane PLS. N.S.W. 1920. 131; Andr. MB. 33.

Hispalis Ramb. Fn. Ent. Andalousie 1838. 135.

Artizoum Gistl Acht hundert und zwanzig neue oder unbeschriebene Thiere, Vacuna ii. 1857. 603.

Megaristerus Nietn. AM. (3). ii. 1858. 427.

Thenarotidius Sloane PLS. NSW. 1898. 461.

M **aenescens** *(Hispalis)* Motsch. Et. Ent. 1858. 23; Andr. TES. 1928. 21. — "East Indies."

M **biguttatus** *(Hispalis)* Motch. Et. Ent. 1858. 25; Bates AF. 1889. 271; Andr. TES. 1928. 21. — Tranquebar. Cochin-China, Cambodia and Annam (all Bates).

E* **bivittatus** × Andr. AM. (9). iii. 1919. 474; id. MB. 34. *vittatus* || Bates (not Gestro) AB. 1892. 231. — Throughout India.

B* **convexus** *(Trechus)* Macl. Ann. Jav. 1825. 20; Andr. TES. 1919. 149. — Java.

X* **dromioides** Bates AF. 1889. 272. — Cochin-China. Annam.

M **femoralis** *(Hispalis)* Motch. Et. Ent. 1858. 24; Bates AG. xxxii. 1892. 334; id. AB. 1892. 231; Andr. TES. 1928. 21. — "East Indies." Chota Nagpur (Bates) Rangoon and Palon (Bates).

M	**flavipes** *(Hispalis)* Motch. Et. Ent. 1858. 24; Bates AG. xxxii. 1892. 333; Andr. TES. 1928. 21.	"East Indies."
M	**fuscescens** *(Hispalis)* Motch. Et. Ent. 1858. 23; Bates AG. xx ii. 1892. 334; Lesne Miss. Pav. 73; Andr. TES. 1919. 198; id. TES. 1928. 21.	"East Indies." Ceylon., Bhamo. Pegu., Rangoon. Siam.
O*	**guttatus** Bates TES. 1873. 327; id AG. xxxii. 1892. 336; id. AB. 1892. 231; Sloane PLS. NSW. 1920. 321.	Chota Nagpur. Rambha (Ganjam). Barkuda. I. (Lake Chilika). Palni Hills., Bhamo. Palon., Rangoon. Victoria Point. China., Cambodia. Philippine Is. Australia.
O*	**guttula** *(Acupalpus)* Dej. Sp. v. 1831. 854; Bates AF. 1889. 271.	"East Indies." Cochin-China and Annam (Bates).
J*	**indicus** *(Megaristerus)* Nietn. AM. (3). ii. 1858. 428; Bates AM. (5) xvii. 1886. 76; id. AF. 1889. 271; id. CRB. 1891. 331; id AG. xxxii. 1892. 336; id. AB. 1892. 231; Sloane PLS. NSW. 1920. 321; Andr. AM. (9). xix. 1927. 103.	Ceylon: Colombo. [Chota Nagpur: Tetara. Ceylon: Kitugallé. Palon. Cochin-China. Annam (all Bates). Australia (Sloane)]
G*	**magnus** Bates AG. xxxii. 1892. 333.	"India" (Brit. Mus.). Mandalay., Prome. Rangoon. Tenasserim.
J*	**mandibularis** *(Megaristerus)* Nietn. AM. (3). ii. 1858. 428; Bates AG. xxxii. 1892. 335; Andr. AM. (9) xix. 1927. 103.	Colombo.
B*	**metallicus** *(Thenarotes)* Blackb. Trans. Roy. Soc. S. Austral. X. 1887. 185; Sloane PlS. NSW. 1920. 320.	Sumbawa. Australia.
O*	**minimus** *(Acupalpus)* Dej. Sp. iv. 1829. 483.	
M	**pallipes** *(Hispalis)* Motch. Et. Ent. 1858. 24; Andr. TES. 1928. 21.	"East Indies."

G*	**punctatus** Bates AG. xxxii. 1892. 335; id. AB. 1892. 231; Andr. RIM. xxii. 1921. 344; id. MB. 34.	Chota Nagpur. Koka (Bhandara). Belgaum., Dharwar. Barkuda I. (Lake Chilka.) Surada., Madura. Bhavani Valley (Nilgiri Hills). Calicut., Malabar. Madras. Vizagapatam. Nilambur., Mysore: Nandidrug, Chikkaballapura. Ceylon: Hambantota, Mihintale, Horawupatane, Anuradhapura, Kandy, Colombo, Kotte, Badulla. Peradeniya. Mandalay.
M	**quadriguttatus** *(Hispalis)* Motch. Et. Ent. 1858. 24; Bates CRB. 1891. 331; id. AG. xxxii. 1892. 336; Andr. TES. 1928. 21.	"East Indies."
A*	**resectus** Andr. MB. 91.	Surada., Mysore: Chikkaballapura.
E*	**rotundiceps** Bates AB. 1892. 231.	Chota Nagpur.
L	**stenolophoides** *(Megaristerus)* Nietn. AM. (3), ii. 1858. 428; Sloane PLS. NSW. 1920. 321; Andr. AM. (9) xix. 1927. 103.	Colombo. Australia (Sloane).
G*	**tetrastigma** Bates AG. xxxii. 1892. 335; Andr. MB. 34.	India., Ceylon. Burma., Cambodia.
O*	**vulneratus** *(Acupalpus)* Dej. Sp. v. 1831. 852; Schaum SEZ. 1847. 49; Motch. Et. Ent. 1855. 43; Bates AG. xxxii. 1892. 336; Andr. MB. 34.	India., Ceylon. Burma., Laos.

AMOLOPS* Tchitcherin.

Amolops Tchitch. Ab. xxix. 1898. 107.

O* **depilatus** *(Molops)* Bates PZS. 1889. 214; Tchitch. Ab. xxix. 1898. 109. — Kashmir: Goorais Valley Skardo.

I* **piliferus** *(Molops)* Bates PZS. 1878. 718; id. PZS 1889. 213; id. SYM. 13; Tchitch. Ab. xxix. 1898. 108. — Kashmir: Goorais Valley, Skardo. Murree., ? Sikkim.

ANATRICHIS* Leconte.

Anatrichis Lec. Trans. Am. Phil. Soc. x. 1853. 391; Lac. 394; Chaud., Monographie des Oodides (1), AF. 1882. 318; Sloane PLS. NSW. 1910. 442.

G **birmanica** Bates AG. xxxii 1892. 325. — Rangoon. Tenasserim.

O* **indica** Chaud. Mon. 320. — Dacca. Sunderbans. Pusa., Prome.

O* **pedinoides** Chaud. Mon. 321; Bates CRB. 1891. 328. — Calcutta., Munser and Bhiwapur (Nagpur). Tharrawaddy.

ANAULACUS* Macleay.

Anaulacus Macl. Ann. Jav. 1825. 22; Lac. 309; Schaum BEZ. 1863. 76; Chaud., Etude monographique des Masoréides, des Tetragonodérides, et du genre Nematotarsus, BM. 1876. ii. 25.

P* **fasciatus** *(Aephnidius)* Schm. Goeb. FCB. 89; Schaum. BEZ. 1863. 78; Chaud. Mon. 25; Bates AG. xxxii. 1892. 405; Andr. TES. 1923. 52; id. RIM. xxvi. 1924. 117. — Assam: Siju Cave. North Kanara. Nilgiri Hills. Ceylon: Matale. Batu Caves (F. M. S.) Karen Hills. Andaman. Is. Indo-China.

X* var. **basalis** Fleut. AF. 1887. 59. t. 4. f. 1; Bates AF. 1889, 282; Andr. TES. 1923. 53. — Annam. Sumatra.

Y* var. **philippinensis** Heller PJS. xxiii. 1923. 302. t. 1. f. 2; Andr. AM. (9). xvii. 1926. 258. — Philippine Is.

P* **quadrimaculatus** *(Aephnidius)* Schm. Goeb. FCB. 90. t. 3. f. 7; Schaum BEZ. 1863. 78; Chaud. Mon. 25; Bates AG. xxxii. 1892. 405; Andr. TES. 1923. 53. — Karen Hills.

B* **sericipennis** Macl. Ann. Jav. 1825. 23. t. 1. f. 4; Schaum BEZ. 1863. 78; Chaud. Mon. 25; Andr. TES. 1919. 158; id. TES. 1923. 53. — Java.

O* **siamensis** Chaud. Mon. 25. — Siam.

ANCHISTA* Nietner.

Anchista Nietn. JASB. xxv. 1856, 523; id. AM. (2). xix. 1857. 374.

Paraphaea Bates. TES. 1873. 312.

O* **binotata** *(Plochinus)* Dej. Sp. i. 1825. 252; Bates AF. 1889. 284. id. AG. xxvii. 1889. 111; id. AG. xxxii. 1892. 423; Lesne, Miss. Pav. 81; Andr. TES. 1923. 20; id. RIM. xxvi. 1924. 117.

O* *discophora (Calleida)* Chaud. BM. 1852. i. 48; id. BM. 1877. i. 236.

B* *signifera (Paraphaea)* Bates TES. 1873. 312; id. TS. 1876. 5. t. 1. f. 5; id. TES. 1883. 208.

— South Eastern Asia, including Japan, the Malay Islands, and the Marianne Islands, but excluding Central and Southern India.

C* **brunnea** *(Lebia)* Wied. Zool. Mag. ii. 1. 1823. 59; Andr. TES. 1921. 172. — Dacca., Pusa. Fyzabad.

S* *modesta* Nietn. JASB. xxv. 1856. 523; id. AM. (2). xix. 1857. 375; Chaud. BM. 1877. i. 239; Andr. Spol. Zeyl. 1924. 140; id. AM. (9). xix. 1927. 103. — Nagpur. Colombo.

O* *picea* Chaud. BM. 1877. i. 238; Andr. TES. 1921. 172.

O* **eurydera** Chaud. BM. 1877. i. 236. — "East Indies."

P* **fenestrata** *(Plochnionus)* Schm. Goeb. FCB. 42; Bates AG. xxxii. 1892. 424; Andr. TES. 1923. 20. — Burma.

O* **glabra** Chaud. BM. 1877. i. 237. — Singhbhum. Dehra Dun., Baihar. Pondicherry. Ceylon. Horawupatana.

O* **subpubescens** Chaud. BM. 1877. I. 238. — "North India." Ajmere.

ANCHOMENUS Stephens.

Anchomenus Steph. Ill. Brit. Ent. i. 67 and 81; Dej. Sp. iii. 1828. 102; Lac. 349; Motch. BM. 1864. ii. 316; Andr. TES. 1919. 90 and 91; id. MB. 48.

Platynus Steph. Ill. Brit. Ent. i. 67 and 83; Dej. Sp. iii. 1828. 96; Lac. 350; Motch. BM. 1864. ii. 316; Sloane PLS. NSW. 1910. 453; Andr. TES. 1919. 92.

Agonum Steph. Ill. Brit. Ent. i. 85 and 182; Dej. Sp. iii. 1828. 132; Lac. 350; Bedel Fn. med. 1902. 216 note (1); Andr. TES. 1919. 90 and 91.

Promecoptera Dej. Sp. v. 1831. 443; Lac. 131; Chaud. Bm. 1871. ii. 281.

Agonothorax Motch. BM. 1864. ii. 317.

A* **benardi** Andr. MB. 102.	Lahoul: Gandla, Keiling, Sisu. Spiti: Pulga, Tehary-Djoni Manikarn. Kulu : Koty, Parbatti Valley. Seraj (Jalori (Pass).
A* **caesitius** Andr. MB. 104; id., EMM. 1926. 69.	Lahoul : Sisu, Sumdeo. Kulu: Kandi, Koty. Sunderdhunga Valley, Gori Valley and Gorge, and Burphu (Almora).
M* **ceylonicus** *(Agonothorax)* Motch. Et. Ent. 1859. 36; Bates AM. (5). xvii. 1886. 146; Andr. TES. 1928. 18.	Ceylon: Nuwara Eliya, Dikoya.
N* **chinensis** *(Agonum)* Boh. ER. 15; Bates CRB. 1891. 335; Andr. MB. 49.	Chota Nagpur; Tetara. Motinala, South Mandla. Ceylon., China. Annam., Tonkin. Singapore.
A* **comatus** Andr. AM. (9) xii. 1923. 219.	Almora., Tanakpur. Ranikhet. Dehra Dun. Assam : Kabo, Bank of Dihong below Pasighot. Eastern Duars. Tonkin.

A*	var **nigritulus** Andr. AM. (9) xii. 1923. 220.	Localities as in type form.
O	**dorsostriatus** *(Agonum)* Fairm. AB. 1888. 13.	Yunnan.
A*	**eurous** Andr. MB. 105; id. EMM. 1926. 69.	Spit: Pulga. Kulu: Kandi, Koty, Manali. U.P.: Dehra Dun, Mundali, Bagar Kote, Nainital, Almora, Kali, Valley, Sunderdhunga Valley, Gori Valley and Gorge
B*	**illocatus** Walk. AM (3). ii. 1858. 203; Bates AM. (5). xvii. 1886. 146; Andr. TES. 1919. 188.	Ceylon : Nuwara Eliya, Bogawantalawa, Pattipola, Galle, Hakgalla,
B*	*degener (Argutor)* Walk. AM. (3). ii. 1858. 204; Bates AM. (5). xvii. 1886. 146; Andr. TES. 1919. 189.	Kikilamane Mt.
A*	**incisus** Andr. Eos. iii. 1927. 77. f. 10.	Kashmir : Sintan.
A*	**indicus** Andr. AM. (9) x. 1922. 165.	Bhiwapur., Khawasa. Mhow., Belgaum.
I*	**ladakensis** Bates PZS. 1878. 718; id. SYM. 11; Andr. MB. 48.	Kashmir: Leh, Kargil, Pangong Valley between Tangtze and Chagra. Tibet : Gyangtse. Pamir.
O*	**lissopterus** Chaud. BM. 1854. i. 136; id. MB. 49; id. EMM. 1926. 69.	Himalayas from Punjab to Sikkim.
I*	*politissimus* Bates PZS. 1878. 719; id. SYM. 12. t. 1. f. 16; Andr. AM. (9). iii. 1919. 475.	
B*	**luteus** Andr. AM. (9). xii. 1923. 218.	Laos.
B"	**magnus** Bates TES. 1873. 278; Putz. CRB. 1875. 50; Har. Abh. Nat. Ver. Bremen. iv. 1875. 286; id. DEZ. 1877. 340.	Yunnan. China. Japan.

C* **marginalis** *(Lebia)* Wied. Zool. Mag. ii. 1. 1823. 60; *(Promecoptera)* Dej. Sp. v. 1831. 444; Brullé HN. iv. 1834. 227; Andr. TES. 1921. 172.	Bengal. Kurseong. Shillong.
O* **mesostictus** *(Agonum)* Bates PZS. 1889. 215.	Kashmir: Goorais Valley.
O **nuceus** Fairm. AB. 1887. 96.	Yunnan.
N **quadripunctatus** *(Carabus)* Deg. Mém. Ins. iv. 1774. 102; Dej. Sp. iii. 1828. 170; Dej. et. Boisd Ic. et HN. ii. 384. t. 122. f. 5; Schaum, Nat. Ins. Deutschl. (1) 1858. 411; Moraw. Beitr. 43; Bates TES. 1873. 281; Reitt. WEZ. 1907. 63; Andr. EMM. 1926. 69.	Kumaon: Milam and Burphu (Gori Valley), Girthi Valley., Simla. Narkanda. Tibet : Lhasa and Gyangtse., Tonkin. Philippine Is.
J *foveolatus (Carabus)* Illig. Mag Ins. i. 1802. 61.	Palaearctic and Nearctic regions.
L *cupratus (Agonum)* Sturm Deutschl. Faun. Ins. v. 1824. 218. t. 137. ff. b. B.	
H' *octocolus* Mann. BM. 1853. ii. 144.	
C' *stigmosus (Platynus)* Lec. Proc. Ac. Phil. 1854. 58.	
Y* *nigrosericans* Heller PJS. xxiii. 1923. 298; Andr. AM. (9). xvii. 1926. 258.	
A* **relueens** Andr. AM. (9). xii. 1923. 681.	Sikkim: Tonglu, Gopaldhara, Lachung, Darjiling.
N* **scintillans** *(Agonum)* Boh. ER. 16.	Northern India. Burma., Laos. Tonkin., China.
O* *aeneotinctus (Agonum)* Bates TES. 1873. 330; id. AG. xxxii. 1892. 371; Andr. TES. 1921. 180.	
O **semicupreus** *(Agonum)* Fairm. AB. 1887. 97.	Yunnan.
B* **viridicans** Andr. EMM. 1926. 69 and 78.	Kumaon: Gori Valley.
A* **ustus** Andr. Eos. iii. 1927. 75. f. 9	Kashmir : Kishtawar, Panjal, Lidar Valley.
A* var. **belli** Eos. iii. 1927. 77.	Kashmir.

ANCUS* Putzeys.

Ancus Putz., Révision Générale des Clivinides, AB. x. 1867. 197.

H*O* **excavaticeps** Putz. Rév. 199; Andr. TES. 1919. 210; id. AM. (9). xvii. 1926. 381. — Siam. Cochin-China. "Malay."

ANENTMETUS* Andrewes.

Anentmetus Andr. AM. (9) xiv. 1924. 585; Bänn, DEZ. 1927. 190 and 191; Andr. FBI. 166. — Haldwani. Madras.

B* **pluto** Andr. AM. (9). xiv. 1924. 586; id. FBI. 166. f. 29.

X* **spissicornis** *(Pseudozaena)* Fairm. AF. 1888. 335; (?Pachyteles) Lesne Miss. Pav. 62 t. 8. f. 13; Andr. AM. (9). iv. 1919. 295 (note); id. AM. (9). xiv. 1924. 586. — Laos

ANISODACTYLUS Dejean.

Anisodactylus Dej. Sp. iv. 1829. 132; Lac. 278; Tchitch. HER. xxxiv. 1900. 370; id. HER. xxxv. 1901. 251; id. HER. xxxvii. 1906. 289.

B* **sjoestedti** Andr. AM. (9) xviii. 1926. 276. — Sumatra. ? Java., Borneo. Philippine Is.

ANOMOSTOMUS* Laferté.

Anomostomus Laf. RMZ. 1853. 376; Andr. Spol. Zeyl. 1923. 235.

A* **orientalis** Andr. Spol. Zeyl. 1923. 237. — Puri., Madras. Madura. Trichinopoly. Ramandrug. Mysore: Chikkaballepura. Ceylon : Hambantota, Vavuniya.

ANOMOTARUS* Chaudoir.

Anomotarus Chaud., Genres aberrants du Groupe des Cymindides, BM. 1875. ii. 48; Sloane PLS. NSW. 1917. 435; id. PLS. NSW. 1920. 170.

Uvea Fauvel BF. 1881. 118; id. RE. 1882. 257; Andr. AM. (9). iii. 1919. 483.

B* **decoratus** Andr. Spol. Zeyl. 1924. 139. — Ceylon : Wirawila.

O* **stigmula** *(Cymindis)* Chaud. BM. 1852. i. 57; id. BM. 1875. ii. 60 and 61; Fauvel RE. 1882. 257; Sloane PLS. NSW. 1917. 436. — Simla. Dehra Dun. Gauhati. Nagpur.

geophila (Cymindis) Montr. AF. 1860. 235. — Belgaum., Khandesh.

B* *pictula (Cymindis)* Bates TES. 1873. 310; id. TES. 1883. 208; Andr. AM. (9). iii. 1919. 483. — Madras. Mysore: Nandidrug. Ceylon : Nalanda, Hambegamuwa. Tharrawaddy. Formosa., Celebes. New Guinea. New Caledonia. Japan.

ANOPLOGENIUS* Chaudoir.

Anoplogenius Chaud. BM. 1852. i. 88; Lac. 304; Tchitch HER. xxxiv. 1900. 364; id. HER. xxxv. 1901. 246; id. HER. xxxvii. 1906. 287.

Loxoncus Schm. Goeb. in litt. FCB. (cover); Motch. BM. 1864. ii. 204.

Lepithrix Nietn. JASB. 1857. ii. 151; id. AM. (2). xx. 1857. 374.

Megrammus Motch. Et. Ent. 1857. 26.

B* **cyanescens** *(Harpalus)* Hope TES. iv. 1845. 15; Andr. TES. 1919. 177. — Yunnan., China. Japan.

M *circumcinctus (Megrammus)* Motch. Et. Ent. 1859. 27. Bates TES. 1873. 269; Putz. CRB. 1875. 49; Tchitch. Ab. xxix. 1897. 61; id. HER. xxxvii. 1906. 288; Andr. TES. 1928. 21.

O* **discophorus** Chaud. BM. 1852. i. 90. — Simla., Haldwani. Pusa., "South India."

J*S* *foliolosus (Lepithrix)* Nietn. JASB. xxvi. 1857. 152; id. AM. (2). xx. 1857. 374; Andr. AM. (9). iii. 1919. 474; id. AM. (9). xix. 1927. 103. — Colombo., Bhamo. Toungoo., Prome. Tharrawaddy. Indo-China. ? Japan.

I. **elevatus** *(Laxoncus)* Schm. Goeb. FCB. t. 3. f. 9; Motch. B.M. 1864. ii. 204; Bates AG. xxxii. 1892. 347; Andr. TES. 1923. 59.	? Burma.
B* **incisus** Andr. AM. (9). xviii. 1926. 279.	Kedah. Lankawi. I. Java., Sumatra. Lombok., Celebes. New Guinea.
B* **microgonus** Bates AM. (5). xvii. 1886. 78; id. AF. 1889. 272; id. CRB. 1891. 333; id. AG. xxxii. 1892. 345.	Bengal : Calcutta, Surjabuggi, Sardah, Sunderbans, Diamond Harbour.
X* *renitens* † Bates (not Bates (1886) AF. 1889. 272; Andr. TES. 1921. 177.	Chota Nagpur. Tetara., Malabar. Barkuda I. (Lake Chilka). Colombo., Bhamo. Palon and Tikekee (Pegu)., Shwegyin. Prome., Tharrawaddy. Insein., Rangoon. Tenasserim. Victoria Point., Penang. Wellesley Provinces. Siam., Cochin-China. Laos., Annam., Tonkin. Sumatra., Philippine Is.
E* **nagpurensis** Bates CRB. 1891. 333.	Chota Nagpur. Tetara., Nagpur. Bilaspur., Madras.
G* **patinalis** Bates AG. xxxii. 1892. 346; Andr. AM. (9). iii. 1919. 474; id. AM. (10). ii. 1928. 587.	Saran., Pusa., Fyzabad. Haldwani., Bhamo. Toungoo.
G* **planicollis** Bates AG. xxxii. 1892. 345.	Sardah., Bhamo., Palon. Tharrawaddy.
B* **renitens** Bates AM. (5). xvii. 1886. 79; id. AG. xxxii. 1892. 346; Andr. TES. 1921. 177; id. TES. 1923. 60.	Ceylon : Colombo, Anuradhapura, Kotte. Mandalay., Toungoo. Shwegu., Palon. Rangoon., Tenasserim. Cochin-China. Cambodia., Tonkin.

G* **rutilans** Bates AG. xxvii. 1889. 103; id. AF. 1889. 272; id. AG. xxxii. 1892. 345; Lesne Miss. Pav. 76. — Pusa. Goalundo to Gauhati (Brahmaputra River). Bhamo., Mandalay. Katha. Tharrawaddy. Prome., Tenasserim. Siam., Cochin-China., Cambodia.

ANTHIA* Weber.

Anthia Weber Obs. Ent. 1801. 17; Bon. Obs. Ent. ii, Mém. Acad. Sci. Turin. xx. 1813. 451; Latr. et Dej. HN. et Ic. 93; Dej. Sp. i. 1825. 338; Lequien Mag. Zool. Cl. ix. 1832. t. 34-41; Chaud. BM. 1850. i. 41; Lac. 177; Obst Synopsis der Col. Gen. Anthia, Arch. Naturg. lxvii. 1901. Beiheft. 265; Rousseau Gen. Ins. Fasc. 38, Anthiinae, 1905. 4; Andr. MB. 14.

Pachymorpha Hope Col. Man. ii. 1838. 51; Lac. 176; Obst Arch. Naturg. 1901. 286; Rousseau Gen. Ins. Fasc. 38, Anthiinae 1905. 5.

B* **sexguttata** *(Carabus)* F. Syst. Ent. 1775. 236; Oliv. Ent. 35. 15. t. 1. f. 6; Dej. Sp. i. 1825. 341; Brullé HN. iv. 1834. 370. t. 9. ff. 2. 2 (larva); Motch. BM. 1845. i. 12. t. 1. f. 3; id. Et. Ent. 1855. 41; Chaud. BM. 1861. i. 562; Motch. BM. 1864. ii. 216; Redt. RN. 8; Bates CRB. 1891. 336; Heyne Tasch. E. K. 17. t. 3. f. 21; Obst Synop. 286; Maindr. BF. 1905. 134 and 161; Rouss. Gen. Ins. Fasc. 38, Anthiinae, t. 1. f. 4; Andr. TES. 1919. 121; id. TES. 1921. 153; id. TES. 1923 (1924). 462; id. MB. 15. — Throughout India (Plains).

O* *indica* Chaud. BM. 1861. i. 562.

L *elliptica* Motch. BM. 1864. ii. 216; Heyne-Tasch. E. K. 17; Andr. TES. 1928. 21.

var. **thunbergi** Fisch. Mem. Soc. Nat. Mosc. i. 1806. 18; Bedel Fn. med. 1914. 306 (Obs).

O* *mannerheimi* Chaud. BM. 1842. ii. 810; Motch. Mag. Zool. 1844. t. 149. f. 1; id. BM. 1845 i. 11. t. 1. f. 2; Chaud. BM. 1861 i. 563; Motch. BM. 1864. ii. 216; Obst Synop. 286; Andr. MB. 14. — Punjab., Persia. Turkestan.

H* var. **orientalis** *(Pachymorpha)* Hope Col. Man. ii. 1838. 163. t. 3. f. 4; Chaud. BM. 1861. i. 563; Motch. BM. 1864. ii. 216; Bates SYM. 19; Heyne-Tasch. E. K. 17; Andr. TES. 1919. 200. — Himalayas. Jhelum Valley. Poona. Bangalore.

APOTOMUS* Illiger.

Apotomus Illig. Mag. Ins. vi. 1807. 348; Dej. Sp. i. 1825. 449; Lac. 172; Reitt., Synopsis, WEZ. 1892. 137 (European species); Dupuis Gen. Ins. Fasc. 117, Apotominae, 1911. 1; Andr. MB. 15.

M* **atripennis** Motch. Et. Ent. 1858. 22; Andr. TES. 1928. 15. — Sikkim : Namsoo. Calcutta.

M* *fuscus* Motch. Et. Ent. 1858. 22; Bates AF. 1889. 264; Dupuis AB. 1911. 36. — Eastern Duars. Pusa., Haldwani.

O* *xanthotelus* Bates EMM. xi. 1874. 95; id. AG. xxvii. 1889. 100; id. AG. xxxii. 1892. 286; Bouch. AF. 1903. 170; Dupuis Gen. Ins. Fasc. 117, Apotominae, 1911. t. f. 11. — Nagpur., Bandra. Colombo. Tharrawaddy. Toungoo., Pegu.

Y* *sumbawanus* † Heller (not Dupuis) DEZ. 1916. 275; Andr. AM. (9) xvii. 1926. 254. — Rangoon. Tenasserim., Kedah. Malacca. Cochin-China. Cambodia., Java. Sumatra., Celebes. Philippine Is. New Guinea

G var. **pallidior** Bates AG. xxx ii. 1892. 286. — Bhamo., Rangoon.

G* **hirsutulus** Bates AG. xxxii. 1892. 287; Andr. MB. 15. — Throughout India and Burma.

E* **sumbawanus** Dupuis AB. 1911. 36; id. Gen. Ins. Fasc. 117, Apotominae, 1911. t. f. 12. — Sumbawa.

APRISTUS Chaudoir.

Apristus Chaud., Enumération des Carabiques et Hrocanthares recueillis pendant un voyage an Caucase et dans les provinces transcaucasiennes, 1846. 62; Lac. 123.

P* **aeneipennis** *(Lionychus)* Schm. Goeb. FCB. 37; Chaud. BM. 1850. i. 67; Motch. Et. Ent. 1855. 50 (note 1); Fairm AF. 1888. 335; Andr. TES. 1923. 15. — Lonavla., Mysore : Teppukadur. Martaban., Tonkin.

O*	**aeneomicans** Chaud. BM. 1850. i. 66; Andr. TES. 1923. 16.	Punjab : Seraj (Jibhi). Simla. U. P. : Almora, Swal River basin, Ranikhet, Dehra Dun, Kali Gad, Kosi River, Haldwani, Arni Gad (Mussoorie), Binal Gad (Chakrata). Sikkim : Gopaldhara, Birik, Tista Valley. Eastern Duars.
B*	**cupreus** Andr. Spol. Zeyl. 1924. 140.	Ceylon: Tissa
M	**subtransparens** Motch. BM. 1861. i. 104; Bates AM. (5). xvii. 1886. 206; id. AB. 1892. 233; Andr. TES. 1928. 21.	Throughout India and Ceylon.

ARHYTINUS* Bates.

Arhytinus Bates AF. 1889. 278; id. AG. xxxii. 1892. 378.

X*	**bemdidioides** Bates AF. 1889. 279; id. AG. xxxii. 1892. 378.	Sikkim., Karen Hills. Perak., Annam. Sumatra.

ARISTOCHROA Tchitcherin.

Aristochroa Tchitch. HER. xxxii. 1898. 70.

O	**latecostata** *(Platysma)* Fairm. AB. 1887. 94; Bates Entom. 1891. Suppl. 70; Tchitch. HER. xxx. 1897. 291; id. HER. xxxii. 1898. 77.	China. Yunnan.

ARISTOLEBIA* Bates.

Aristolebia Bates AG. xxxii. 1892. 428; Andr. TES. 1923. 36; id. AM. (9). xvii. 1926. 259.

Hoplomenes Heller PJS. xix. 1921. 528; Andr. AM. (9). xvii. 1926. 259.

Y*	**davaonis** (*Hoplomenes*) Heller PJS. xix. 1921. 529. t. 1. f. 3.	Philippine Is.

O* **prattiana** *(Lebia)* Bates PZS. 1889. 218; Andr. AM. (9). xiv. 1924. 593.	China.
G* **quadridentata** Bates A. G. xxxii. 1892, 428; Andr. TES. 1923. 36.	Assam. Bhamo.

ASAPHIDION* Des Gozis.

Asaphidion × Des Gozis, Recheche de l' espèce typique de quelques anciens genres, rectifications synonymiques, et notes diverses, 1886. 6; Bedel Fn. med. 1896. 55; Andr. MB. 18; id. EMM. 1925. 49.

Tachypus || Dej. (not Weber) Sp. v. 1831. 35; Cast. Hist. Nat. i. 1840. 152; Lac. 381.

B* **championi** Andr. MB. 84. t. 2. f. 2; id. EMM. 1925. 52. t. 1. f. 4.	Doiphang Valley (Assam-Bhutan frontier). Kalimpong. Spiti: Manikarn. U. P : Almora, Kaldbunga (Haldwani) Binal Gad, Bodyar, Tadi Gad, Manjgaon.
B* **cuprascens** Andr. EMM. 1925. 53. t. 1. f. 5.	Kaldbunga (Haldwani).
A* **cupreum** Andr. EMM. 1925. 51. t. 1. f. 2.	Yunnan. China.
A* **fragile** Andr. EMM. 1925. 55. t. 1. f. 8.	Yunnan.
A* **granulatum** Andr. EMM. 1925. 51. t. 1. f. 3.	China.
A* **griseum** Andr. EMM. 1925. 54. t. 1. f. 6.	Eastern Duars.
O* **indicum** *(Tachypus)* Chaud. BM. 1850. ii. 189 (note); Andr. EMM. 1925. 55. t. 1. f. 9.	Sitamarhi., Kurseong. U. P. : Mussoorie, Haldwani, Almora, Bodyar, Kanasar, Manjgaon, Binal Gad, Sijla Gad, Dodora Khud, Chulli Khud. Siwalik Hills. Simla Hills., Kotgarh. Mandi., Dharmsala.

B* **obscurum** Andr. EMM. 1925. 52. t. 1. f. 1. — Garhwal:, Dudhatoli. ?Dehra Dun.

B* **ornatum** Andr. EMM. 1925. 57. t. 1. f. 11. — U. P. : Almora, Mussoorie, Binal Gad, Chakrata.

L* **semilucidum** *(Tachypus)* Motch. Et. Ent. 1862. 24; id. BM. 1864. ii. 180; Bates TES. 1873. 300; Sem. WEZ. 1889. 60; Andr. EMM. 1925. 57 t. 1. f. 12. — China. Japan.

Q *nubiferum (Tachypus)* Moraw. Mel. Biol. Acad. Sci. Petr. iv. 1862. 246; id. BAP. v. 1863. 327; id. Beitr. 82.

B* **substriatum** Andr. EMM. 1925. 56. t. 1. f. 10. — Punjab : Seraj. Almora.

A* **viride** Andr. EMM. 1925. 54. t. 1. f. 7. — Sikkim: Dikchu in the Tista Valley.

AXONYA* Andrewes.

Axonya Andr. AM. (9). xii. 1923. 679.

A* **championi** Andr. AM. (9). xii. 1923. 680; id. MB. 22. t. 2. ff. 5a. 5b. 5.; id. EMM. 1926. 68. — Kangra., Nalagarh. Almora., Haldwani. Kathgodam., Bijnor. Nainital. Dehra Dun. Lucknow., Sikkim. Birik (Tista Valley). Namsoo. Assam : between Pasighat and Kobo, Dihong River, Mangaldai.

BADISTER Schellenberg.

Badister Schell. Helv. Ent. ii. 1806. 90/91; Dej. Sp. ii. 1826. 405; Lac. 234.

Amblychus Gyll. Ins. Suec. i. (2). 1810. 74.

Trimorphus Steph. Ill. Brit. Ent. i. 1828. 180.

A* **sundaicus** Andr. AM. (9). xviii. 1926. 275. — Kuala Lampur. Java., Sumatra.

C* **thoracicus** Wied. Zool. Mag. ii. 1. 1823. 57; Andr. TES. 1921. 171 and 187. — Calcutta. Berhampur.

BARYSOMUS* Dejean.

Barysomus Dej. Sp. iv. 1829. 56; Lac. 290.

Oosoma, Nietn. JASB. xxvi. 1857 144; id. AM. (2). xx. 1857. 368.

O* **gyllenhali** Dej. Sp. iv. 1829. 59; Bates AM. (5). xvii. 1886. 77.	Calcutta. Gopkuda I.
J*S* *arenarius (Oosoma)* Nietn. JASB. xxvi. 1857. 146; id. AM. (2). xx. 1857. 370; Andr. AM. (9). xix. 1927. 103.	(Lake Chilka). Madras. North Kanara. Malabar., Ceylon. Colombo. Trincomali. Puttalam. Niroddumunai.
C* **semivittatus** *(Carabus)* F. Suppl. Ent. Syst. 1798. 59; Dej. Sp. iv. 1829. 60; Motch. Et. Ent. 1855. 44; Redt. RN. 14; Bates AF. 1889. 269; Andr. RIM. xxii. 1921. 344; id. TES. 1921. 160; id. TES. 1923 (1924). 467.	Madhupur., Pusa. Balighai. Chandipur., Baihar. Madras : Barkuda I. (Lake Chilka).
J&S* *gerstaeckeri (Oosoma)* Nietn. JASB. xxvi. 1857. 147; id. AM. (2). xx. 1857. 370; Andr. AM .(9). xix. 1927. 103.	Madras. Trichinopoly. Devaraya Drug. Ceylon : Colombo, Kanthalai, Niroddummnai, Anuradhapura. Cochin-China. Annam., Hongkong.

BEMBIDION Latreille.

Bembidion, Latr. Hist. Nat. Crust. et Ins. iii. 1802. 82.

Bembidium, Dej. Sp. v. 1831. 31; Lac. 382; Andr. MB. 16.

Ocydromus, Schell. Helv. Ent. ii. 1806. 16/17.

P. *Peryphus,* Steph. Ill. Brit. Ent. ii. 2. and 11.

N. *Notaphus,* Steph. Ill. Brit. Ent. ii. 2 and 17.

Pr. *Princidium,* Motch. BM. 1864 ii. 181.

C. *Chlorodium,* Motch. BM. 1864. ii. 182.

Pl. *Plataphus,* Motch. BM. 1864. ii. 184.

E. *Emphanes,* Motch. BM. 1864. ii. 185.

B. *Bracteon,* Bedel Fn. Seine i. 1879. 27.

Ps. *Pseudolimnaeum,* Kr. DEZ. 1888. 365.

T. *Testediolum,* Gangl. Käf. Mitteleur. i. 1892. 170.

P. *Daniela,* Net. WEZ. 1910. 210.

Tr. *Trichoplataphus,* Net. Ent. Blätt. 1914. 51 = *Triporus* Andr. EMM. 1921. 251.

Nc. *Notaphocampa,* Net. Ent. Blätt. 1914. 167.

Pa. *Pamirium,* Net. Ent. Mitt. ix. 1920. 112.

M. *Microserrullula* × Net. Arch. Naturg. 1921. A. H. 7. 185. = Serrula || Net. (not Mörch) WEZ 1910. 209.

A* P. **aquilum** Andr. MB. t. 1. f. 6. — Lahoul : Patseo. Kashmir : Gond, Baltal. Ladak : Lah, Mulbeck.

B* P. **babaulti** Andr. MB. 80. t. 1. f. 7; id. EMM. 1926. 67. — Dharmsala. Jibhi (Seraj)., Mandi. Simla Hills: Fagu, Gahan. U.P. : Chakrata, Binsar, Bhowali, Dehra Dun, Binal Gad, Almora, Sunderdhunga Valley, Gori Valley, R. Sarda Gorge, Nainital.

A* P. **bakeri** Andr. EMM. 1924. 197. — Philippine Is.

A* P. **betegara** Andr. MB. 82. — Bashahr : Sangla. Spiti : Pulga. Kulu : Koty. Mandi.

A* **bolivari** Andr. Eos. iii, 1927. 67. f. 2. — Sikkim.

O* P. **bracculatum** Bates PZS. 1889. 212; Andr. EMM. 1924. 194; id. EMM. 1926. 67. — Kashmir : Goorais Valley. Kumaon : Gori River Gorge, Milam, Laptel.

I* T. **braminum** Andr. AM. (9). xii. 1923. 442.	Kodaikanal.
A* P. **bryanti** Andr. EMM. 1921. 249.	Java.
N. **callipygum** Boh. ER. 17; Andr. TES. 1928. 5.	Hongkong.
B* Ps. **cameroni** Andr. EMM. 1922. 177; id. EMM. 1926. 68.	Kulu : Monali. Simla Hills : Gahan, Almora : Gori River Gorge.
B* P. **charon** Andr. EMM. 1926. 67 and 74.	Kumaon : Burphu and Gori River Gorge.
O. P. **chloreum** Bates TES. 1873. 332; id. TES. 1883. 275.	China.
B* P. **cimmerium** Andr. EMM. 1922. 238.	Simla Hills: Matiana, Gahan. Chulli Khud (Chakrata).
A* **clarum** Andr. EMM. 1923. 98; id. EMM. 1926. 67.	Tista Valley (Sikkim). Assam : Naga Hills, Yambung, Janakhmukh. Manipur : Shugnu. U.P. : Almora, Chakrata, Khaula, Ranikhet, Tanakpur, Nainital, Upper Gumti Valley, R. Sarda Gorge. Garhwal : Siuni Gad.
O* P. **collutum** Bates TES. 1873. 332.	China.
B* Pr. **compactum** Andr. EMM. 1922. 240.	Manipur : Shugnu. U. P. : Almora, Ranikhet, Bhatkot, Swal River basin, R. Sarda Gorge, Dehra Dun, Haldwani.
O* P. **dardum** Bates PZS. 1889. 212; Andr. EMM. 1924. 194.	Kashmir : Skardo.

B* **eurydice** Andr. EMM. 1926. 67 and 73.	Kumaon : Gori River Gorge.
A* P. **eutherum** Andr. EMM. 1923. 96.	Eastern Duars. Tista Valley (Sikkim). Naga Hills. Pasighat., Kobo. U. P. : Nainital, Almora. R. Sarda Gorge. Tanakpur, Ranikhet, Dehra Dun, Mussoorie, Siwaliks Chulli Khud, Haldwani. Simla Hills : Matiana. Shahpur (Kangra).
A* Ps. exaratum Andr. MB. 81.	Spiti : Mankaran. Simla Hills : Gahan. U. P. : Chulli Khud, Muktesar, Almora. Almora.
B* **exquisitum** Andr. EMM. 1923. 99.	Simla Hills: Kotagarh. Formosa.
D* B. **fusiforme** Net. Ent. Mitt. 1914. 168.	
A* T. **gagates** Andr. MB. 72. t. 1. f. 4.	Kashmir : Gond, Gilgit, Jhelum Valley, Lidar Valley, Pohru Valley, Lolab Valley, Gulmarg, Batote. Bashahr : Daran. Punjab : Seraj (Jalori Pass), Dharmsala. Spiti : Mankaran, Pulga Kangra: Shahpur and Baijnath. Lahoul : Keiling,

T. **gagates** Andr.—	Patseo, Sumdeo. Kulu : Koty, Naggar, Parbath Valley., Chamba. Simla Hills: Gahan, Jubbul, Kotkhai and Tharoch. U.P. : Sunderdhunga Valley, Binal Gad, Chulli Khud, Sainj Khud.
A* P **hasurada** Andr. MB. 77.	Kashmir : Ramsu, Batote. Ladak : Mulbeck. Bashahr : Poo. Spiti : Pulga. Naggar. Simla Hills : Matiana, Kotgarh. U. P. : Almora Nainital, Bimal Gad, Bodyar.
A* T. **himalayanum** Andr. MB. 74; id. EMM. 1926. 67.	Lahoul: Patseo. Dharmsala. Baijnath (Kangra), Seraj. (Jalori Pass). Kurram Valley : Parachinar. Tibet : Supi River.
B* **incisum** Andr. EMM. 1921. 250; id. EMM. 1923. 101.	Indo-China.
A* Tr. **kara** Andr. EMM. 1921. 250; id. EMM. 1923. 101.	Garo Hills. Naga Hills. Eastern Duars. Sikkim : Namsoo, Gopaldhara. U. P. : Almora Swal River basin, Khairna, Dehra Dun, River Song,, Ranikhet, Haldwani Siwaliks., Tavoy. Indo-China.

A* P. **kaschmirense** Net. Ent. Mitt. 1920. 117.	Bashahr : Poo.
B* P. **kempi** Andr. EMM. 1922. 178.	Simla Hills : Matiana, Theog, Kotgarh, Fagu. Bashahr : Daran. U. P. : Dehra Dun, Mussoorie, Mohna, Almora, Upper Gumti Valley, Nun Nadi, Dodora Khud, Binal Gad, Manjgaon, Khedar Khud, Sainj Khud, Chakrata.
A* P. **ladakense** Andr. EMM 1924. 195 I* *tibiale* † Bates (not Dufts.) SYM. 19; Andr. MB. 17. *bracculatum* † Andr. (not Bates) MB. 17 (part).	Kashmir : Gond, Baltal. Ladak : Mulbeck. Lahoul : Sumdeo.
A* P Ladas Andr. MB. 79. t. 1. f. 5.	Ladak : Leh, Mulbeck. Rukshu : Nima-Mud, Tsho-Morari, Salt Lake.
O* E. **latiplaga** Chaud. BM. 1850. ii. 185; Andr. RIM. xviii. 1919. 100. *quadrispilotum* Schauf. Verh. Zool-bot. Ges. Wien. xxxi. 1881 (1882). 620; Net. Ent. Blätt. 1914. 55 *(tetraspilotum).*	Seistan : Labi Baring, Nasratabad. Mediterranean region. S. Russia. Turkestan.
A* **leptaleum** Andr. EMM. 1922. 242.	Sikkim : Lachung.
A* P. **leve** Andr. MB. 78.	Ladak : Mulbeck.
A* P. **limatum** Andr. MB. 77.	Kangra.
B* C. **loricatum** Andr. EMM. 1922. 239.	Fyzabad., Lyallpur. Pusa., Kumaon : Ranikhet, Almora, Bhatkot.
B* M. **luniferum** Andr. EMM. 1924. 192.	Fyzabad., Nainital.
A* Pc. **luntaka** Andr. MB. 75.	Bashahr : Poo, Sangla. Kashmir : Baltal, Sonemarg. Ladak : Leh,

Pc. **luntaka**—	Mulbeck, Nuria. Kulu : Koty. Rukshu : Nima-Mud. Lahoul : Keiling, Sumdeo, Sisu, Zingzinbar.
J* M. **luridipenne** Schaum BEZ. 1860. 199; Andr. AM. (9). xix. 1927. 108.	Calcutta. Sunderbans., Pusa.
O* Nc. **niloticum** Dej. Sp. v. 1831. 73; Bate. TES. 1873. 301; Har. DEZ. 1877. 342; Net. Ent. Blätt. 1914. 167; Müll. Kol. Rundsch. 1918. 61; Andr. AM. (9). iii. 1919. 472; id. RIM. xviii. 1919. 100; id. EMM. 1921. 249; id. MB. 18. E* *hamatum* Kol. Mel. Ent. i. 1845. 75. *batesi* Putz. CRB. 1875. 52; Bates TES. 1876. 5; id. TES. 1883. 269.	Scistan : Nasratabad. Kalka. Daphar Plantation (Upper Jhelum). Almora., Ranikhet. Mussoorie. Lyallpur. Dhappa (Calcutta). Naga Hills., Egypt. Mesopotamia. Turkestan., China. Indo-China, Japan.
B* Pl. **nivicola** Andr. AM. (9). xi. 1923. 275.	Mount Everest Expeditions base Camp.
B* **notatum** Andr. EMM. 1922. 175; id. EMM. 1926. 67.	U. P. : Almore Nainital Gori Valley, Upper Gumti Valley, Binsar, Dodora Khud, Sainj Khud, Chulli Khud, Jadi Gad. Pusa., Kurseong. Garo Hills.
J* Nc. **opulentum** Nietn. AM. (3). ii. 1858. 420; Har. DEZ. 1877. 341; Bates TES. 1883. 269; Net. Ent. Blätt. 1914. 167; Andr. EMM. 1921`. 249; id. AM. (9). xix. 1927. 104.	India., Ceylon. Burma., Siam. Cambodia. Java.
O *? hamiferum* ‖ Fvl. (not Chaud.) RE. 1882. 231; id. RE. 1903. 218.	Australia. ? New Caledonia.

B* *europs* Bates AM. (5). xvii. 1886. 156; Andr. AM. (9). iii. 1919, 472.	
A* *riverinae* Sloane PLS. NSW. 1894 (1895). 405; id. PLS. NSW. 1920. 321.	
? *caledonicum* × Csiki Col. Cat. 97. Carab. 1928. 69.	
B* P. **orinum** Andr. EMM. 1922. 239; id. EMM. 1926. 67.	Kumaon : Almora, Sunderdhunga Valley, Gori River Gorge, Burphu. Kulu : Parbatti Valley.
I* P. **pamirense** Bates PZS. 1878. 718; id. SYM. 18; Müll. Kol. Rundsch. 1918. 90.	Pamir. Gilgit.
B* Pl. **persephone** Andr. EMM. 1926. 67 and 72.	Kumaon : Shelshel
A. Pa. **petrimagni** Net. Ent. Mitt. 1920. 63 and 113.	Pamir.
B* P. **phaedrum** Andr. EMM. 1923. 97.	Kumaon : Almora. Bhatkhot, Swal River. basin, Ranikhet. Garhwal : Siuni Gad.
M Pa. **platypterum** Solsky FT. 120; Mars. Ab. xix. 1880. 482; Net. WEZ. 1910. 48; id. Ent. Mitt. 1920. 112.	? Pamir. Turkestan.
I* *punctulipenne* Bates PZS. 1878 718; id. SYM 18; Net. WEZ 1910. 48.	
A* Pl. **pluto** Andr. MB. 72 t. 1. f. 3 ; Id. EMM. 1926. 67.	Kumaon : Sagcha, Laptel, Shelshel. Rukshu : Nima-Mud. Tsho-Morari, Salt Lake. Leh.
B* **psuchrum** Andr. EMM. 1922. 241.	Sunderdhunga Valley (Almora).
M. **quadricolle** Motch. Ins. Sib. 1844, 270; Müll. Kol. Rundsch. 1918, 55; *(Bembidium* sp.) Andr. RIM. xviii. 1919. 99; id. AM. (9). xiv. 1924. 588.	Seistan: Nasratabad. S. E. Europe. Mesopotamia.
O* *inserticeps* Chaud. BM. 1850. il. 173.	

A* P. **radians** Andr. EMM. 1922. 177; id. EMM. 1926. 67.	Sikkim : Lachung, Gopaldhara, Karponang, Tonglu. U. P. : Almora, Sunderdhunga Valley, Pindar Valley, Gori river Gorge, Nainital, Ranikhet, Bahtkot. Bashahr : Khadvala. Simla Hills : Baghi, Kufri, Narkanda, Suttoo.
B* P. **regale** Andr. EMM. 1922. 174; id. EMM. 1926. 67.	U. P. Tanakpur, River Sarda Gorge, Gori River Gorge. Kulu : Parbatti Valley.
A* P. **rubidum** Andr. EMM. 1924. 196.	Bashahr : Poo, Sangla.
A* P. **satanas** Andr. EMM. 1924. 193.	Kashmir : Goorais Valley.
B* M. **splendens** Andr. EM. 1923. 95.	Kumaon: Almora, Ranikhet R. Sarda Gorge.
A* **tambra** Andr. EMM. 1923. 100.	Yunnan.
B* **trimaculatum** Andr. Eos. iii. 1927. 68. f. 3.	Punjab : Upper Jhelum. Lyallpur. Dehra Dun.
P **turanicum** × Csiki Col. Cat. 97. Carab. 1928. 97.	Ladak : Leh, Mulbeck, Nuria.
?O *pictum* ‖ Fald. (not Dufts.) Mem. pres. Acad. Petersb. ii. 1835. 357; Motch. Ins. Sib. 1844; 243; Müll Kol. Rundsch. 1918. 90; Andr. MB. 16.	Kashmir : Baltal. Rukshu : Nima-Mud. Tsho-Morari, Salt Lake. Turkestan., Siberia.
A* P. **uvidum** Andr. MB. 83. *bracculatum* † Andr. (not Bates) MB. 17 (part); id. EMM. 1924. 196.	Kangra. Nainital. Ranikhet.

N. **variam** *(Carabus)* Oliv. Enc. Méth. v. 1790. 358; id. Ent. 35. 110. t. 14. f. 165; Bates TES. 1883. 269; Müll. Kol. Rundsch. 1918. 60 ; Andr. RIM. xviii. 1919. 100.	Seistan : Jellalabad, Labi Baring. Kashmir : Jhelum Valley. Palaearctic region.
B* P. **vitalisi** Andr. EMM. 1921. 251.	Laos.
O* N. **xanthacrum** Chaud. BM. 1850. ii. 175 (note); Andr. EMM. 1921. 249.	India. Burma., Siam.
G* *xanthotelum* Bates AG. xxxii. 1892. 287; Andr. AM. (9). iii. 1919. 472.	Indo-China.
A* **xanthochiton** Andr. EMM. 1922. 176.	Sikkim : Rangit Valley, Gopaldhara. U. P. : Almora, Upper Gumti Valley, Ranikhet, Mussoorie Haldwani. Simla Hills : Theog.
B* P. **xestum** Andr. EMM. 1923. 97.	Nainital.
A* P. **yunnanum** Andr. EMM. 1923. 100.	Yunnan.

BOTHYNOPTERA* Schaum.

Bothynoptera Schaum. Journ. Ent. ii. 1863. 75; Chaud., Mon. des Callidides, AB. xv. 1872. 181.

J* **dorsigera** Schaum Journ. Ent. ii. 1863. 76. t. 4. f. 3'; Chaud. Mon. 181; Andr. AM. (9). xix. 1927. 108.	Allahabad. Sikkim : Gopaldhara, Pashok., Cherrapunji. Tonkin.
B* **perforata** Bates TES. 1873. 313.	Japan.
B* **tripunctata** Bates TES. 1873. 314.	Japan.

BRACHICHILA* Chaudoir.

Brachichila Chaud. Mémoire sur les Thyréoptérides, AB. xii. 1869. 123; Dupius, AB. 1214. 29.

O* **hypocrita** Chaud. Mém. 123; Bates AF. 1889. 283; Dupuis AB. 1914. 29.	Assam: Yambung. Kobo., Tonkin. Annam., Hongkong. Formosa.

G* **rugulipennis** Bates AG. xxxii. 1892. 406. — Thagata (Tenasserim). Palon.

BRACHIDIUS* Chaudoir.

Brachidius Chaud. BM. 1852. i. 78; Lac. 264; Chaud., Essai monographique sur les Cratocérides, AB. xv. 1872. 18.

O* **crassicornis** Chaud. BM. 1852. i. 80; id. Mon. 19; id. in Oberthür Col. Nov. i. 1883. 39; Heller DEZ. 1916. 276. — Bhamo. Karen Hills, Dawna Hills. Andaman Is. Malay Peninsula. Siam., Laos., Java. Sumatra., Borneo. Buru., Timor. Tenimber. Philippine Is.

O* *corpulentus* Chaud. Mon. 20; Bates AG. xxxii. 1892. 351.

BRACHINUS Weber.

Brachinus Weber. Obs. Ent. 1801. 22; Latr. et. Dej. HN. et. lc. 104; Dej. Sn. i. 1825. 297; . . . Lac. 99; (Biology) Wickham. Bull. Lab. Iowa 1893. 330. t. 9. ff. 1. 2. ; id. Canad. Ento. 1894. 39.

Aploa Hope. TZS. i. 1834. 91.

Brachynus Chaud. Mon. des Brachynides, AB. 1876. 49; Andr. MB. 56.

O* **atripennis** Chaud. Mon. 71. — Simla.

O* **bigutticeps** Chaud. Mon. 52. — Sumatra.

Z* *hageni* Oberth. NLM. v. 1883. 217; Andr. AM. (10). ii. 1928. 589. — Boreno.

O* **charis** × Andr. TES. 1923. 41. scitulus † Chaud. (not Schm. Goeb) Mon. 59. — Tranquebar.

O* **chinensis** Chaud. BM. 1850. i. 81; Redt. RN. 5; Chaud. Mon. 52; Andr. TES. 1923 (1924). 460. — Myawadi (Burmo Siamese frontier). Siam

L *confucius* Boh. ER. 3. — Annam. Hongkong.

O* **cinctellus** Chaud. Mon. 70. — Decca., "North West Provinces."

G* **circumtinctus** Bates AG. xxxii. 1892. 396. — Burma : Senmigion. Laos.

A	**costulipennis** Liebke Ent. Anz. 1928. 19.	Yunnan.
G*	**evanescens** Bates AG. xxxii. 1892. 395.	Karen Hills.
G	**exquisitus** Bates AG. xxxii. 1892. 394.	Burma : Teinzo.
G*	**flavicapillus** Bates AG. xxxii. 1872. 393.	Burma : Bhamo, Palon, Rangoon, Theinzeik, Pyaung-gyaung.
O*	**flaviventris** Chaud. Mon. 68.	Coromandel. Nagpur.
O	**frontalis** Chaud. in Oberthur Col. Nov. i. 1883. 18.	New Guinea.
O*	**hexagrammus** Chaud. Mon. 55.	Throughout India.
O*	**illotus** Chaud. Mon. 58; Vuillet Ins. ii. 1912. 17.	Dacca., Mhow. ? Cochin-China.
G*	**intactus** Bates AG. xxxii. 1892. 396.	Rangoon., Palon.
X*	**lesnei** × Andr. AM. (9). xii. 1923. 222. *sexmaculatus* † Lesne (not Dej.) Miss. Pav. 79.)	Siam., Annam.
O*	**limbellus** Chaud. Mon. 70; Andr. TES. 1923. 42.	Assam., Calcutta. Dacca., Puri., Bandra. Trichinopoly. Ceylon : Murunkam. Rangoon. Senmigion., Palon. Yenangyaung (Magwe).
G*	*clarescens* Bates AG. xxxii. 1892. 395; Andr. AM. (10). ii. 1928. 589.	
O*	**limbicoills** Chaud. Mon. 67.	Dacca., Pusa.
C*	**longipalpis** Wied. Germ. Mag. iv. 1821. 118; Dej. Sp. i. 1825. 314; Chaud. Mon. 87; Andr. TES. 1921. 167.	Sikkim : Tista Valley. Almora., Tanakpur. Haldwani.
G*	*concinnus* Bates AG. xxxii. 1892. 393; Andr. AM. (9). xiv. 1924. 592.	Dehra Dun. Garhwal:Jumnagwar, Boxar. Dikhala., Madras. Burma: Senmigion, Myenkyan, Rangoon.
O*	**luzonicus** Chaud. Mon. 68; Heller DEZ. 1916. 269.	Philippine Is.
P*	**Melancholicus** *(Aptinus)* Schm. Goeb. FCB. 71; Andr. TES. 1923. 40.	Bhamo.

G* *caligatus* Bates AG. xxvii. 1889. 109; id. AG. xxxii. 1892. 393.

P* **modestus** Schm. Goeb. FCB. 73; Chaud. Mon. 69; Bates AG. xxxii. 1892. 396; Andr. TES. 1923. 42. — Calcutta., Bhamo. Tharrawaddy. Prome., Toungoo. Rangoon. Moulmein. Cochin-China.

O* **orientalis** Chaud. Mon. 67. — Hongkong.

F* **pallidipes** Reitt. Ent. Blätt. 1919. 137. — Kashmir : Goorais Valley. Dehra Dun., Almora. Bashahr., Poo. Kangra. R. Sarda Gorge. Tanakpur.

stenoderus ‡ Andr. (not Bates) MB. 57; id. AM. (10). ii. 1928. 589.

X* **paviei** Lesne Bull. Mus. Paris. ii. 1896. 245; id. Miss. Pav. 79. t. 8. f. 12. — Cambodia.

O* **piceus** Chaud. Mon. 53. — Philippine Is.

L **pictus** *(Aploa)* Hope TZS. i. 1834. 92. t. 13. f. 1; Brullé HN. iv. 1834. 223; Chaud. Mon. 54; Bedel. Fn. med. 1914. 311. (note); Andr. TES. 1919. 202; id. MB. 57. — Daltonganj., Pusa. Siwaliks., Delhi. Lyallpur.

O* *figuratus* Chaud. BM. 1852. i. 41. — Nagpur., Belgaum. Poona., Trichur. Madras., Udayagiri. Bangalore. Ceylon. : Hambantota.

P* **puncticollis** Schm. Goeb. FCB 72; Chaud. Mon. 69; Bates AG. xxxii. 1892. 395; Andr. TES. 1923. 41. — Karen Hills., Palon. Kawkareik., Tonkin. Laos.

A* **reyi** Andr. MB. 108. t. 4. f. 1. — Igatpuri. North Kanara. Bandra. Madras : Surada

P* **scitulus** Schm. Goeb. FCB. 72; Andr. TES. 1923. 41. — Burma.

V* **scotomedes** Redt. RN. 5; Bates TES. 1873. 305; Chaud. Mon. 53; Heyne-Tasch. E. K. 15; Andr. TES. 1923 (1924). 460. — China. Japan.

O* **scutellatus** Chaud. Mon. 69.	Dacca.
O* **sexmaculatus** Dej. Sp. i. 1825. 312; Chaud. Mon. 63.	Throughout India and Ceylon. Tharrawaddy.
L* **sexpustulatus** (Carabus) F. Syst. Ent. 1775. 825; Andr. AM. (10). ii. 1928. 588.	Ind. Or.
A* **stevensi** Andr. MB. 107. t. 4. f. 2.	Assam : Kobo, Dihong River below Pasighat. Sikkim : Tista Valley Kalimpong. Kangra. Garhwal : Dhikala. U. P. : R. Sarda Gorge, Ranikhet. Dehra Dun.
O* **suturellus** Chaud. Mon. 69; Bates AG. xxxii. 1892. 394; (*"suturiger"*) Vuillet Ins. ii. 1912. 17.	Almora., Lucknow. Nagpur (Kangra). Chota Nagpur : Barway., Mhow. Bhowali., Belgaum. Bangalore. Tharrawaddy. Rangoon. ? Cochin-China.
O* **tetracolon** chaud. Mon. 61; Bates CRB. 1891. 336.	Dacca., Asansol. Ceylon : Vavuniya, Hambantota.
O* **tetragrammus** Chaud. Mon. 54; Andr. MB. 117.	Bengal., Pusa. Nagpur., Surada. Malabar., Belgaum. Bandra.,
O* **vigilans** Chaud. Mon. 68; Bates CRB. 1891. 336.	Chota Nagpur : Tetara Nagpur., Madras.
O* **vitticollis** Chaud. Mon. 56.	Tharrawaddy. Rangoon.

BRACHYCTIS* Chaudoir.

Brachyctis Chaud. Mémoire sur les Coptodérides, AB. xii. 1869. 252.

O* **rugulosa** Chaud. Mém. 252. — Penang., Singapore. Sumatra., Borneo. Philippine Is.

BRACHYONYCHUS* Chaudoir.

Brachyonychus Chaud.; Mon. sur les Panagéides, AB. 1878. 86.

O* **andersoni** Bates Journ. Linn. Soc. xxi. 1887. 135. — Mergui Archipelage : Elphinstone I. Kedah., Langkawi I.

O* **laevipennis** Chaud. Mon. 87. — Siam., Cochin-China.

G* var. **parumpunctatus** Bates AG. xxxii. 1892. 298. — Malewoon (Tenasserim). Siam. Malay Peninsula.

X* **perraudieri** Bates AF. 1889. 264. — Cochin-China.

G* **punctipennis** Gestro AG. 1882. 305. — Laos.

O* **sublaevis** *(Epicosmus)* Chaud. RMZ. 1869. 67; id. Mon. 89; Lesne Miss. Pav. 69. — Cochin-China. Cambodia.

O* *humeratus (Epicosmus)* Chaud. RMZ. 1869. 69; id. Mon. 89; Andr. AM. (9). xiv. 1924. 588.

BRADYBAENUS* Dejean.

Bradybaenus Dej. Sp. iv. 1829. 160; Lac. 292.

Calodromus Nietn. AM. (3). ii. 1858. 181.

O* **festivus** Dej. Sp. iv. 1829. 163; Bates AM. (5). xvii. 1886. 77; id. CRB. 1891. 333. — Panighata. Chota Nagpur : Konbir., Ajmere. Madras., Madura. Coimbatore. Mysore : Chikkaballapura. Ceylon : Colombo, Kandy, Kotte.

J* *exooratus (Calodromus)* Nietn. AM. (3). ii. 1858. 181; Heyne-Tasch. E. K. 24; Andr. AM. (9). xix. 1927. 104.

V* *ornatus* Redt. RN. 14; Andr. TES. 1923 (1924). 467.

BROSCOSOMA* Rosenhauer.

Broscosoma Rosenh., Broscosoma und Laricobius, zwei neuen Käfergattungen (Erlangen), 1846, 8 pp. 1 t.; Putz. Carab. gen. nov. 1846; Lac. 242; Putz., Less Broscides, SEZ. 1868. 353; Sem. HER. xxxiv. 1900. 77 and 80.

A* **gracile** Andr. Eos. iii. 1927. 72. f. 6. — Sikkim : Gnatong.

S* **ribbei** Putz. SEZ. 1877. 100; Andr. TES. 1922. 249. — Sikkim: Gnatong. Dzalong, Darjiling. Mungphu, Tonglu. Tonkin.

BROSCUS* Panzer.

Broscus Panz. Index Ent. 1813. 62; Lac. 239; Putz., Les Broscides, SEZ. 1868. 307; Andr. MB. 20.

Cephalotes Dej. Sp. iii. 1828. 426; Andr. TES. 1919. 91.

A* **bipilifer** Andr. Eos. iii. 1927. 71. f. 5. — Sikkim : Gnatong.

O* **punctatus** *(Cephalotes)* Dej. Sp. iii. 1828. 431; Putz. Mon. 309; Bates SYM 4; (Biology) Fletch. Agr. Res. Inst. Pusa. Bull. 89. 1919. 34; Andr. MB. 20; Sem. RRE. xx. 1926. 41. — India north of the Ganges, including the whole Himalayan tract. Egypt., Sinai. Agrabia. Mesopotamia. Turkestan., China.

B* *nepalensis (Percus)* Hope Zool. Misc. 1831. 21; Dohrn SEZ. 1879. 458; Andr. TES. 1919. 170.

K' ***limbatus*** Ball. BM. 1870. ii. 327; Chaud. BM. 1878. ii. 2; Andr. TES. 1919. 171.

Q* *batesi* Sem. HER. xxv. 1891. 276 (note); Andr. TES. 1919. 171.

O* var **davidianus** Fairm AB. 1888. 7; Andr. TES. 1919. 171. — Hongkong. Yunnan.

A* **taurulus** Andr. Eos. iii. 1927. 70. f. 4. — Sikkim : Gnatong.

BROTER* Andrewes.

Broter Andr. AM. (9). xii. 1923. 449.

A* **ovicollis** Andr. AM. (9). xii. 1923. 449. — Kodaikanal. Shembaganur.

CAELOSTOMUS* Macleay.

Caelostomus Macl. Ann. Jav. 1825. 23.

D *Drimostoma* Dej. Sp. v. 1831. 745; Lac. 313; Chaud., Essai monographique sur les Drimostomides, AB. xv. 1872. 9; Tchitch. HER. xxxii. 1898. 530 and 536.

Stomonaxus Motch. Etud. Ent. 1859. 34; Chaud., Essai monographique sur les Drimostomides, AB. xv. 1872. 13; Tchitch. HER. xxxii. 1898. 530 and 537; id. HER. xxxv. 1902. 506. *Coelostomus* Andr. MB. 44.

O	**abruptus** *(Stomonaxus)* Jord. NZ. 108.	Tenimber.
O*	**celebensis** *(Stomonaxus)* Chaud. Mon. 14.	Celebes.
O	**convexior** *(Stomonaxus)* Jord. NZ. 109.	Sumbawa.
G	**dilaticollis** *(Stomonaxus)* Bates AG. xxxii. 1892. 351.	Karen Hills.
G*	**inermis** *(Stomonaxus)* Bates AG. xxxii. 1892. 352.	Patkai Hills. Burma : Teinzo, Thagata, (Tenasserim)., Siam.
Q	**japonicus** *(Stomonaxus)* Tchitch. HER. xxxii. 1898. 14; id. HER. xxxiv. 1900. 263.	Japan.
	striaticollis (Stomonaxus) † Bates (not Dej) TES. 1873. 283.	
B*	**laeviventris** *(Stomonaxus)* † Bates TES. 1883, 290.	Japan.
O	**minor** *(Stomonaxus)* Jord. NZ. 108.	Sumbawa.
B*	**picipes** Macl. Ann. Jav. 1825. 24; Hope. Col. Man. ii. 1838. t. 2. f. 6 Andr. TES. 1919. 160; Sloane PLS. NSW. 1920. 321; Andr. MB. 44.	The whole of South Eastern Asia, excluding Japan. Australia.
N*	*rufipes (Drimostoma)* Boh. ER. 13; Tchitch. HER. xxxiv. 1900. 262; Lesne Miss. PAV. 77; Andr. TES. 1921. 180.	
O*	*striaticollis (Stomonaxus)* † Chaud. (not Dej.) Mon. 13; id. in Oberthür Col. Nov. i. 1883. 39; Bates AF. 1889. 277; id. CRB. 1891. 334 (*as strigicollis;* id. AG. xxxii. 1892. 351; Tchitch. HER. xxxiv. 1900. 264.	
Q	var. **borneensis** *(Stomonaxus)* Tchitch. HER. xxxii. 1898. 13; id., HER xxxiv. 1900. 264.	Borneo.
O*D	**rectangulus** *(Drimostoma)* Chaud. Mon. 11; id. in Oberthür Col. Nov. i. 1883. 39; Bates AG. xxvii. 1889. 106.	Buram : Bhamo, Teinzo, Shwegu, Tenasserim. Singapore., Java.
A*	**ruber** Andr. AM. (9). x. 1992. 164.	Nagpur. New Guinea.
L	**sculptipennis** *(Stomonaxus)* Motch. Etud. Ent. 1859. 35. f. 1 f. 6. (as *sculpticollis* in fig.); Tchitch. HER. xxxiv. 1900. 263 (note); Andr. TES. 1928. 22.	Ceylon : Nuwara Eliya, ? Vavuniya, ? Anuradhapura.
O	**similis** *(Stomonaxus)* Jord. NZ. 109.	Buru.

O D **subsinuatus** *(Drimostoma)* Chaud. in Oberthür Col. Nov. i. 1883. 38. — New Guinea.

CALATHOMIMUS* Bates.

Calathomimus Bates AM. (5) xvii. 1886. 77.

B* **consors** Bates AM. (5). xvii. 1886. 78. — Ceylon : Bogawantalawa.

B* **maculatus** Bates AM. (5). xvii. 1886. 77. — Ceylon : Bogawantalawa.

B* **vittatus** Andr. AM. (9). xvii. 1926. 254. — Ceylon : Woodside, Urugalla.

CALATHUS Stephens.

Calathus Steph. Ill. Brit. Ent. i. 67 and 1828. 97; Dej. Sp. iii. 1928. 62; Lac. 342; Gautier des Cottes, Mon. du genre Calathus (espèces européennes et méditerranéenn-es), SEG. ii. 1867. 235-86; Putz., Mon. des Calathides, AB. 1873. 19-96; Andr. TES. 1919. 91; id. MB. 46.

Odontonyx Steph. Ill. Brit. Ent. i. 96.

Pristosia Motch. BM. 1864. ii. 311; Andr. TES. 1928.

O* **aeneocupreus** Fairm. AF. 1886. 314. — Yunnan.

S* **amaroides** Putz. SEZ. 1877. 103; Andr. TES. 1922. 250. — Sikkim : Darjiling, Gopaldhara.

B* **atrema** Andr. EMM. 1926. 69 and 77. — Kumaon : Burphu (Gori Valley) and Gori River Gorge.

A* **clarus** Andr. MB. 101. t. 3. f. 4. — Spiti : Pulga. Kulu : Kandi, Koty. Simla Hills : Gahan, Fagu, Kotgarh, Seraj.

O **delavayi** Fairm. AB. 1887. 96. — Yunnan.

O **falsicolor** Fairm. AF. 1886. 315. — Yunnan.

O* **himalayae** Bates Entom. 1891. Suppl. 9; Andr. MB. 47; id. EMM. 1926. 69. — Kulu : Koty. Rotang Valley. Gori River Gorge, Kali Valley and Dhauli Ganga (Almora).

V*	**kollari** × Putz. Mon. 72; Bates CRB. 1891. 334; Andr. MB. 46; id. EMM. 1926. 69. *angustatus* || Redt. (not Ramb) HK. 500; Bates SYM. 11.	The Whole Himalayan tract. Chota Nagpur : Tetara. Baluchistan: Nushki.
O	**lateritius** Fairm. AF. 1886. 314.	Yunnan.
W*	**melanocephalus** *(Carabus)* L. Syst. Nat. Ed. x. 1758. 415; Bates SYM. 11.	Pamir., Europe. Siberia., Persia. Turkestan.
O	**nubilipennis** Fairm. AB. 1888. 13.	Yunnan.
O*	**pectiniger** Putz. Mon. 86.	
M*	*piceus (Pristosia)* Motch. BM. 1864. ii. 312; Putz. AB. 1873. 91; Bates TES. 1883. 253; Andr. TES. 1928. 11.	"North India."
O	**strigipennis** Fairm. AB. 1888. 12.	Yunnan.

CALLEIDA* Latreille et Dejean.

Callcida Latr. et. Dej. HN. et Ic. 132; Dej. Sp. i. 1925. 220; Schm Goeb. FCB. 32; Lac. 105.

Callida Chaud., Mon. des. Callidides, AB. xv. 1872. 103.

O*	**chloroptera** Dej. Sp. v. 1831. 340; Schm. Goeb. FCB. 33; Chaud. Mon. 112; Andr. TES. 1923. 11.	Tenasserim., Java. Sumatra.
V*	**var.** lepida Redt. RN. 6; Chaud. Mon. 112; Bates TES. 1873. 317; Andr. TES. 1923. 12; id. TES. 1923 (1924). 461.	China. Japan.
S	**cyclops** Matsumura Mém. Soc. Ent. Belg. xviii. 1911. 147.	Formosa.
Y*	**discoidalis** Heller PJS. xix. 1921. 529.	Philippine Is.
G*	**doriae** Bates AG. xxxii. 1892. 422.	Bhamo. Assam : Noa Dehing.
O*	**femoralis** Chaud. Mon. 112.	Dacca.
O*	**lativittis** Chaud. Mon. 113.	Dacca.
G*	*excelsa* Bates AG. xxxii. 1892. 422; Andr. TES. 1921. 150.	Sibsagar., Naga Hills. Purnea., Nainital. Haldwani. Burma : Karen Hills, Rangoon, Senmigion.

B*	**onoha** Bates TES. 1873. 317.	Japan.
O*	**permunda** Chaud. Mon. 113.	Celebes.
X*	**propinqua** Fleut. AF. 1887. 59; Bates AF. 1889. 283.	Annam., Tonkin.
? C*	**splendidula** *(Carabus)* F. Syst. Eleuth. i. 1801. 184; *(Lebia)* Macl. Ann. Jav. 1825. 26; Dej. Sp. v. 1831. 341; Schm. Goeb. FCB. 32; Motch. Et. Ent. 1855. 51; Chaud. Mon. 113; Bates AF. 1889. 283; id. AB. 1892. 233; Lesne Miss. Pav. 81; Maindr. AF. 1905. 334; Andr. TES. 1919. 164; id. TES. 1921. 161; Heller PJS. xix. 1921. 530; Andr. TES. 1923. 11.	South East Asia, excluding Japan and New Guinea.
M.	var. **rnbricata** Motch. BM. 1864. ii. 238; Chaud. Mon. 113; Bates AF. 1889. 283; Gardn. Ind. For. Rec xiii. 1927. 65. t. 5. ff. 1-5 (larva); Andr. TES. 1928. 22.	As type form.
G	**sultana** Bates AG. xxxii. 1892. 421.	Garo Hills. Karen Hills.
B*	**terminata** Waterh. TES. 1876. 11; Andr. TES. 1919. 197.	Borneo.

CALLISTOMIMUS* Chaudoir.

Callistomimus Chaud. BM. 1872. i. 382; Andr. PZS. 1921. 233; id. MB. 23.

Pristomachaerus Bates TES. 1873. 323; id. AG. xxxii. 1892. 304; Chaud. AB. 1878. 85.

O*	**acuticollis** *(Callistus)* Fairm. AF. 1889. 6; Andr. PZS. 1921. 237 and 242.	Yunnan.
X*	*yunnanus (Pristomachaerus)* Maindr. BF. 1909. 275.	
A*	**belli** Andr. PZS. 1921. 237 and 241. t. 1. f. 8.	Kadra (North Kanara).
G*	**cauliops** *(Pristomachaerus)* Bates AG. xxxii. 1892. 306; Andr. PZS 1921. 236. t. 1. f. 3.	Palon.
C*	**chalcocephalus** *(Panagaeus)* Wied. Zool. Mag ii. 1. 1823. 57; Chaud. BM. 1861. ii. 351; id. AB. 1878. 84; Bates SYM. 4; Andr. PZS. 1921. 236 and 238. t. 1. f. 1. ; id. TES. 1921. 171; id. EMM. 1926. 68.	Jhelum Valley. Kangra : Baijnath. Sarju and East. Ramganga Valleys. Almora. Haldwani. Dehra Dun.

O*	*messii (Pristomachaerus)* Bates TES. 1873. 324.	Sikkim: Gopaldhara. Nilgiri Hills. Maymyo., Java. Krakatau., China.
G*	var. **quadristigma** *(Pristomachaerus)* Bates AG. xxxii. 1892. 303; Andr. PZS. 1921. 236 and 238.	Bhamo., Palon. Shwegu., Rangoon. Cochin-China., Laos.
V	**chlorocephalus** *(Panagaeus)* Koll. Ann. Wien. Mus. i. 1835. 335. t. 31. f. 4; Chaud. BM. 1861. ii. 351; id. AB. 1878. 84; Andr. PZS. 1921. 238; id. TES. 1921. 171.	? India.
O*	**coarctatus** *(Callistus)* Laf. AF. 1851. 230; Andr. PZS. 1921. 237 and 246. t. 1. f. 12. a-b; id. TES. 1921. 146; id. MB. 23; id. EMM. 1926. 68.	Kulu : Manali U. P. Almora, Ranikhet, Upper Gumti Valley, Sarju Valley, Khairna, Malwa Tal, Siwalik Hills, Kanarichina. Eastern Duars. Sikkim : Gopaldhara, Namsoo, Darjiling, Kalimpong, Pashok, Mungphu, Kurseong. Spiti : Manikarn.
A*	**d'abreui** Andr. PZS. 1921. 237 and 242 t. 1. f. 9.	Sitapur (U. P.) Pusa., Mandla.
B*	**dicksoni** Waterh. AM. (5). xiv. 1884. 429; Andr. TES. 1919. 197; id. PZS. 1921. 236.	Formosa.
A*	**dux** Andr. PZS. 1921. 237 and 244. t. 1. f. 10.	Prome.
A*	**jucundus** Andr. PZS. 1921. 236 and 239. t. 1. f. 2.	North Kanara. Nilgiri Hills.
G*	**lebiodes** Bates AG. xxxii. 1892. 306; Andr. PZS. 1921. 237.	Naga Hills. Karen Hills.
L	**littoralis** *(Callistus)* Motch. Et. Ent. 1859. 33; Andr. PZS. 1921. 237 and 245. t. 1. f. 11; id. TES 1921. 146; id. TES. 1928. 22.	Chota Nagpur : Tetars, Barway. Bhopal., Nagpur.
J*	*westwoodi (Callistus)* Schaum BEZ. 1863. 85; Andr. AM. (9). xix. 1927. 108.	Jubbulpur., Raipur. Bhandara.

O* *coarctatus* Chaud. † (not Laf.) BM. 1872. i. 383; Bates CRB. 1891. 327.	Hoshangabad. Mandla., Belgaum. Mysore : Nandidrung. Madras : Malabar. Tranquebar, Madura, Genji, Trichinopoly.
D* var. **ceylonicus** Dupuis AB. 1913. 6; Andr. PZS. 1921. 237 and 245.	Ceylon : Anuradhapura.
J* **modestus** *(Callistus)* Schaum BEZ. 1863. 85; Bates TES. 1873. 246; id. AG. xxxii. 1892. 303 (part); Andr. PZS. 1921. 237 and 243; id. TES. 1923 (1924). 468; id. AM. (9). xix. 1927. 108.	Gokteik., Bhamo. Kawkareik., Palon. China. Japan.
A* var. **humeralis** Andr. PZS. 1921. 237 and 243.	Haldwani., Calcutta.
modestus † Fairm. AF. 1888. 336; Bates AG. xxxii. 1892. 303. (part).	Sardah., Palon. Tonkin., Laos.
A* var. **insularis** Andr. PZS. 1921. 237 and 244.	Java.
X* **nair** *(Pristomachaerus)* Maindr. BF. 1909; 276; Andr. PZS. 1921. 237 and 244.	Mahe., Nilgiri Hills.
A* **var.** nilgirinus Andr. PZS. 1921. 237 and 244.	Nilgiri. Hills.
S* **quadricolor** *(Pristomachacrus)* Putz. SEZ. 1877. 101; Andr. PZS. 1921. 247; id. TES. 1922. 249.	Sikkim : Darjiling. Gopaldhara. Karen Hills.
G* *eucharis (Pristomachaerus)* Bates AG. xxxii. 1892. 305; Andr. PZS. 1921. 237 and 247.	Bhamo. Palon.
S* **quadriguttatus** *(Pristomachaerus)* Putz. SEZ. 1877. 101; Andr. PZS. 1921. 238; id. TES. 1922. 249.	Nainital. Dehra Dun.
G* **rubellus** *(Pristomachaerus)* Bates AG. xxxii. 1892. 305; Andr. PZS. 1921. 236. t. 1. f. 4.	Naga Hills. Karen Hills.
A* **sikkimensis** Andr. PZS. 1921. 236 and 238. t. 1. f. 5.	Sikkim : Gopaldhara.
B* **subnotatus** Andr. PZS. 1921. 236 and 239.	Cambodia.
X* **suturalis** Fleut. AF. 1887. 61. t. 4. f. 2; Bates AF. 1889. 265; Andr. PZS. 1921. 236.	Annam.
B* **venustus** Andr. PZS. 1921. 236 and 240. t. 1. f. 6.	Laos.
A* **virescens** Andr. PZS. 1921. 236 and 241. t. 1. f. 7.	Gokteik., Kyanktan. Tharrawaddy.
B* **vitalisi** Andr. PZS. 1921. 237 and 246.	Annam., Tonkin.

CALOSOMA* Weber.

Calosoma Weber Obs. Ent. i. 1801. 20; F. Syst. Eleuth. i. 1801. 211; De. Sp. ii. 1826. 190; Lac. 58; Reitt., Bestimmungs Tabelle der eur. Col., xxxiv. Carabini 1896. 46; Andr. MB. 6; Breuning, Mon. der Gattung Calosoma (i), Kol. Rundsch. 1927. 129; Andr. FBI. 49.

Charmosta Motch. BM. 1865. ii. 301.

Caminara Motch. BM. 1865. ii. 303.

Campalita Motch. BM. 1865. ii. 304.

Ctenosta Motch. BM. 1865. ii. 306.

Callipara Motch. BM. 1865. ii. 308.

B*	**beesoni** Andr. AM. (9). iv. 1919. 290; Beeson Ind. For. 1925. 570. t. 31. ff. 1. and 6; Gardn. Ind. For. Rec. xiii. 1927. 63 t. 5. ff. 6—10 (larva); Breun. Mon. 139, 147 and 175; Andr. FBI. 21. f. 6a, 27. f. 7A, 51 and 52. f. 9.	Jaunsar; Molta, Mondali, Kathian, Bodyar. Chakrata., Simla. Chamba.
B*	**chinense** Kirby TLS. 1818. 379; Dej. Sp. v. 1831. 563; Putz. CRB. 1875. 45; Bates TES. 1883. 232; id. Entom 1890. 212; Reitt. Best. Tab. 51; Roeschke Ent. Nachr. 1900. 58 and 60; Maindr. BF. 1906. 225; Lapouge Bull. Soc. Sci. Med. Ouest 1908. 159 (larva); Andr. TES. 1919. 130; Sem. RRE. xx. 1926. 38; Breun. Mon. 140 and 217.	Manchuria. Siberia. Korea. Japan. China. Yunnan.
L	*aeneum* Motch. BM. 1859. ii. 489; Andr. TES. 1928. 22.	
O	**davidis** Géhin Cat. Carab. 1885. 61; Breun. Mon. 140, 149, and 226.	China. Yunnan.
O*	*thibetanum* Fairm. AB. 1887. 92; Bates Entom. 1890. 212; Heyne-Tasch. E. K. 12. t. 2. f. 18.	
G*	**himalayanum** Gestro AG. vii. 1875. 851; Breun. Mon. 139 and 174; Andr. FBI. 51 and 55.	Srinagar., Ladak. Sindh Division (Kashmir). Chamba., Khalatse. Bashahr : Poo.
J*	**imbricatum** Klug in Symb. Phys. Decas iii. 1832. t. 23. f. 11; Heyne Tasch. EK. 11; Roeschke Ent. Nachr. 1900. 61; Andr. TES. 1921. 145; id. AM. (9). xix. 1927. 100; id. Ent. Mitt. 1927. 142; Breun.	Karachi., Basra. Ashtola Island (Persian Gulf). Hejaz.

	Mon. 140, 149 and 221; Alluaud AG. lii. 1928. 69; Andr. FBI. 51 and 63.	Abyssinia., Teneriffe. Cape Verde Is.
? M	*arabica (Caminara)* Motch. BM. 1865. ii. 304.	
	var. **andrewesi** Breun. Mon (iii), Kol. Rundsch. 1928. 95; Andr. FBI. 51 and 64.	Assam. Bengal.
	orientale † Chaud. (not Hope) AF. 1869. 368; Breun. Mon. 140 and 223.	
B*	[**maderae** *(Carabus)* F. Syst. Ent. 1775. 237; Dej. Sp. v. 1831. 563; Mayet BF. 1887. 171 (larva); Reitt. Best. Tab. 50; Roeschke Ent. Nachr. 1900. 57 and 60; Lapouge Bull. Soc. Sci. Med. Ouest. 1908. 156 (larva); Andr. TES. 1919. 171; id. TES. 1921. 153; Breun. Mon. 140, 149 and 204; Andr. FBI. 56.	Palaearctic region.
C*	*indagator (Carabus)* F. Mant. Ins. i. 1787. 197; Andr. TES. 1921. 154; Breun. Mon. 140 and 210.]	
L	var. **auropunctatum** *(Carabus)* Herbst in Arch. Ins.-gesch. (Fuessly) v. 1784. 131; Payk. Mon. Carab. 1790. 68; Dej. et Boisd. Ic. et HN. ii. 51. t. 70. f 4; Reitt. Best. Tab. 49; Roeschke Ent. Nachr. 1900. 60; Lapouge Bull. Soc. Sci. Med. Ouest 1908. 158 (larva); Burgess and Collins U. S. Dept. Agric. Bull. 417. 1917. 107 (biology); Andr. MB. 6; id. AM. (9). xix. 1927. 98; Breun. Mon. 140 and 211; Andr. FBI. 51 and 58.	Palaearctic region. Kashmir: Sonemarg. Konain (Chakrata). Bannu., Chamba. Rotang Valley. Lahoul: Samdeo, Patseo, Sisu, Keiling, Cimur.
K	*sericeum (Carabus)* F. Ent. Syst. i. (1). 1792. 147; Schiaedte Nat. Tidsskr. 1867. 480. t. 16. ff. 15—18 (larva); Bates Entom. 1891. Suppl. 8; Andr. TES. 1921. 156.	
I*	*orientale* † Bates (not Hope) SYM. 3.	
L	var. **indicum** Hope Zool. Misc. 1831. 21; Roeschke Ent. Nachr. 1900. 59. and 60; Fletch. Agr. Res. Inst. Pusa. Bull. 89. 1919. 33. f. 26 (biology) Andr. TES. 1919. 171; id. MB. 6; Gardn. Ind. For. Rec. xiii. 1927. 64. t. 5. f. 11 (larva); Breun. Mon. 140 and 215; Andr. FBI. 29. f. 8a, 51 and 61. f. 10.	Kashmir ; Srinagar. Yusimarg, Markand, Gandarbal, Upper Munda, Gulmarg, Lidar Valley, Jhelum Valley., Peshawar. Taru. Kohat., Kangra. Phulga (Kulu). Spiti : Tsho-ti. Kumaon., Bhatkot. Dehra Dun., Calcutta. Chapra., Pusa.
O*	*scabripenne* Chaud. AF. 1869. 371; Bates Entom. 1891. Suppl. 8; Andr. TES. 1921. 145.	
A	*nivale* Breun. Mon. 140 and 216.	

A	var. **kashmirense** Breun. Mon. 140 and 216; ? Lapouge Bull. Soc. Sci. Med. Ouest. 1908. 158 (larva); Andr. FBI. 51 and 60.	Kashmir.
A	var. **yunnanense** Breun. Mon. 140 and 217.	Yunnan.
Q	**maximowiczi** Moraw. Beitr. 20. Bates TES. 1873. 234; id. TES. 1883. 232; Heyd. DEZ. 1886. 282; Kraatz. DEZ. 1896. 331; Reitt. Best. Tab. 48; Breun. Mon. 139, 147, and 176.	Siberia., Korea. Japan. China. Formosa.
B*	*mikado* Bates TES. 1873. 235; id. TES. 1883. 232.	
A	var. **sauteri** Born. Soc. Ent. xxiv. 1909. 99; Breun. Mon. 139 and 178.	Formosa.
A	var. **touzalini** Lapouge Misc. Ent. xxviii. 34 (42 in text).	Yunnan.
L	**nigrum** Parry TES. iv. 1845. 85; Breun. Mon. 140, 215 and 218; Andr. FBI. 69.	Khasi Hills.
L	**oceanicum** Perr. Ann. Soc. Linn. Lyon. xi. 1864. 49. t.1.f.1; Heyne-Tasch. EK. 11; Heller Nov. Caledon Zool. ii. 1916. 237; Breun. Mon. 139. 146, and 157.	Australasia.
O*	var. **timorense** Chaud. AF. 1869. 367; Breun. Mon. 139 and 358.	Timor. Wetar.
A	var. **klynstrai** Breun. Mon. 139 and 158.	Tenimber.
O*	**olivieri** Dej. Sp. v. 1831. 559; Mayet BF. 1887. 173 (larva); Bedel Fn. med. 1895. 18 and 20; Lapouge Bull. Soc. Sci. Med. Ouest 1908. 162 (larva); Alluaud BF. 1918. 251; Andr. TES. 1921. 145; id. Ent. Mitt. 1927. 143; Breun. Mon. 140, 149, and 224; Alluaud. AG. Lii. 1928. 68; Andr. FBI. 52 and 65.	Peshawar., Taru. Baluchistan : Nushki, Quetta. Persia., Irak. Transcaspia. Asia Minor. North Africa. Cape Verde.
L	*azoricum* Heer, Ueber fossil. Calos. in Leonhard and Bronn N. Jahrb. Mineral. (Stuttgart). 1861. 55; Woll. Cat. Col. Canar. 1864. 4; Reitt. Best. Tab. 49.	Cape Verde Is. Azores., Canaries.
I*	*orientale* ‡ Cotes (not Hope) BNHS. vi. 1891. 416.	
L	**orientale** Hope TZS. i. 1834. 92; Roeschke Ent. Nachr. 1900. 61; Andr. TES. 1919. 202; id. TES. 1921. 145; Breun. Mon. 184; id. Mon. (iii). Kol. Rundsch. 1928. 95; Andr. FBI. 52 and 67. f. 11.	Bengal., Chapra. Khandwa., Poona. Nasik., Coimbatore.

O* *squamigerum* Chaud. AF. 1869. 368; Breun. Mon. 139, 147, 184 and 223. — Kodaikanal. Madura., Manaparai.

CAPHORA* Schmidt-Goebel.

Caphora Schm. Goeb. FCB. 91; Lac. 309; Chaud. Etude monographique des Masoréides, des Tetragonodérides et du genre Nematotarsus, BM. 1876. ii. 8.

P* **humilis** Schm. Goeb. FCB. 91. t. 3. f. 8; Chaud. Mon. 8; Andr. TES. 1923. 53. — Bengal., Purnea. Dehra Dun. Moulmein. Tenasserim. Siam., Java.

CARABUS Linnaeus.

C. *Carabus* L. Syst. Nat. Ed. X. 1758. 413; Dej. Sp. ii 1826. 30; Lac. 54; Reitt., Bestimmungs-Tabelle Eur. Col., xxxiv. Carabini. 1896. 54; Lapouge, Carabes nouveaux ou mal connus, Misc. Ent. xxi. xxvi. xxviii. xxx. 1913-28. (sep. pag.) ; Andr. FBI. 70.

A. *Apotomopterus* Hope Col. Man. ii. 1838. 48; Reitt. Best. Tab. 170.

Co. *Coptolabrus* Solier in Baudi et Truqui Stud. Ent. i. 1848. 58; Reitt. Best. Tab. 124.

P. *Paraplesius* Moraw. MAP. xxxiv. 9. 1886. 51; Reitt. Best. Tab. 125.

I. *Imaibius* Bates PZS. 1889. 211; Kraatz DEZ 1895. 363; Reitt. Best. Tab. 94; Roeschke Ent. Nachr. 1900. 58; id. DEZ. 1907. 542; Andr. MB. 7. *Tropidocarabus* Kraatz DEZ. 1895. 366; Reitt. Best. Tab. 95; Roeschke Ent. Nachr. 1900. 58.

M. *Meganebrius* Kraatz. DEZ. 1895. 366; Reitt. Best. Tab. 192; Andr. AM. (9). vii. 1921. 409.

Ps. *Pseudocoptolabrus* Reitt. Best. Tab. 95.

L. *Laocarabus* Lapouge Misc. Ent. 1916. 86 (sep. pag.).

Ind. *Indocarabus* Lapouge Misc. Ent. 1921. 119 (sep. pag.).

O*I **barysomus** Bates PZS. 1889. 210; Kraatz. DEZ. 1895. 364. f. 369; Reitt. Best. Tab. 94; Roeschke DEZ. 1907. 544 and 547; Lapouge Bull. Soc. Sci. Med. Ouest. 1907. 37 (larva); Andr. FBI. 73 ad 87. ff. 14 and 14a. — Kashmir : Goorais VAlley, Batote, Pir Panjal. Hazara : Lower Kagan Valley. Agra (Ind. Mus.)

A*	var **hügeli** Roeschke DEZ. 1907. 548; Lapouge Bull. Soc. Sci. Med. Ouest. 1907. 39 (larva); Andr. FBI. 73 and 90.	Mountains north east of Islamabad.
	var. **heroicus** × Sem. RRE. 1906. 395; Andr. FBI. 73 and 91. f. 15.	Mountains west of Pir Panjal and north of Punch.
Q	*colossus* || Sem. (not Blenze) RRE. 1906. 266.	
A*	*caroli* Roeschke DEZ. 1907. 530, 549 and 559.	
B*	**birmanus** Andr. FBI. 73 and 85.	Ruby Mines.
B*	**I boysi** Tatum AM. (2). viii. 1851. 51; Bates Entom. 1891. Suppl. 8; Roeschke Ent. Nachr. 1900. 62; id. DEZ. 1907. 546 and 553; Andr. TES. 1919. 181; id. MB. 7; id. EMM. 1926. 66; id. FBI. 74 and 95.	Southern Kashmir. Kulu : Kandi. Spiti : Tsho-ti, Tchary Djoni, Pulga. Chamba., Simla. Bashahr.
D*	*wallichi* † Kraatz (not Hope) DEZ. 1895. 366; Reitt. Best. Tab. 95.	Garhwal : Balcha Tehri, Lohba. U. P. : Almora, Samgong, Mundali, Deoban, Bodyar, Dudhatoli, Sunderdhunga, Valley.
V*I.	**caschmirensis** Redt. HK. 499. t. 23. f. 4; Bates SYM. 3; Kraatz. DEZ. 1895. 366; Reitt. Best. Tab. 94; Roeschke Ent. Nachr. 1900. 58 and 62; id. DEZ. 1907. 544 and 549; Lapouge Bull. Soc. Sci. Med. Ouest. 1908. 168 (larva); Fowl. Fauna. Br. India. Col. Gen. Intr. etc., 1912. 54. f. 24; Andr. FBI. 74 and 91. f. 16.	Southern Kashmir. Murree., Simla Kulu., Garhwal. U. P. : Mussoorie, Nainital Kathian, Kanasar, Binsar, Deola.
B*	*lithariophorus* Tatum AM. xx. 1847. 14; Andr. TES. 1919. 181.	
O*I	**dardiellus** Bates PZS. 1889. 211; Roeschke DEZ. 1907. 546 and 556; Andr. FBI. 75 and 97.	Kashmir : Goorais Valley, Radjawar, Gulmarg, Nagmarg (Lolab Valley), Songam, Dusu, Batote.
D*	*opacus* Kraatz. DEZ. 1895. 367; Reitt. Best. Tab. 95.	
O*	var. **granulisparsus** Bates PZS. 1889. 211; Roeschke DEZ. 1907. 559; Andr. FBI. 75 and 99.	Kashmir: Goorais Valley Pir Panjal, Gulmarg.

D* var. **klaegeri** Kraatz DEZ. 1895. 368; Reitt. Best. Tab. 95; Roeschke. DEZ. 1907. 558; Andr. FBI. 75 and 98.	Kashmir: Radjawar, Pir Panjal.
A* var. **subpunctulus** Roeschke DEZ., 1907. 558; Andr. FBI. 75 and 98.	Kashmir: Eastern Kishtwar, Panjal, Tanin (Lidar Valley).
O* A **delavayi** Fairm. AF. 1886. 308; Reitt. Best. Tab. 169; Born Soc. Ent. xxxiii. 1918. 11.	Yunnan.
A *talianus* Hauser Soc. Ent. xxxiv. 1919. 25; Breun. Kol. Rundsch. 1926. 69.	
V P. **dokhtouroffi** Gangl. HER. xx. 1886. 268; Reitt. Best. Tab. 126.	Pamir.
D* A. **eccoptopterus** Kraatz DEZ. 1894. 138. t. 1. ff. 17—19; Reitt. Best. Tab. 170; Maindr. BF. 1906. 217; Born. Soc. Ent. xxxiii. 1918. 10.	China. Yunnan.
Q I. **epipleuralis** Sem. RRE. 1906. 268; Roeschke DEZ. 1907. 546, 554 and 559; Andr. FBI. 74 and 96.	Mountains west of Pir Panjal and north of Punch.
B* **everesti** Andr. FBI. 72 and 76.	Tibet: Rongshar Valley
G*A. **feae** Gestro AG. 1888. 106; Bates AG. xxxii. 1892. 269; Breun. Kol. Rundsch. 1926. 70; Andr. FBI. 73 and 84.	Kachin.
O Co. **gemmifer** Fairm. AB. 1887. 91; Meyer. Darcis SEG. 1902. 407; Born. Soc. Ent. xxv. 1910. 74.	Yunnan.
A var. **cyaneofemoratus** Born: Soc. Ent. xxiv. 1910. 185; id. Soc. Ent. xxv. 1910. 74.	Yunnan.
A var. **yunnanus** Born. Soc. Ent. xxv. 1910. 74.	Yunnan.
W*C. [**granulatus** L. Faun. Suec. 1761. 217; Moraw. Beitr. 12.]	Palaearctic region.
A var. **luctuosus** Lapouge Misc. Ent. 1924. 188.	Tonkin.
O*M. **indicus** Fairm. BF. 1889. 15; Reitt. Best. Tab. 192; Roeschke Ent. Nachr. 1900. 59; Andr. AM. (9). vii. 1921. 409; id. FBI. 4. f. 1, 7. ff. 1 and 2, 12. f. 3a, 72 and 78.	Sikkim : Darjiling, Mungphu, Tonglu, Gopaldhara.
A A. **infirmior** Hauser Soc. Ent. xxxix. 1924. 6.	Yunnan.
X A. **kouanping** Maindr. BF. 1906. 217; Born. Soc. Ent. xxxiii. 1918. 10.	Yunnan.

O	A. **protenes** Bates PZS. 1889. 217; id. Entom. 1890. 210; Maindr. BF. 1906. 216; Born. Soc. Ent. xxxiii. 1918. 10.	China Yunnan.
	protensus Reitt. Best. Tab. 170.	
X	Co. **pustulifer** Lucas BF. 1869. 10; id. AF. 1872. 293. t. 14. f. 12; Bates PZS. 1889. 217; id. Entom. 1890. 211; Heyne-Tasch. EK. 10. t. 2. f. 10; Reitt. Best. Tab. 124; Maindr. BF. 1906. 225.	China. Yunnan.
A	var. **guerryi** Born. Soc. Ent. xxv. 1910. 73.	
X	*rectiperlatus* Maindr. BF. 1906. 225; Born. Soc. Ent. xxxiii. 1918. 10.	Yunnan.
Q	I. **rostianus** Sem. RRE. 1906. 267; Roeschke DEZ. 1907, 545, 550, and 559; Andr. FBI. 74 and 93.	Kashmir : Kishtwar, Songam, Dusu, Tanin (Lidar Valley).
A*	Ind. **sanchari** Andr. AM. (9). vii. 1921. 407; id. FBI. 73 and 82. f. 13.	Sikkim : Gnatong and "beyond Jalep."
A	*wagae* † Lapouge (not Fairm.) Misc. Ent. 1921. 118 (sep. pag.); Andr. EMM. 1926. 66; Breun. Kol. Rundsch. 1926. 67.	
D*	A. **sauteri** Roeschke Suppl. Ent. Berl. i. 1912. 4.	Formosa.
I*	I. **stoliczkanus** Bates PZS. 1878. 713; id. SYM. 3. t. 1. f. 13; Reitt. Best. Tab. 94; Roeschke Ent. Nachr. 1900. 62; id. DEZ. 1907. 546 and 552; Andr. FBI. 74 and 94.	Murree. Kashmir : West of Pir Panjal.
O	Ps. **taliensis** Fairm. AF. 1886. 308; Reitt. Best Tab. 95.	Yunnan.
A	L. **vitalisi** Lapouge Misc. Ent. 1916. 85 (sep. pag).	Laos.
O*	**wagae** Fairm. AF. 1882. 65; Andr. EMM. 1926. 66. and 70; id. FBI. 73 and 80. f. 12.	North India (in description), Tibet (on label). Kumaon : Girthi Valley.
B*M.	**wallichi** Hope Zool. Misc. 1831. 21; Bates CRB. 1891. 324; Reitt. Best. Tab. 95; Roeschke Ent. Nachr. 1900. 59; Andr. TES. 1919. 171; id. AM. (9). vii. 1921. 409; id. FBI. 72 and 75.	"Nepal."
A	A. **yunnanensis** Born Soc. Ent. xx. 1905. 33.	Yunnan.
G'	*konanti* Maindr. BF. 1906. 215; Born. Soc. Ent. xxxiii. 1918. 10.	

G' var. **modestior** Maindr. BF. 1906. 216 (note). — Yunnan.

O A. **yunnanus** Fairm. AF. 1886. 309; Reitt. Best. Tab. 169; Hauser Arch. Naturg. Abt. A. 1922. H. 7. 109; Breun. Kol. Rundsch. 1926. 69. — Yunnan.

G' *albrechti* ‡ Maindr. (not Moraw). BF. 1906. 216; Born. Soc. Ent. xxxiii. 1918. 10.

A var. **bornianus** Hauser Arch. Naturg. Abt. A 1922. H. 7. 108; Breun. Kol. Rundsch. 1926. 69. — Yunnan.

CATADROMUS* Macleay.

Catadromus Macl. Ann. Jav. 1825, 18; Dej. Sp. iii. 1828. 186; Tchitch. Ann. Mus. Zool. St. Petersb. 1896. 141; id. HER. xxxv. 1902. 507; Sloane PLS. NSW. 1920. 152.

V **malayanus** Tchitch. Ann. Mus. Zool. St. Petersb. 1896. 146. — Singapore.

L **tenebrioides** *(Carabus)* Oliv. Enc. Méth. V. 1790. 324; id. Ent. 35. 17. t. 6. f. 67; Mac. Ann. Jav. 1825. 18. t. 1. f. 5; Dej. Sp. iii. 1828. 187; Dej. Et Boisd. Ic. et HN. ii. 397. t. 124. f. 2; Brullé HN. iv. 1834. 331; Heyne-Tasch. EK. 24; Tchitch. Ann. Mus. Zool. St. Petersb. 1896. 148; Andr. TES. 1919. 148. — Java. Amboina. Australia.

C* *rajah (Harpalus)* Wied. Anal. Ent. 1824. 7; Andr. TES. 1921. 174.

B* *elseyi* White PZS. 1859. 117. t. 58. f. 3; Sloane PLS. NSW 1920. 321.

CATASCOPUS* Kirby.

Catascopus Kirby TLS. xiv. 1825. 94; Latr. et Dej. HN. et. Ic. 115; Macl. Ann. Jav. 1825. 14; Dej. Sp. i. 1825. 328; Schm. Geob. FCB. 80; Lac. 145; Chaud. BEZ. 1861. 116; id. RMZ. 1872. 244; Andr. MB. 62.

O* **aculeatus** Chaud. BEZ. 1861. 121; Schauf. HER. xix. 1885. 184. — Celebes.

L *rugicollis* Saund. TES. 1863. 464. t. 18. f. 6; Chaud. RMZ. 1872. 249; Andr. AM. (9). iii. 1919. 482; id. AM. (9). xvii. 1926. 259; id. AM. (9). xix 1927. 107 (malformation).

J*	**aeneus** Saund. TES. 1863. 467. t. 17. f. 2; Chaud. RMZ. 1872. 249; Heyne-Tasch. EK. 15; Andr. TES 1921. 149; id. AM. (9). xvii. 1926. 259; id. AM. (9). xix. 1927. 107.	Sarawak.
O*	**aequatus** Dej. Sp. v. 1831. 452; Chaud. BEZ. 1861. 121; Putz. CRB. 1875. 45; Heyne-Tasch. EK. 16; Heller DEZ. 1916. 273.	Philippine Is.
O*	**agnatus** Chaud. RMZ. 1872. 244.	Moluccas.
O*	**andamanensis** Chaud. BM. 1877. i. 200.	Andaman and Nicobar Is.
O*	**aruensis** Saund. TES. 1863. 458. t. 17. f. 5; Chaud. RMZ. 1872. 249. *cupricollis* || Chaud. (not Waterh) in Oberthür Col. Nov. i. 1883. 24; Andr. AM. (9). xiv. 1924. 593.	Aru Is. New Guinea
	aeneicollis × Andr. AM. (9). iii. 1919. 481.	
A*	**bellus** Andr. AB. 1921. 207.	North Kanara. Nilgiri Hills.
O*	**brachypterus** Chand. BEZ. 1861. 119; Saund. TES. 1863. 468.	Sarawak., Malacca. Singapore.
A*	**bryanti** Andr. AB. 1921. 209.	Penang.
B*	**cingalensis** Bates AM. (5). xvii. 1886. 203; Andr. MB. 117.	Chota Nagpur: Tetara., Mhow.
O*	*reductus* † Chaud. (not Walk.) BEZ. 1861. 117; id. RMZ. 1872. 245; Andr. TES. 1919. 185; id. TES. 1921. 149.	Surada., Nilgiri Hills. Mysore : Chikkaballapura.
E*	*severini* Bates CRB. 1891. 339; Andr. AB. 1921. 202.	Ceylon ; Kandy, Balangoda, Bentote, Kotte, Anuradhapura, Galgamuwa, Katupota,[1] Digandenia, Polpitya.
B*	**cupripennis** *(Pericalus)* Thoms. Arch. Ent. i. 1857. 282; Chaud. BEZ. 1861. 122; Heyne-Tasch. EK. 15. t. 2. f. 34; Bouch. AF. 1903. 175; Andr. TES. 1919. 182.	Penang., Perak. Malacca. Singapore. Sumatra. Borneo. Celebes.

O*	**cyanellus** Chand. BM. 1848. i. 113; id. BEZ. 1861. 118.	"Nepal." Dehra Dun., Dapoli. North Kanara. Coimbatore.
O*	**diffinis** Chaud. RMZ. 1872. 244.	Celebes., Khasi Hills.
	elegans *(Elaphrus)* Weber Obs. Ent. 1801. 45; *(Carabus)* F. Syst. Eleuth. i. 1801. 184; Hope Col. Man. ii. 1838. 78; Schaum SEZ. 1847. 45; Chaud. BM. 1850. i. 354; Motch. Etud. Ent. 1855. 47; Chaud. BEZ. 1861. 120; id. RMZ. 1872. 247; Bates AG. xxxii. 1892. 411; Heller DEZ. 1916. 273; Sloane PLS. NSW. 1920. 322; Andr. TES. 1921. 161; id. TES. 1923. 48.	Sunderbans., Insein. Moulmein. Victoria Point. Andaman and Nicobar Is. Indo-China., Siam. The entire Malay region., N. Australia. Solomon Is. New Britain.
L	*nitidulus* Cast. Etud. Ent. 1834. 60.	
P*	*elevatus* Schm. Geob. FCB. 84; Chaud. BEZ. 1861. 122; Andr. TES. 1923. 48 (malformation).	
O*	*amoenus* Chaud. BEZ. 1861. 120; Saund. TES. 1863. 468; Chaud. RMZ. 1872. 247.	
X*	var **lateralis** Brullé HN. iv. 1834. 233. f. S. f. 3; Chaud. RMZ. 1872. 250.	Celebes. Australia.
B*	*celebensis (Pericalus)* Thoms. Arch. Ent. i. 1857. 282; Chaud. BEZ. 1861. 120; Schauf. HER. xix. 1885. 184; Andr. TES. 1919. 182.	N. Australia.
H*	var. **australasiae** Hope AM. ix. 1842. 426; Chaud. RMZ. 1872. 247; Solame PLS. NSW. 1920. 322.	N. Australia.
O*	**var. cyaneus** Chaud. RMZ. 1872. 247.	N. Australia.
G*	**var. scintillans** Bates AG. xxxii. 1892. 411.	Sikkim., Khasi Hills.
P*	*elegans* † Schm. Goeb. (not Weber) FCB. 83; Chaud. BEZ. 1861. 120; Andr. TES. 1923. 48.	Rajabhatkawa. Sunderbans. Thagata. (Tenasserim). Andaman Is.
C*	**facialis** *(Carabus)* Wied. Zool. Mag. i. 3. 1819. 165; Dej. Sp. i. 1825. 329; Brullé HN. iv. 1834. 232; Schm. Goeb. FCB. 81; Chaud. BM. 1850. i. 352; id. BEZ. 1861. 116; Saund. TES. 1863. 468; Bates AF. 1889. 283; id. AG. xxvii. 1889. 111; id. CRB. 1891.	Throughout Northern India, Assam, and Burma.

339; id. AG. xxxii. 1892. 409; Heyne-Tasch. EK. 16; Bouch. AF. 1903. 175; Lesne Miss. Pav. 80; Dupuis AB. 1914. 119; Andr. TES. 1921. 165; id. TES. 1923. 47; id. MB. 62.

O* *hardwicki* † Dej. (not Kirby) Sp. i. 1825. 329; Latr. et Dej. H. N. et Ic. 116. t. 7. f. 8; Schm. Goeb. FCB. 81.

goebeli × Mun. Cat. 153; Andr. TES. 1923. 47.

O* var. **angulatus** Chaud. BEZ. 1861. 117; id. RMZ. 1872. 244; Schauf. HER. xix. 1885. 184; Heyne-Tasch. EK. 16. t. 2. f. 38. — Malay Peninsula and Archipelago.

B* *elegans* † Macl. (not Weber) Ann. Jav. 1825. 15; Chaud. BEZ. 1881. 117; Andr. TES. 1919. 141.

O* *facialis* † Dej. Sp. v. 1831. 452.

O* *oxygonus* Chaud. BEZ. 1861. 117; id. RMZ. 1872. 244; Saund. TES. 1863. 468.

M* *excisus* Motch. BM. 1864. ii. 303; Chaud. RMZ. 1872. 249; Andr. TES. 1928. 10.

O* var. **goebeli** † Chaud. (not Mun. Cat.) RMZ. 1872. 245; Andr. AM. (9). xiv. 1924. 593. — Malacca.

O* var. **basalis** Chaud. RMZ. 1872 245; Lesne Miss. Pav. 80; Dupuis AB. 1914. 119; Andr. AM. (9). xiv. 1924. 593. — Darjeeling. Burma., Indo-China. Siam., Malay region. Yunnan.

E* var. **poultoni** Andr. AB. 1921. 203. **fuscoaeneus** × Chaud. RMZ. 1872. 247; Bates AG. xxxii. 1892. 410. (var.) — Sarawak. Burma : Bhamo, Thagata. Andaman Is.

M* *aeneus* || Motch. (not Saund). BM. 1864. ii. 303; Andr. TES. 1928. 10. — Indo-China. Malay Peninsula. Borneo., Sumatra.

B* var. **cupreicollis** Waterh. TES. 1877. 1; Bates AG. xxxii. 1892. 410; Andr. TES. 1919. 197. — Andaman. Is.

B* **hardwicki** Kirby TLS. xiv. 1825. 98. t. 3. f. 1; Andr. TES. 1919. 130. — Singhbhum. Angul.

O* *aeneipennis* Chaud. BEZ. 1861. 118; Andr. AM. (9). xiv. 1924. 593. — U. P. Haldwani, Jubliganj., Baihar.

B*	**ignicintus** Bates TES. 1883. 280.	Japan.
O*	**laevigatus** Saund. TES. 1863. 458. t. 18. f. 2; Chaud. RMZ. 1872. 249.	Batchian., Ternate. Aru Is.
J*	**laevipennis** Saund. TES. 1863. 465. t. 18. f. 3; Chaud. RMZ. 1872. 249; Andr. AM. (9). xvii. 1926. 259; id. AM. (9). xix. 1927. 107.	Tonda. Moluccas. Celebes.
B*	**laotinus** Andr. AB. 1921. 205.	Laos.
	schaumi † Dupuis (not Saund). AB. 1914. 119; Andr. AB. 1921. 202.	Tonkin. Annam.
B*	**lissonotus** Andr. AB. 1921. 208.	Wandei (South Kanara). Sidapur, Savira Butti. Coorg; Cochin State.
G*	**mirabilis** Bates AG. xxxii. 1892. 409; Dupuis AB. 1913. 419.	Assam : Garo Hills, Rotung. Karen Hills. Ruby Mines., Laos. Tonkin., Formosa.
X*	**perignitus** Maindr. BF. 1906. 152.	Sumatra. Negri Sembilan.
A*	**phlogops** Andr. AM. (9). xviii. 1926. 283.	Johore., Sumatra.
B*	**presidens** *(Pericalus)* Thoms. Arch. Ent. i. 1857. 281; Chaud. BEZ. 1861. 122; Andr. TES. 1919. 182; id. AM. (9). xvii. 1926. 259.	Malay Peninsula. Sumatra., Boreno. Celebes.
O*	*costulatus* Chaud. RMZ. 1862. 489; id. RMZ. 1872. 249; Andr. TES. 1921. 149.	
O*	*splendidus* Saund. TES. 1863. 459. t. 17. f. 1.	
O*	**punctipennis** Saund. TES. 1863. 464. t. 18. f. 4; Chaud. RMZ. 1872. 249.	Malay Peninsula. Java., Borneo., Siam. Annam., Tonkin.
B*	var. **salvazae** Andr. AB. 1921. 210.	Laos., Tonkin.
B*	**reductus** Walk. AM. (3) ii. 1858. 203; Bates AM. (5) xvii. 1886. 210; Heyne-Tasch. EK. 16; Andr. TES. 1919. 185.	Ceylon.

P*	**regalis** Schm. Goeb. FCB. 84; Chaud. BEZ. 1861. 122; Bates AG. xxxii. 1892. 410; Andr. TES. 1923. 49.	Assam : Naga Hills, Khasi Hills, Dikrang, Silonibari, Dejoo, Rotung, Kobo., Sylhet. Sikkim, Bengal: Lower Tonda. Gopaldhara. Burma ; Karen Hills, Tharrawaddy, Thagata, Tavoy. Andaman and Nicobar Is., Laos. Tonkin.
B*	**saphyrinus** Andr. AB. 1921. 206.	Laos.
D*	**sauteri** Dupuis AB. 1913. 419.	Formosa.
O*	**schaumi** Saund. TES. 1863. 457. t. 17. f. 3; Chaud. RMZ. 172. 248; Heyne-Tasch. EK. 16.	Sarawak.
O*	**simplex** Chaud. RMZ. 1872. 246.	Philippine Is.
O*	**smaragdulus** Dej. Sp. i. 1825. 331; Schaum. SEZ. 1847. 45; Chaud. BEZ. 1861. 119.	Bengal : Buxa, Rajabhatkawa, Jalpaiguri, Kuntimari, Sunderbans. Tenasserim. Malay Peninsula. Tonkin., Java. Sumatra., Borneo. Samar. I.
P*	*pauper* Schm. Goeb. FCB. 84; Andr. TES. 1923. 49.	
M*	**subquadratus** Motch. BM. 1864. ii. 302 Chaud. RMZ. 1872. 249; Andr. TES. 1928. 9.	"East indies."
O*	**versicolor** Saund TES. 1863. 463. t. 18. f. 1; Chaud. RMZ. 1872. 248.	Sumatra.
P*	**violaceus** Schm. Goeb. FCB. 82; Chaud. BEZ. 1861. 122; Bates AG. xxxii. 1892. 410; Andr. TES. 1923. 48.	Rajabhatkawa. Throughout Sikkim, Assam, Burma, and Malay Peninsula.
O*	*cyanipennis* Chaud. BM. 1854. i. 130; id. BEZ. 1861. 118.	
O*	**virens** Chaud. RMZ. 1872. 245.	Celebes.

E* **vitalisi** Dupuis AB. 1914. 119. — Tonkin.

O* **vollenhoveni** Chaud. RMZ. 1872. 248. — Sumatra.

O* **wallacei** Saund. TES. 1863. 462. t. 17. f. 4; Chaud. RMZ. 1872. 249; Heyne-Tasch. EK. 16; Arrow TZS. xx. 1915. 499; Andr. TES. 1919. 205. — Waigiou., Mysol. New Guinea. Aru Is.

H* **whithilli** Hope Col. Man. ii. 1838. 164. t. 3. f. 2; Chaud. BEZ. 1861. 116; Bates CRB. 1891. 339; Heyne-Tasch. EK. 16; Andr. TES. 1919. 200. — "Nepal.", Kheri. Mhow., Dapoli. Coimbatore.

D* *grandis* Dupuis AB. 1913. 428; Andr. AB. 1921. 202; id. Spol. Zeyl. 1924. 139. — Mysore : Aynur. Ceylon : Anuradhapura.

CELAENEPHES* Schmidt-Goebel.

Celaenephes Schm-Goeb. FCB. 77; Lac. 138.

Taromorpha Blckb. PLS. NSW. 1894. 85; Andr. AM. (9) xx. 1927. 272.

P* **parallelus** Schm-Goeb. FCB. 78. t. 2. f. 5; Bates AF. 1889. 286; id. AG. xxxii. 1892. 420; Bouch. AF. 1903. 176; Lesne Miss. Pav. 80; Vuillet Ins. ii. 1912. 17; Sloane PLS. NSW. 1920. 322; Andr. TES. 1923. 46; id. Ins. Samoa vi. 1927. 11.

B* *linearis (? Leistus)* Walk. AM. (3). ii. 1858. 203; Bates AM. (5). xvii. 1886. 211; Andr. TES. 1919. 188.

B* *allernatus (Taromorpha)* Blckb. PLS. NSW. 1894. 85; Andr. AM. (9) xx. 1927. 272.

G* *forsteri* Bouch. AF. 1903. 176.

V* *rechingeri* Csiki Denkschr. Akad. Wiss. Wien Math. nat. Kl., 1915. xci. 164.

— Chittagong. Ruby Mines. Tenasserim. Ceylon., Andaman Is. Malay Peninsula. Siam., Indo China. Malay Archipelago. Australia. New Caledonia, Samoa.

CHAETOBROSCUS* Semenow.

Chaetobroscus Sem. HER. xxxiv. 1900. 44; Andr. MB. 21.

O* **anomalus** *(Broscus)* Chaud. BM. 1878. ii. 1; Sem. RRE. x. 1910. 215; Andr. MB. 21. t. 2. f. 3. — Kashmir : Nagaberau, Tragbul Pass, Killanmarg, Lidar, Valley, Rotang Valley. Lahoul : Sisu. Spiti : Pulga. Seraj (Jalori Pass). ? Sikkim.

CHLAEMINUS* Motchulsky.

Chlæminus Motch. BM. 1864. ii. 350; Chaud; Essai monographique sur le genre Abacetus, BM. 1869 i. 401.

Distrigodes Motch. BM. 1864. ii. 353 (part).

Q	**annamensis** Tchitch. HER. xxxvi. 1903. 58.	Annam.
M*	**biguttatus** Motch. BM. 1864. ii. 351; Chaud. Mon. 401; Bates AF. 1889. 277; id. AG. xxxii. 1892. 363; Andr. TES. 1928. 14.	India., Burma. Ceylon., Cambodia. Siam., Java.
O*	**biplagiatus** Chaud. Mon. 402; Bates AG. xxxii. 1892. 363.	Pusa., Sylhet. Burma., Rangoon. Teinzo.
O*	**cruciatus** Chaud. Mon. 403.	Bengal.
L	**flavoguttatus** *(Distrigodes)* Motch. BM. 1864. ii. 354; Chaud. Mon. 404; Bates AG. xxxii. 1892. 363; Andr. TES. 1928. 22.	Bhamo. Tenasserim. Siam.
O*	**quadriplagiatus** Chaud. Mon. 403; Bates AG. xxxii. 1892. 363.	? Dacca (Chaudoir). ? Bhamo (Bates). ? Saigon.

CHLAENIUS Dejean.

Chlaenius Dej. Sp. ii. 1826. 297; Laf. AF. 1851. 238; Lac. 224; Chaud. BM. 1856. in 192; id., Mon. des Chléniens, AG. viii. 1876. 10; Sloane PLS. NSW. 1910. 437; Andr. TES. 1919. 91; id. MB. 24.

Lissauchenius Macl. Ann. Jav. 1825. 13; Lac. 217; Chaud. Mon. 12.

Epomis Dej. Sp. ii. 1826. 368; Lac. 223; Chaud. Mon. 16; Andr. TES. 1919. 91.

Indalma Schm. Goeb. in litt. FCB. (cover); Andr. TES. 1923. 58.

Diaphoropsophus Chaud. BM. 1850. i. 407; Lac. 221; Chaud. Mon. 11.

Barynorphus Laf. AF. 1851. 213 and 235.

Amblygenius Laf. AF. 1851. 212 and 263.

Ocybatgus × Laf. AF. 1851. 293; Lac. 219; Chaud. Mon. 12.

Ocydromus Laf. || (not Schell). AF. 1851. 212.

Homalolachnus × Laf. AF. 1851. 293; Lac. 220; Chaud. Mon. 11.

Omalotrichus|| Laf. (not *Homalotrichus* Sol.) Af. 1851. 233.

Chlaenites Motch. BAP. 1860. ii. 515.

Callistoides Motch. BM. 1864. ii. 334.

Poeciloistus Motch. BM. 1864. ii. 347.

Chlaenioctenus Bates AG. xxxii. 1892. 309.

E*	**abjectus** Andr. AB. 1920. 23.	Sumatra.
G*	**acroxanthus** Chaud. Mon. 112; Bouch. AF. 1903. 171.	Singapore., Siam Jave, Borneo. Sumatra.
A*	**adonis** Andr. AM. (9). xii. 1923. 215.	Annam.
O*	**agilis** Chaud. BM. 1856. ii. 246; id. Mon. 193; Bates Entom. 1891. Suppl. 8; Andr. MB. 26.	Kangra., Naggar. U. P. : Almora, Ranikhet, Kathgodam, Nainital, Siwaliks, Dehra Dun, R. Song, Nim Nadi, Lucknow. Bashahr : Poo. Darjiling, Eastern Duras. Assam between Pasighat and Kobo. Indo-China.
G*	**amabilis** Chaud. Mon. 279.	Siam., Tonkin.
B*	**amarae** Andr. EMM. 1920. 239; id. Ent. Mitt. 1927. 143.	Arabia. Irak., Seistan.
E*	**amplipennis** Chaud. Mon. 252.	Java.
B*	var. **niasanus** Andr. AB. 1920. 24.	Sumatra., Nias. Is.
O*	**anchomenoides** Bates PZS. 1889. 212.	Kashmir : Goorais Valley.
A*	**andamanensis** Andr. EMM. 1920. 236. and 237.	Andaman Is.
G	**annulipes** Bates AG. xxxii. 1892. 312.	Karen Hills.
C*	**apicalis** *(Carabus)* Wied. Zool. Mag. i. 3. 1819. 166; Dej. Sp. ii. 1826. 324; Laf. AF. 1851. 245; Chaud. BM. 1856. ii. 244; id. Mon. 89; Gestro.	Assam., Dacca. Kurseong., Pusa. Almora., Ranikhet.

	AG. 1882. 297; Bates AG. xxxii. 1892. 311; Andr. TES. 1921. 165.	Tanakpur., Haldwani. Burma Bhamo, Teinzo. Minhla, Senmigion, Rangoon, Prome. ? Java.
A*	**apollo** Andr. AM. (9). iv. 1919. S.	Nilgiri Hills.
O*	**atripes** Chaud. Mon. 160.	Kashmir., Assam. Dacca., Calcutta. Rangoon.
O*	**bengalensis** Chaud. BM. 1856. ii. 262; id. Mon. 157; Bates AF. 1889. 266.	Calcutta. Tranquebar., Ceylon.
L	*quadricolor (Poecilöistus)* † Motch. (not Oliv). BM. 1864. ii. 248.	Manipur., Rangoon. Siam., Cochin-China. Annam., Tonkin. China.
O*	**bicolor** Chaud. Mon. 130.	Gowchar (Nepal Valley). Dikrang Valley (Dafla Hills)., Dacca. Laos., Tonkin.
O*	**bimaculatus** Dej. Sp. ii. 1826. 301; Chaud. BM. 1856. ii. 198; id. Mon. 51; Bates AG. xxxii. 1892. 310; Bouh. AF. 1903. 171.	Nepal., Bhutan. Sikkim., Assam.
O*	var. **lynx** Chaud. BM. 1856. ii. 199; id. Mon. 50; Heyne-Tasch. EK. 21; Andr. AM. (9). xiii. 1924. 468.	Manipur., Calcutta. Dacca., Allahabad.
V*	*bimaculatus* † Redt. RN. 9; Andr. TES. 1923 (1924). 462.	Berhampur. Pusa.
J	var. **celebensis** Schauf. HER. xxi. 1887. 105; Andr. Am. (9). xix. 1927. 110.	Chota Nagpur; Barway, Tetara. Ceylon., Karen Hills. Bhamo., Tharrawaddy. Laos., Annam. Yunnan., China. Malay Archipelago.
B*	**binghami** Andr. AM. (9). iv. 1919. 12.	Maymyo.
O*	**bioculatus** Chaud. BM. 1856. ii. 198; id. Mon. 50; Bates CRB. 1891. 327; id. AG. xxxii. 1892. 310; Fletch. Agr. Res. Inst. Pusa Bull. 89. 1919. 31. f. 23. a-c.	Throughout India, Ceylon, and Burma Andaman Is.

B*	*aspericollis* Bates TES. 1873. 248; Chaud. Mon. 50.	Borneo., Indo-China.
O*	**birmanicus** Chaud. Mon. 93.	Sumatra, Hongkong. Japan, Dehra Dun. Darjiling. Assam: Sibsagar, Shillong, Harmatti (Dafla Hills), Garo Hills, Rotung, Sadiya Manipur. Toungoo., Rangoo. Johore
O*	**braminus** Chaud. Mon. 139.	?"Coromandel." Maldah.
M	**caeruleus** *(Carabus)* Stev. Mém. Soc. Nat. Mosc. ii. 1809. 38; Dej. Sp. ii. 1826. 363; Dej. et Boisd. Ic. et HN. ii. 190. t. 95. f. 4; Chaud. Mon. 228; Bates PZS. 1889. 213; Andr. MB. 27.	Kashmir : Goorais Valley, Srinagar. Ladak. Punjab : Seraj. Baluchistan : Nushki, Armenia. Southern Russia.
A*	**caesitius** Andr. AM. (9) xii. 1923. 215.	Annam.
B*	**callichloris** Bates TES. 1873. 250; Chaud. Mon. 198; Vuillet Ins. ii. 1912. 19.	Laos., China. Formosa., Japan.
X*	**cambodiensis** Bates AF. 1889. 266; id. AG. xxxii. 1892. 318.	Khasi Hills. Sibsagar., Sadiya. Calcutta., Karen Hills. Bhamo., Tenasserim. Indo-China.
G*	**camillae** Gestro AG. 1888. 108; Bates AG. xxxii. 1892. 313.	Burma ; Teinzo. Laos.
P*	*elegans (Indalma)* Schm. Goeb. FCB. t. 3. f. 5; Andr. TES. 1923. 58.	Cambodia.
O*	**celer** Chaud. Mon. 201; Bates AG. xxxii. 1892. 319.	Dehra Dun., Pusa. Chapra., Dacca. Between Goalundo and Gauhati. Prome., Bhamo. Shwegu., Katha. Palon.

G* **halcoderus** Chaud. Mon. 139. — Siam., Laos.

C* **chalcothorax** Wied. Zool. Mag. ii. 1. 1823. 51; Dej. Sp. ii. 1826. 304; Laf. AF. 1851. 244; Chaud. BM. 1856 ii. 233; id. Mon. 138; Bates CRB. 1891. 328; id. AG. xxxii. 1892. 315; Andr. TES. 1921. 169. — Calcutta (the other lacalities given by the authors cited are all uncertain).

B* **championi** Andr. AM. (9). xi. 1923. 343. — Almora.

B* **chapanus** Andr. AM. (9) iv. 1919. 5. — Tonkin.

O* **chlorodius** Dej. Sp. iii. 1826. 365; Laf. AF. 1851. 245; Chaud. BM. 1856. ii. 287; id. Mon. 176; Ball. BM. 1869. i. 213; Andr. EMM. 1920. 235. — Tezpur., Jorhat. Malda., Barrackpore. Sardah., Murshidabad. Calcutta., Pusa. Chapra. U.P. : Sitapur, Bareilly, Ramnagar (Fatehpur) Dehra Dun, Almora. Nagpur., "Saugor."

B* **cinctus** *(Carabus)* F. Spec. Ins. I. 1781. 310; Hope Col. Man. ii. 1838. 77; Schaum SEZ. 1847. 44; Motch. Etud. Ent. 1855. 61; Andr. TES. 1919. 122; id. TES. 1921. 154; id. MB. 26. — Surada.

circumdatus × Brullé Silb. Rev. Ent. iii.1835. 283; Chaud. Mon. 114; Bates AM. (5). xvii. 1886. 74; id. CRB. 1891. 328; id. AG. xxxii. 1892. 314; Bouch. AF. 1903. 171; Vuillet Ins. ii. 1912. 19; Fletch. Agr. Res. Inst. Pusa Bull. 89. 1919. 32. f. 24. a-c; Andr. MB. 25. — India., Ceylon. Burma. Malay Peninsula and Archipelago. Siam., Indo-China. Yunnan., China.

O* *limbatus* || Dej. (not Wied) Sp. ii. 1826. 306; Laf. AF. 18151. 243; Chaud. BM. 1856. ii. 227.

B* *cinctus* † Macl. (not F.) Ann. Jav. 1825. 13; Andr. TES. 1919. 137.

O* *javanus* Chaud. BM. 1856. ii. 229; id. Mon. 115; Bates AF. 1889. 265; Bouch. AF. 1903. 171; Andr. TES. 1921. 147 and 176.

S* *cupricollis* Nietn. JASB. xxv. 1856. 387; id. AM. (2). xix 1857. 243; Andr. AM. (9). xix. 1927. 104.

O*	var. **xanthopleurus** Chaud. BM. 1856. ii. 230; id. Mon. 115; Bates TES. 1873. 249; id. AF. 1889. 266; Heyne-Tasch. EK. 22; Lesne Miss. Pav. 70; Andr. TES 1921. 176.	Dehra Dun., Siam. Indo-China. Sumatra., China. Formosa.
M*	*nicanus (Chalenites)* Motth. BM. 1864. ii. 346; Andr. TES. 1928. 12.	Japan. Korea.
Q*	**circumductus** Moraw. BAP. v. 1863. 246; Bates TES. 1873. 249; Chaud. Mon. 210.	China. Japan.
B*	**comans** Andr. AM. (9). iv. 1919. 15.	Tonkin. Cochin-China. China.
A*	var. **exoletus** Andr. MB. 89.	Nepal : Pharping.
O*	*velocipes* Chaud. Mon. 266 (Siamese examples).	Dunsiri Valley (Dafla Hills). Berhampur. Birbhum., Calcutta. Kharagpur., Sardah. Parabatipur., Pusa. Narpur (Kangra). C. P. : Mangli, Khawasa., Surada. Tharrawaddy. Rangoon., Siam. Laos., Tonkin. Yunnan.
O*	**contractus** Chaud. Mon. 202.	Dacca., Pusa. Chapra., Dehra Dun.
O*	**convexus** Fairm. AF. 1886. 310.	Yunnan.
A*	*yunnanus* Andr. EMM. 1920. 236 and 238; id. AM. (9) xiv. 1924. 588.	
A*	**corbetti** Andr. AM. (9). iv. 1919. 13.	Rangoon.
G	**corrosulus** Bates AG. xxxii. 1892. 319.	Burma., Teinzo.
O*	**costiger** Chaud. BM. 1856. ii. 258; Bates TES. 1873. 253; Chaud. Mon. 95; Bates AF. 1889. 265; id. Entom. 1890. 212; Heyne-Tasch. EK. 22; Lesne. Miss. Pav. 70.	Japan., China. Yunnan., Korea. Formosa. Indo-China. Sumatra.
G*	var. **bhamoensis** Bates AG. xxxii. 1892. 311; Andr. EMM. 1920. 239.	Bhamo.

B*	var. **almorae** Andr. EMM. 1920. 239.	Sunderdhunga Valley (Almora).
O*	**crebrepunctatus** Chaud. BM. 1856. ii. 204 ; id. Mon. 55.	Khargpur, Balaghat. Bandra., Bombay. Belgaum., Madras.
G	**cupreolineatus** Chaud. Mon. 158.	Siam.
O*	**cyaniceps** Bates TES. 1873. 325; Chaud. Mon. 278; Andr. AM. (9). xiii. 1924. 468.	Laos., Hongkong. Macao.
B*	**cyanostolus** Andr. Spol. Zeyl. 1924. 133.	Ceylon., Mankulam.
G	**delicatus** Bates AG. xxxii. 1892. 316.	Burma., Teinzo.
M*	**dilatatus** *(Poeciloistus)* Motch. BM. 1864. ii. 348; Chaud. Mon. 157; Bates AG. xxxii. 1892. 319; Lesne Miss. Pav. 70; Andr. TES. 1928. 14.	Halem., Sibsagar. Kaliganj., Calcutta. Port. Canning. Birbhum., Siliguri. Berhampur., Dacca. Pusa., Chapra. "Hurdwar & Roorkee." Burma : Senmigion. Penang., Siam. Camodia.
O*	**distigma** Chaud. Mon. 49.	Singapore. Borneo., Timor.
X*	**djaina** *(Ocybatus)* Maindr. BF. 1899. 251; Andr. AM. (9). iv. 1919. 3.	Karachi., Chapra. Chota Nagpur; Barway.
G	**doriae** Chaud. Mon. 137.	Siam.
X*	**dureli** Maindr. BF. 1899. 252.	Sikkim : Gopaldhara, Maria Basti. Assam : Rotung, Kobo.
O*	**duvauceli** *(Epomis)* Dej. Sp. v. 1831. 668; Laf. AF. 1851. 253; Chaud. BM. 1856. ii. 237; Motch. BM. 1864. ii. 344; Chaud. Mon. 122; Andr. MB. 25.	Assam., Bengal. Pusa., Kalka. Bareilly., Allahabad. Lucknow., Surat. Nagpur. Trichinopoly.
O*	**eneides** *(Chlaenioctenus)* Bates AG. xxxii. 1892. 310 (note).	Assam.
A*	**exilis** Andr. AM. (9). xi. 1923. 337.	Tharrawaddy.

B*	**fastigatus** × Andr. TES. 1921. 182.	Kandy
	frater † Bates (Not Chaud) AM. (5) xvii. 1886. 74; Andr. TES. 1921. 147.	.
G	**feanus** *(Homalolachnus)* Bates AG. xxxii. 1892. 308.	Burma., Teinzo.
O*	**femoratus** Dej. Sp. ii. 1826, 328; Laf. AF. 1851. 251; Chaud. BM. 1856. ii. 244; id. Men. 93.	Java. Sumatra.
O*	*flavofemoratus* † Chaud. (not Cast.) BM. 1856. ii. 244; id. Mon. 93.	
O*	var. **costatus** Laf. AF. 1851. 251; Chaud. BM. 1856. ii. 244; id. Mon. 93.	
B*	**flaviguttatus** Macl. Ann. Jav. 1825. 14; Chaud. Mon. 52; Andr. TES. 1919. 141; Sloane PLS. NSW. 1920. 322.	Java. Sumatra. Australia.
O*	*binotatus* Dej. Sp. ii. 1826. 302; Laf. AF. 1851. 242; Chaud. Mon. 48; Sloane DEZ. 1907. 467; id. PLS. NSW. 1910. 438.	
O*	*punctatus* || Chaud. (not Motch). BM. 1856. ii. 200.	
L	*maculifer* Cast. Trans. Roy. Soc. Vict. viii. 1868. 148.	
	puncticeps × Mun. Cat. 224.	
M*	var. **guttatus** Eschsch. Zool. Atl. v 1833. 26. t. 25. f. 8; Chaud. Mon. 49; Heller DEZ. 1916. 276; Andr. Ins. Samoa iv. Col. 1927. 3; id. TES. 1928. 20.	Philippine Is. New Caledonia. New Guinea. Samoa.
	biguttatus (Lissauchenius) || Montr. (not Motch) AF 1860. 237.	
Q	**flavipes** Mén. Cat. rais. 1832. 114; Fald. Nouv. Mém Soc. Imp. Nat. Mosc. iv. 1835. 36; Chaud. Mon. 263.	Hurmuk (Perso-Baluch-Afghan frontier)., Lenkoran.
B'	*exutus* Friv. A Kir. Mag. Term. társ. Erk. i. 1845. 178. t. 1. f. 1.	Smyrna. Persia.
V	*persicus* Redt. Denkschr. Wien. Akad. 1850. i. 47.	Syria., Crete.
I*	*semicyaneus* † Andr. (not Solsky) RIM. xviii. 1919. 100; id. AM. (9). xiv. 1924. 588.	Dalmatia.
H*	**flavofemoratus** CAt. Etud. Ent. 1834. 81. t. 1. f. 3; Heyne-Tasch. EK. 22. t. 3. f. 31; Andr. TES. 1919. 207.	Shan Hills. Karen Hills. Bhamo., Palon.

M*	*nigricoxis* Motch. BM. 1864.ii. 339; Chaud. Mon. 94; Bates AG. xxxii. 1892. 312; Andr. TES. 1928. 12.	Java., Sumatra. Laos., Annam. Hongkong.
B*	**fletcheri** Andr. AM. (9). iv. 1919. 10.	Ceylon: Diyatalawa, Kotmalie.
O*	**frater** Chaud. Mon. 261; Andr. MB. 27.	Pusa., Singhbhum. Kathiawar: Sasan. Mhow., Nagpur. Hoshangabad. Motinala. Mysore: Teppukadu. Madras., Malabar.
X*	**fraterculus** Maindr. BF. 1899. 250; Andr. AM. (9). xi. 1923. 343.	Sikkim: Maria Basti, Gopaldhara, Tista Valley. Eastern Duars.
A*	var. **atupus** Andr. AM. (9). xi. 1923. 344.	Sikkim : Maria Basti, Gopaldhara, Tista Valley. Naga Hills. Manipur.
O*	**fugax** Chaud. Mon. 266; Andr. MB. 28.	Nurpur (Kangra). Chota Nagpur, Konbir, Tetara, Ranchi., Singhbhum. Nagpur., Seoni. Raipur., Chanda. Belgaum., Poona. Kasara., Surada. Madras., Horseley Konda., Malabar. Bangalore. Rangoon River.
E*	*submarginatus* † Bates (not Chaud.) CRB. 1891. 328; Andr. TES. 1921. 147.	
O*	**germanus** Chaud. Mon. 199.	Rangoon., Laos.
O*	**gestroi** Chaud. Mon. 51.	Malacca., Laos. Tonkin., China.
O*	**guttula** Chaud. BM. 1856. ii. 216; id. Mon. 281.	China., Yunnan. Sumatra.
Y*	var. **cuspidatus** Heller PJS. xxiii. 1923. 301; Andr. AM. (9). xvii. 1926. 254.	Philippine Is.

O* **hamatus** Dej. Sp. v. 1831. 633; Eschsch. Zool. Atl. v. 1833. 26; Laf. AF. 1851. 242; Chaud. BM. 1856. ii. 209; id. Mon. 63.	Philippine Is.
O* **hamifer** Chaud. BM. 1856. ii. 209; id. Mon. 62; Bates AG. xxxii. 1892. 311; id. AB. 1892. 230; Bouch. AF. 1903. 171; Lesne Miss Pav. 69; Andr. TES. 1919. 140; id. MB. 24.	India, Ceylon. Burma., Siam. Singapore. Java.
O* *bihamatus* Chaud. BM. 1856. ii. 210; id. Mon. 62; Bates. AF. 1889. 265.	Sumatra., Borneo. Indo-China.
A* *queenslandicus* Sloane PLS. NSW. 1910. 439; id. PLS. NSW. 1920. 322.	Hongkong. Formosa.
O* **hemichlorus** Fairm AB. 1888. 8.	Yunnan.
A* **henryi** Andr. AM. (9). iv. 1919. 11; id. EMM. 1920. 236; id. RIM. xxii. 1921. 342; id. MB. 26.	Chota Nagpur. Barwa., Mandla. Surada., Barkuda I. (Lake Chilka) Ceylon : Hambegamuwa
B* **himalayicus** Andr. AM. (9). xi. 1923. 335.	Brahmaputra River above Jorbat. Almora., Jaunsar.
O* **impressicollis** Chaud. Mon. 236.	"North India" Nagpur., Udvada.
O* **inops** Chaud. BM. 1856. ii. 239; id. Men. 262 (excluding Siamese examples); Bates TES. 1873. 250; Putz. CRB. 1875. 49; Fairm. RE 1888. 113; Heyne Tasch. EK. 22.	Tonkin. China. Formosa. Japan.
M* *arcuaticollis* Motch. Etud. Ent. 1860. 7; Andr. TES. 1928. 19.	
Q *vestitus* † Moraw. (not Payk.) Beitr. 35.	
B* **jacobsoni** Andr. AM. (9). xviii. 1926. 274.	Sumatra.
B* **junceus** Andr. AM. (9). xi. 1923. 336.	China.
A* **kanarae** Andr. (9). iv. 1919. 3.	North Kanara. Goa., Mormugao.
X* **kolariensis** Maindr. BF. 1898. 130.	Chota Nagpur : Nowafully.

O* **kuluensis** Bates Entom. 1891. Suppl. 9; Andr. MB. 27.	Kashmir : Sonemarg, Srinagar, Martand. Kulu : Koty.
A* **lacunosus** Andr. EMM. 1920. 240.	Yunnan.
O* **laetiusculus** Chaud. BM. 1856. ii. 248; id. Mon. 264; Bates SYM. 5; id. Entom. 1891. Suppl. 8; Andr. MB. 28.	Throughout the Himalayan tract from Abbottabad to Cachar. Lucknow., Ceylon.
O* **laevipennis** Chaud. Mon. 196; Bates AG. xxxii. 1892. 319.	Dacca., Sylhet. Kumaon., Bhatkot. Burma: Teinzo (Bates).
O* **lafertei** Guer. Voy. Deless. ii. 1843. 36; Chaud. Mon. 86; Bates. CRB. 1891. 327.	Throughout India and Ceylon.
O* *diffinis* Laf. AF. 1851. 241; Heyne-Tasch. EK. 21.	
O* *centromaculatus* Chaud. BM. 1856. ii. 218.	
M* *maculipennis* Motch. BM. 1864. ii. 341; Andr. TES. 1928. 12.	
B* **laotinus** Andr. AM. (9). iv. 1919. 14.	Laos., Yunnan.
C* **leucops** *(Harpalus)* Wied. Zool. Mag. ii. 1. 1823. 52; Chaud. Mon. 71; Bates AM. (5). xvii. 1886. 75; id. AF. 1889. 265; Lesne Miss. Pav. 70; Andr. TES. 1921. 170; id. MB. 25.	India., Ceylon. Burma. Andaman Is. Siam., Indo-China.
O* *aeruginosus* Chaud. BM. 1856. ii. 271.	Sumatra., Sinki I. (Sumatra)., Ceram. Philippine Is.
O* **limbicollis** Chaud. Mon. 41.	Dacca (Chaudoir) Formosa (Chaudoir). Laos., Annam.
A* **lioderus** Andr. AM. (9). xi. 1923. 342.	Rangoon.
B* **luculentus** Andr. EMM. 1920. 235. and 236.	Bengal : Barkura. Chapra.
opacipennis † Andr. (not Chaud.) AM. (9). iv. 1919. 11.	Sitapur (U. P.) Nagpur.
O* **luteicauda** Chaud. Mon. 201.	Dacca.
O* **luzonicus** Chaud. BM. 1856. ii. 261; id. Mon. 159.	Philippine Is.

B*	**macleayi** × Andr. TES. 1919. 139. *micans* † Macl. (not F.) Ann. Jav. 1925. 14; Chaud. Mon. 52.	Kuala Lumpur. Java., Sumatra. Borneo.
O*	**macropus** Chaud. Mon. 140.	Pusa., Chapra. Deccan. North Kanara. Ceylon.
O*	**maculatus** Dej. Sp. ii. 1826. 300; Laf. AF. 1851. 241; Chaud. BM. 1856. ii. 220; id. Mon. 99; Bates AG. xxxii. 1892. 310.	Dacca., Sylhet. Mandalay. Tharrawaddy. Rangoon., Insein. Pegu., Tenasserim. Siam.
M*	**malachinus** *(Callistoids)* Motch. BM. 1864. ii. 335; Andr. TES. 1928. 12.	Throughout India and Ceylon.
O*	*pudicus* † Chaud. (not F.) Mon. 280.	
A*	**masoni** Andr. AM. (9). xi. 1923. 341.	Kodaikanal. Shembaganur.
O*	**medioguttatus** *(Lissauchenius)* Chaud. Mon. 35; Bates AG. xxxii. 1892. 309.	Sikkim : Gopaldhara. Dacca., Bharmo.
O*	**melanopterus** Chaud. Mon. 226.	Siam., Laos.
A*	**melanopus** Andr. AM. (9). xi. 1923. 344.	Yunnan.
X*	**micans** *(Carabus)* F. Ent. Syst. i. 1792. 151; Hope Col. Man. ii. 1838 83; Motch. Etud. Ent. 1855. 61; Chaud. BM. 1856. ii. 206; id. Mon. 62; Andr. TES. 1919. 139; id. TES. 1921. 156.	China. Japan.
M*	*biguttatus* Motch. Etud. Ent. 1854. 63; Chaud. Mon. 274; Har. DEZ. 1877. 338; Fairm. RE. 1888. 113; Andr. TES. 1928. 15.	
O*	*subhamatus* Chaud. BM. 1856. ii. 211; Bates TES. 1873. 248; id. TES. 1883. 207.	
V*	*vulneratus* † Redt. (not Dej.) RN. 9; Andr. TES. 1923 (1924). 463.	
A*	**montivagus** Andr. AM. (9). xi. 1923. 338.	Yunnan.
O*	**mouhoti** Chaud. in Oberthür Col. Nov. i. 1883. 33.	Laos.
A*	**multicolor** Andr. AM. (9). iv. 1919. 4.	Belgaum.

B*	**mutatus** × Gemm. & Har. Mun. Cat. 222.	Java.
	apicalis || Macl. Ann. Jav. 1825. 14; Bouch. AF. 1903. 171; Andr. TES. 1919. 137; id. TES. 1921. 165.	
Q	**naeviger** Moraw. Mel. Biol. Acad. Sci. Patr. iv. 1862. 241; id. BAP. v. 1863. 324; id. Beiter. 33; Bates TES. 1873. 246; id. Entom. 1890. 212; Putz. CRB. 1875. 45; Chaud. Mon. 273; Andr. TES. 1928. 15.	China. Japan. Korea.
B*	**nepalensis** Hope Zool. Misc. 1831. 21; Dohrn SEZ. 1879. 458; Andr. TES. 1919. 171 and 202; id. AM. (9) xiii. 1924. 468.	India., Burma. Ceylon. Sumba.
O*	*mellyi (Diaphoropsophus)* Chaud. BM. 1850. i. 407; id. BM. 1856. ii. 213; id. Mon. 30; Bates CRB. 1891. 327; id. AG. xxxii. 1892. 309.	Laos., Combodia. Formosa., Hongkong. Macao.
O*	*planicornis (Barymorphus)* Laf. AF. 1851. 236.	
O*	*concinnus (Barymorphus)* Laf. AF. 1851. 236.	
O*	*swinphoci* Bates PZS. 1866. 342.	
C*	**nigricans** Wied. Germ. Mag. iv. 1821. 110; Dej. Sp. ii. 1826. 371; Laf. AF. 1851. 253; Chaud. BM. 1856. ii. 238; id. Mon. 126; Bates AG. xxxii. 1892. 314; Bouch. AF. 1903. 171; Andr. TES. 1921. 166.	India., Ceylon. Burma., Java. Nias I., Sumatra. Borneo.
O*	*rugicollis (Epomis)* Laf. AF. 1851. 253 (note 1).	Indo-China. Yunnan.
V*	*duvauceli* † Redt. (not Dej.) RN. 9; Andr. TES. 1923 (1924). 462.	Formosa. China.
B*	*culminatus* Bates TES. 1873. 251; Har. DEZ. 1877. 338; Bates TES. 1883. 208.	Japan.
O*	**nigripennis** Chaud. BM. 1856. ii. 250; id. Mon. 225.	Burdwan., Pusa.
B*	*melanopterus* † Bates (not Chaud.) AM. (5). xvii. 1886 75; Andr. Spol. Zeyl. 1924. 133.	Chota Nagpur : Nowatoli. Haldwani., Mandla. Hoshangabad. Peradeniya.
A*	**nilgiricus** Andr. AM. (9). iv. 1919. 9.	Nilgiri Hills. Coimbatore.
O*	**nitidicollis** Dej. Sp. ii. 1826. 314; Laf. AF. 1851. 243; Chaud. BM. 1856. ii. 228; id. Mon. 117; Bates AG. xxxii. 1892. 314; Andr. MB. 25.	Throughout India. Burma : Tharrawaddy, Senmigion.

A*	**ocellatus** Andr. AM. (9). xi. 1923. 339.	Nagpur., Jubbulpur. Mandla.
G*	**Ochroperas** Bates AG. xxxii. 1892. 316.	Bhamo., Rangoon.
B*	**ocreatus** Bates TES. 1873. 252; Chaud. Mon. 268.	Yunnan., Japan.
O*	**opacipennis** Chaud. Mon. 176; Bates CRB. 1891. 328; Andr. EMM. 1920. 236.	Bengal : Barkura. Chota Nagpur : Konbir Kalka. Nagpur., Baihar. Ramandrug.
O*	**orbicollis** Chaud. Mon. 40.	Sardah. Chota Nagpur. Barwa., Chapra. Almora., Nagpur. Kodaikanal., Nilgiri. Hills., Ceylon., Rangoon.
A*	**ostrinus** Andr. MB. 88.	Nepal. : Chitlong. Surada.
O*	**pallipes** *(Epomis ?)* Gebl. Mém. Soc. Nat. Mosc. vi. 1823. 128; Dej. et. Boisd. Ic. et HN. ii. 173. t. 92. f. 1; Moraw. Beitr. 34; Bates TES. 1873. 253; Putz. CRB. 1875. 49; Chaud. Mon. 222; Heyd. DEZ. 1879. 330; Fairm. RE. 1888. 113.	Siberia., China. Japan.
M*	*corpulentus* Motch. Etud. Ent. 1860. 7; Andr. TES. 1928. 19.	
O*	**panagaeoides** *(Omalotrichus)* Laf. AF. 1851. 235; Chaud. Mon. 28.	Malabar. Horsely Konda. Ramandrug. Mysore : Bangalore, Nandidrug, Chikkaballapura. ? Siam.
O*	**parallelus** Dej. Sp. v. 1831. 627; Laf. AF. 1851. 242; Chaud. Mon. 69.	Madras., Tranquebar. North Kanara.
J*	*dohrni* Nietn. JASB. xxvi. 1857. 149; id. AM. (2). xx. 1857. 372; Andr. AM. (9). xix. 1927. 104.	Ceylon : Colombo, Hambantota, Dehiwela.
G*	**pectinipes** *(Chlaenioctenus)* Bates AG. xxxii. 1892. 309.	Karen Hills., Sylhet.

	pericallus × Redt. RN. t. 1. f. 4; Bates TES. 1873. 249; Chaud. Mon. 286; Fairm. RE. 1888. 113.	China. Japan.
V*	*pulcher* || Redt. (not Nietn). RN. 10; Andr. TES. 1923 (1924). 463.	
O*	**phaenoderus** Chaud. Mon. 161; Bates CRB. 1891. 328.	Harmatti (Dafla Hills)., Dacca. Sahibganj. Saidpur (Bengal). Chota Nagpur; Konbir, Barwa. ? Bangalore.
	pictus × Chaud. BM. 1856. ii. 208; id. Mon. 62; Bates CRB. 1891. 327; id. AG. xxxii. 1892. 311; Bouch. AF. 1903. 171; Fletch. Agric. Res. Inst. Pusa. Bull. 89. 1919. 33; Andr. TES. 1921. 156; id. MB. 24.	Khasi Hills. Sibsagar., Calcutta. Saran, Birbhum. Pusa., Chapra.
O*	*schoenherri* || Dej. Sp. v. 1831. 626; Laf. AF. 1851. 242.	Tetara., Kangra. Mhow., Bombay. Deolali. Kathiawar : Sasan. Madura., Coimbatore. Bhamo., Java., Sumatra (Bouchard). Siam. Indo-China., China.
O*	**pleuroderus** Chaud. in Oberthür Col. Nov. i. 1883. 30.	Nasik.
G*	**poecilinus** Bates AG. xxxii. 1892. 317.	"India.", Kawkareik.
O*	**porrectus** Chaud. Mon. 224.	Dacca.
M*	**posticalis** Motch. Etud. Ent. 1853. 44; id. BM. 1864. ii. 340; Chaud. Mon. 273; Fairm. RE. 1888. 113; Heyne-Tasch. EK. 21; Andr. TES. 1928. 15.	China. Japan.
Q	*hospes* Moraw. BAP. v. 1863. 245; id. Beitr. 32; Bates TES. 1873. 247; id. TES. 1883. 207; Heyd. DEZ. 1879. 330.	
C*	**posticus** *(Carabus)* F. Suppl. Ent. Syst. 1798. 57; Hope Col. Man. ii. 1838. 82; Schaum SEZ. 1847. 46; Motch. Etud. Ent. 1855. 61; Chaud. BM. 1856. ii. 198; id. Mon. 55; Andr. TES. 1921. 159; id. MB. 24.	Throughout India and Ceylon.

O* *neelgheriensis* Guér. RZ. 1840. 38; Laf. AF. 1851. 242; Chaud. BM. 1856. ii. 206; id. Mon. 54; Bates AB. 1892. 230.

O* *bilunatus* Guér Voy. Deless. ii. 1843. 36.

O* *formosus* Chaud. BM. 1856. ii. 206.

J* *maleolens* Nietn. JASB. xxvi. 1857. 148; id. AM. (2). xx. 1857. 371; Andr. AM. (9). xix. 1927. 104.

M* *binotulatus* Motch. BM. 1864. ii. 341; Andr. TES. 1928. 12.

O* **postscriptus** Bates TES. 1873. 326; Chaud. Mon. 156. — Hongkong.

O* *extremus* Chaud. Mon. 112; Andr. Am. (9). xiv. 1924. 588.

B* **Praefectus** Bates TES. 1873. 253; Chaud. Mon. 156; Bates AG. xxxii. 1892. 314. — Brahmaputra River above Jorhat. Dacca., Calcutta. Pusa., Dehra Dun. Puri., Jhelum., Kalsi. Bhandara. Burma : Rangoon, Teinzo., Annam. Java., Sumatra. China., Japan.

O* *ducalis* Chaud. Mon. 155.

O* **pretiosus** Chaud. BM. 1856. ii. 288; id. Mon. 178; Andr. EMM. 1920. 235. — Delhi. U. P. : Dehra Dun, Moghal Sarai, Sitapur, Lucknow. Rawalpindi. "Saugor.", Ajmere. Ceylon (Stettin Mus.)

G* **privatus** Bates AG. xxxii. 1892. 315. — Karen Hills. Tharrawaddy.

O* **prostenus** Bates TES. 1873. 325; Chaud. Mon. 259; Bates TES. 1883. 235. — China. Japan.

O* *pratensis* Chaud. Mon. 210; Andr. AM. (9). xiv. 1924. 588.

O* **proximus** Chaud. Mon. 113. — Decca.

A* **ptuchodes** Andr. AM. (9). xi. 1923. 340. — Tharrawaddy.

O* **pubipennis** Chaud. BM. 1856. ii. 233; id. Mon. 138; Bates AG. xxxii. 1892. 315; Andr. TES. 1921. 169. — Pusa., Raipur. Bhandara., Balaghat. Tranquebar.

C* **pudicus** *(Carabus)* F. Syst. Eleuth. i. 1801. 193; Hope. Col. Man. ii. 1838. 83 (No. 123). ; Schaum SEZ. 1847. 46; Motch. Etud. Ent. 1855. 61; Chaud. BM. 1856. ii. 217; Andr. TES. 1921. 162. — Bengal. Karen Hills.

G* *caeruleiceps* Bates AG. xxxii. 1892. 320.

T* **pugni** Cam. Att. R. Aead. Sci. Tor. xiv. 1879. 146; Gestro AG. 1882. 306. — Mandalay. Minhla.

S* **pulcher.** Nietn. JASB. xxv. 1856. 387; id. AM. (2). xix. 1857. 242; Andr. Spol. Zeyl. 1924. 132; id. AM. (9). xix. 1927. 104. — Throughout India and Ceylon.

O* *marginatus* || Dej. (not Rossi) Sp. ii. 1826. 305; Laf. AF. 1851. 243; Chaud. BM. 1856. ü. 227.

murginifer × Chaud. Mon. 118.

O* **punctatostriatus** Chaud. BM. 1856. ii. 244; id. Mon. 91; Andr. MB. 25; id. EMM. 1926. 68. — Himalayan tract from Simla to Bhutan., Ceylon. Laos., Yunnan.

O* *cyaneonitens* Fairm. AF. 1886. 310.

O* **puncticollis** Dej. Sp. ii. 1826. 315; Laf. AF. 1851; 245; Chaud. BM. 1856. ii. 243; id. Mon. 196; Bates AG. xxxii. 1892. 319; Andr. MB. 27. — Throughout India and Ceylon. Burma. : Mandalay, Senmigion.

L* **quadricolor** *(Carabus)* Oliv. Enc. Méth. v. 1790. 344; id. Ent. 35. 77. t. 10. f. 111; F. Syst. Eleuth. i. 1801. 180; Macl. Ann. Jav. 1825. 14; Hope. Col. Man. ii. 1838. 75; Schaum. SEZ. 1847. 44; Motch. Etud. Ent. 1855. 61; Chaud. BM. 1856. ii. 287; Motch BM. 1864. ii. 348; Chaud. Mon. 154; Bates AG. xxxii. 1892. 314; Vuillet Ins. ii. 1912. 19; Andr. TES. 1919. 139; id. TES. 1921. 161; id. TES. 1928. — Throughout India and Ceylon Burma. Java. Indo-China.

O* *orientalis* Dej. Sp. ii. 1826. 339; Laf. 1851. 246.

O* *chlaenioides (Amblygenius)* Laf. AF. 1851. 263.

J*S* *princeps* Nietn. JASB. xxvi. 1857. 147; id. AM. (2). xx. 1857. 371; Andr. Spol. Zeyl. 1924. 133; id. AM. (9). xix. 1927. 104.

M* *laevicollis (Poeciloistus)* Motch. BM. 1864. ii. 348; Chaud. Mon. 154; Andr. TES. 1928. 13.

E* **rayotus** Bates CRB. 1891. 327; Andr. MB. 116. — Throughout India Ceylon., Laos. Cambodia.

E* **rotundus** Andr. AB. 1920. 24. — Cambodia. Cochin-China.

G* **rudesculptus** Chuad. Mon. 136. — Siam.

B* **rufifemoratus** *(Lissauchenius)* Macl. Ann. Jav. 1825. 13 t. i. f. 1; Chaud. BM. 1856. ii. 198; id. Mon. 35; Heyne-Tasch. EK. 22; Andr. TES. 1919. 136. — Java.

C* **rufithorax** Wied. Germ. Mag. iv. 1821. 112; Dej. Sp. ii. 1826. 322; Laf. AF. 1851. 245; Chaud. Mon. 259; Andr. TES. 1921. 167. — Chilla (Hardwar). Bengal., Noa Dehing Valley. (Assam).

S* **rugulosus** Nietn. JASB. xxv. 1856. 388; id. AM. (2). xix. 1857. 243; Chaud. Mon. 90; Bates AM. (5). xvii. 1886. 74; Andr. MB. 25; id. AM. (9). xix. 1927. 104.

G* *pachysomus* Chaud. Mon. 117; Andr. AM. (10). ii. 1928. 587. — Chota Nagpur. Nowatoli. Singhbhum Kathiawar : Sasan. Satara. C. P. : Basatar, Khawasa, Hoshangabad, Kanpa, Motinala, Subkar. Mysore : Teppukadu. Ceylon., Negombo. Kuala Lumpur. Laos., Tonkin. Siam.

O* **scapularis** Chaud. Mon. 98. — Bengal., Pusa. Nainital., Almora. Dehra Dun. Nurpur (Kangra). Simla., Delhi. Raipur., Saugor. Coimbatore. Madras. Bangalore. Ceylon : Anuaradhapura, Murunkam.

E* **semiviridis** Andr. AB. 1920. 22. — Java.

O* **semperi** Chaud. Mon. 92. — Philppine Is.

O* **sericimicans** Chaud. Mon. 235. — Tonkin., Hongkong.

A* **sexguttatus** *(Homalolachnus)* Andr. AM. (9) iv. 1919. 2. — Maymyo.

O* **sinensis** Chaud. BM. 1856. ii. 263; id. Mon. 162; Heyne-Tasch. EK. 22; Andr. AM. (9). xiii. 1924. 468. — Hongkong. Macao.

O*	**sinuatus** Dej. Sp. ii. 1826. 321; Laf. AF. 1851. 244; Chaud. BM. 1856. ii. 242; id. Mon. 194.	"East Indies."
O*	**sobrinus** Dej. Sp. ii. 1826. 316; Laf. AF. 1851. 245; Chaud. BM. 1856. ii. 243; id. Mon. 197.	Almora., Ranikhet. Nainital. Garhwal: Motisal. Dehra Dun.
J*	**spoliatus** *(Carabsu)* Rossi Faun. Etrusc. Mant. 1792. 79; F. Ent. Syst. Suppl. 1798. 54; Dej. Sp. ii. 1826. 312; Dej. et. Boisd. Ic. et. HN. ii. 168. t. 90. f. 4; Brullé HN. iv. 1834. 417; Chaud. Mon. 88; Fairm. RE. 1888. 113; Bedel Fn. med. 1897. 95; Andr. RIM. xviii. 1919. 100; id. TES. 1921. 157; id. AM. (9). xix. 1927. 99; id. Ent. Mitt. 1927. 143.	Seistan: Labi Baring., Irak. Arabia. Mediteranean basin. Central Asia. China., Korea. Japan.
O*	**stenoristus** Chaud. Mon. 265.	"North India" Malabar.
M'	**subcostatus** W. Macl. TES. NSW. 1864. i. 114; Chaud. Mon. 116; Sloane PLS. NSW. 1910. 437 and 439.	Java. Australia
O*	**submarginatus** Chaud. Mon. 235; Bates AG. xxxii. 1892. 320; Vuillet Ins. ii. 1912. 19.	Chapra, Karen Hills. Prome. Tharrawaddy. Rangoon. Cambodia.
H*	**sykesi** Hope TZS. i. 1834. 93. t. 13. f. 2; Bates NLS. xi. 1889. 207; Andr. TES. 1919. 203.	Poona., Bombay. "South India."
K'	**tenuelimbatus** Ball. BM. 1870. ii. 326; Solskey FT. 62; Chaud. Mon. 263; Bates SYM. 5; Andr. MB. 27.	Kashmir: Goorais. Valley., Leh. Turkestan.
O*	**tetragonoderus** Chaud. Mon. 68; Bates AG. xxxii. 1892. 311.	Pusa., Palon. Tenasserim. Andaman Is. Malay Peninsula. Sumatra., Sumbawa. Tonkin., Formosa. China.
	trachys × Andr. AM. (9). xii. 1923. 679.	Indo-China. Sumatra.
B*	*asper* || Andr. (not Chaud.) TES. 1921. 190.	China.
X*	*cinctus* ‡ Bates (not F.) AF. 1889. 266; Andr. TES. 1921. 176.	

B*	var. **fumatus** Andr. Spol. Zeyl. xiii. 1924. 133.	Murshidabad. Calcutta., Chapra. Pusa., Bhiwapur.
L	*cinctus* ‡ *(Carabus)* Herbst (not F.) Arch. Ins.-gesch. (Fuessly) v. 1784. 135. t. 9. f. 7. (?); Oliv. Ent. 35. 87. t. 3. f. 28. (part) (?); Illig. Mag. für. Ins. i. 1802. 351 (?); Dej. Sp. ii. 1826. 307; Hope Col. Man. ii. 1838. 77; Laf. AF. 1851. 243; Chaud. BM. 1856. ii. 232; id. Mon. 135; Bates AM. (5). xvii. 1886. 74; Andr. TES. 1919. 122; id. AM. (9). xix. 1927. 98.	Ramandrug., Trichur. Ennur., Bangalore. Ceylon : Hambegamuwa, Murunkam, Hambantota, Anuradhapura.
B*	*pulcher* † Bates (not Nietn). AM. (5). xvii. 1886. 74.	
Y*	**trigonotomoides** van Emden DEZ. (1928) 1929. 378.	Java.
O*	**trinototus** Chaud. Mon. 279.	Pusa., Chapra. Purnea., Dehra Dun.
A*	**tudicus** Andr. AM. (9). iv. 1919. 7.	Belgaum. North Kanara.
B*	**uninotatus** Andr. AM. (9). iv. 1919. 16.	Naga Hills. Assam. Valley. Manipur. Upper Chindwin.
Q	**variicornis** Moraw. Beitr. 35; Bates TES. 1873. 252; Chaud. Mon. 267.	China. Japan.
O*	**variipes** Chaud. BM. 1856. ii. 268; id. Mon. 87; Bates CRB. 1891. 328.	Chota Nagpur : Tetara, Barwa. Chapra., Allahabad. Delhi., Rawalpindi "Saugor"., Baihar.
O*	**velocipes** Chaud. Mon. 266 (part); Bates AM. (5). xvii. 1886. 74; Andr. MB. 28.	Dacca., Purulia. U. P.: Someswar, Nainital, Almora, Bhimtal, Haldwani, Kangra. C. P. : Mangli, Nagpur., Belgaum. Kasara., Nilgiri Hills. Kodaikanal. Ceylon : Dikoya. Manipur.
O*	**virgulifer** Chaud. Mon. 61.	China.
B*	*pictus* † Bates (not Chaud) TES. 1873. 247.	Japan.

B* **vitalisi** Andr. AM. (9). iv. 1919. 6.
Laos.

O* **vividus** Andr. Chaud. Mon. 176; Andr. EMM. 1920. 235.
Pusa., Mhow. Nagpur., Buldana. Bombay., Nasik. Belgaum., Mysore. Nandidrug., Madras. Trichinopoly. Manaparai. Udayagiri.

O* **valneratus** Dej. Sp. v. 1831. 624; Laf. AF. 1851. 242; Chand. BM. 1856. ii. 203; id. Mon. 52; Bates Entom. 1891. Suppl. 8.
Bengal., Pusa. Chapra., Delhi., Mhow. Ajmere., Nagpur. Bombay. Kathiawar : Gogo. Karachi.

O* **wallacei** Chaud. Mon. 134.
Celebes.

C* **xanthospilus** *(Carabus)* Wied. Germ. Mag. Ent. iv. 1821. 115; Chaud. Mon. 285; Bates AF. 1889. 266; id. CRB. 1891. 328; Heyne-Tasch. EK. 21. t. 3. f. 46; Lesne Miss. Pav. 70; Andr. TES. 1921. 167.
Throughout India and Ceylon. Siam. Indo-China.

J* *quinquemaculatus* Nietn. JASB. xxv. 1856. 386; id. AM. (2). xix. 1857. 242; Andr. AM. (9). xix. 1927. 104.

CHYDAEUS* Chaudoir.

Chydaeus Chaud. BM. 1854. ii. 343; Andr. MB. 29.

Acrogeniodon Tchitch. Ab. xxix. 1897. 65; id. HER. xxxiv. 1900. 370; id. HER. xxxv. 1901. 251; Andr. AM. (9). iii. 1919. 473.

A* **bakeri** Andr. AM. (9). xviii. 1926. 277.
Philippine Is.

X* **bedeli** *(Acrogeniodon)* Tchitch. Ab. xxix. 1897. 66; Andr. MB. 29. t. 2. ff. 6 (♂) and 7 (♀); id. EMM. 1926. 68.
Sikkim : Darjiling Mungphu Gopaldhara, Lachung. Gori River Gorge, Sunderdhunga and Pindar. Valleys (Almora). Spiti : Pulga. Kulu: Kandi, Koty. China.

O* **obscurus** Chaud. BM. 1854. ii. 344; Andr. EMM. 1926. 68. — Sikkim : Darjiling Gopaldhara, Tonglu, Kurseong.

O* *semenowi (Acrogeniodon)* Tchitch. HER. xxxii. 1899. 660; Andr. AM. (9). xiv. 1924. 590. — Deoban and Kalsi (Chakrata). Gori Valley and Gori River Gorge (Almora). Bengal.

CLIVINA* Latreille.

Clivina Latr. Hist. Nat. Crust. et. Ins. iii. 1802. 96; Bon. Obs. Ent. ii., Mém. Acad. Sci. Turin xx. 1813. 480; Dej. Sp. i. 1825. 411; Putz., Mon. des Clivina et genres voisins, Mém. Liége. ii. 1846. 523 and 577; Lac. 204; Putz., Postscr. ad Cliv Mon., Mém. Liége. xviii. 1863. 29; id., Rév. Gén des Clivinides, AB. x. 1867. 107-94; id. Suppl. Réy. Gén. A. xi. 1868. 10-20; id., Suppl. ii. Rév. Gén., AB. xvi. 1873. 15-17; Fleisch. Best. Tab. Eur. Col. xxxix. Scaritini. 1899. 33; Tchitch. RRE. 1904. 258; Andr. MB. 11; id. AM. (9). xvii. 1926. 372; id. FBI. 344 and 351.

Eupalamus Schm. Goeb. in litt. FCB. (cover); Motch. BM. 1861. i. 101.

Dacca Putz., Postscr ad Cliv. Mon., Mém. Liége xviii. 1863. 68; Andr. AM. (9). iii. 1919. 470.

X* **alutacea** Lesne Bull. Mus. Paris ii. 1896. 241. f. 3; id. Miss. Pav. 67 t. 8. f. 5. and text fig. 3. — Cambodia. Cochin-China.

H* **assamensis** Putz. Mon. 584; id. Postscr. 31; id. Rév. 108; Andr. TES. 1919. 205; id. AM. (9). xvii. 1926. 373; id. FBI. 354 and 364. — Assam., Sunderbans. Sardah., Pusa. Chapra. Bhubaneswar. Rambha (Ganjam).

J* **attenuata** *(Scarites)* Herbst Nat. Syst. Ins. Käf. x. 1806. 264. t. 176. f. 7; Putz. Mon. 626; id. Rév. 110; id., AB. xi. 1868. t. 1. f. 2; Bates AF. 1889. 262; id. CRB. 1891. 325; Andr. RIM. xxii. 1921. 340; id. MB. 11; id. AM. (9) xvii. 1926. 374; id. AM. (9). xix. 1927. 98; id. FBI. 353 and 355 f. 51. — Throughout North India and as far south as the Central Provinces. Tenasserim. F. M. S. Indo-China. Sumatra., Bushire.

T* *picipes* Bon., Obs. Ent. ii., Mém. Acad. Sci. Turin. xx. 1813. 481; Dej. Sp. i. 1825. 416; Putz. Mon. 623; id. Postscr. 51; Andr. AG. xlix. 1922. 393.

H*	*melanaria* Putz. Mon. 586; id. Postscr. 51; Andr. TES. 1919. 206.	
O*	var. **sulcigera** Putz. Rév. 110; Bates AF. 1889. 262; Andr. AM. (9). xvii. 1926. 374; id. FBI. 357.	Siam. Cochin-China.
G*	var. **bhamoensis** Bates AG. xxxii. 1892. 275; Andr. FBI. 357.	Bhamo.
X*	**bacillaria** Bates AF. 1889. 261; Andr. TES. 1921. 175.	Cochin-China.
O*	**bengalensis** Putz. Mon. 603; id. Rév. 137; id. AB. xi. 1868. t. 1. f. 11; Andr. Am. (9). xvii. 1926. 377; id. FBI. 354 and 370.	Calcutta. Pusa. Ragoon.
L(E)*	*pluridentata* Putz., CRB. 1877. 42.	
G*	*scissa* Bates AG. xxxii. 1892. 278.	
O	**biplagiata** Putz. SEZ. 1866. 43; id. Rév. 191; id. AB. xi. 1868. 19; Sloane PLS. NSW. 1896. 155.	Celebes. Australia.
O*	**brevior** Putz. Rév. 126; Bates AG. xxxii. 1892. 277; Andr. AM. (9). xvii. 1926. 375; id. FBI. 355 and 378.	Pusa. Burma : Rangoon, Senmigion, Tharrawaddy, Teinzo, Kawkareik, Zigon., Kedah.
A*	**bullata** Andr. AM. (9). xx. 1927. 270.	Timor.
B*	**castanea** Westw. PZS. 1837. 128; Putz. Rév 131 (note); Andr. TES. 1919. 179; id. AM. (9). xvii. 1926. 377; id. FBI. 355 and 374. f. 54.	The whole of South Eastern Asia, including Japan in the north and New Guinea in the south.
L*	*clivinoides (Eupalamus)* Schm. Goeb. FCB. t. 3. f. 4; Andr. TES. 1923. 58.	
E*	*parryi* Putz. Postscr. 60; id. Rév. 130; id. AB. xi. 1868. t. 1. f. 9; id. CRB. 1875. 49; Bates TES. 1876. 3; id. AM. (5). xvii. 1886. 72; id. AF. 1889. 262; id. AG. xxvii. 1889. 100; id., AG. xxxii. 1892. 279.	
O*	*lata* Putz. Rév 131; Bates TES. 1873. 238; Vuillet. Ins. ii. 1912. 18.	
G*	**javanica** † Bouch (not putz.) AF. 1903. 169.	

H*	var. **agona** Putz. Rév. 131; Andr. TES. 1919. 208; id. AM. (9). xvii. 1926. 377.	Siam.
A*	**costata** Andr. FBI. 354 and 364.	Nilgiri Hills.
G*	**costulipennis** Bates AG. xxxii. 1892. 281; Andr. AM. (9). xvii. 1926. 373; id. FBI. 354 and 366.	Shillong., Kobe. Salween Div. Palon., Dawna Hills. Perak., Penang. Selangor., Sumatra.
E*	**dolens** Putz. Suppl. ii.15; Andr. AM. (9). xvii. 1926. 375.	China.
O*	**elongatula** Nietn. JASB. xxv. 1856. 390; id. AM. (2). xix. 1857. 245; Putz. Rév. 123; Bates AM. (5). xvii. 1886. 72; Andr. AM. (9). xvii. 1926. 375 and *elegantula* in error) 376; id. AM. (9). xix. 1927. 105; id. FBI. 355 and 378.	Ceylon : Colombo. Murunkam. Horawupotana.
H*	**ephippiata** Putz. Mon. 602; id. Postscr. 39; id. Rév. 185; Andr. TES. 1919. 209; id. AM. (9). xvii. 1926. 377.	Java. Celebes. Australia.
A	*felix* Sloane PLS. NSW. 1896. 213; id. PlS. NSW. 1920. 320.	
A	*eximia* Sloane PLS. NSW. 1896; 214; id. PLS. NSW. 1920. 320.	
	extensicollis Putz. Mon. 601; id. Rév. 115; Andr. AM. (9). xvii. 1926. 375 and 376; id. FBI. 354 and 372.	? Calcutta. Dehra Dun., Java. Sumatra., Lombok. Philippine Is.
G*	**extensicollis** † Bouch. (not Putz.) AF. 1903. 169.	Sumatra.
E*	**forcipata** *(Dacca)* Putz. Postscr. 68. t. i. f. 41; id. Rév 25; Andr. FBI. 355 and 380. f. 56.	Srimangal., Dacca. Calcutta., Sardah. Rajmahal. Mertlik (Bengal). Pusa. Barkuda I. (Lake Chilka). Chalakudi (Cochin State). Ceylon : Kotte.
B*	**gamma** Andr. FBI. 354 and 368.	"India."

L (E)*	**helferi** Putz. Rév. 126; Andr. AM. (9). xvii. 1926. 375; id. AM (9) xix. 1927. 109; id. FBI. 354 and 367.	Calcutta., Khargpur. Parbatipur. Sunderbans.
S*	*grammica* Putz. CRB. 1877. 41.	Pusa.
G*	*debilis* || Bates AG. xxxii. 1892. 278.	Dinapur., Balugaon.
	invalida × Andr. AM. (9). iii. 1919. 470; id. MB. 12.	Barkuda Is. (Lake Chilka). Tharrawaddy. Prome., Mandalay. Rangoon., Toungoo. Kondan (Insein).
	javanica Putz. Mon. 592; id. Rév. 124; Andr. TES. 1919. 209; id. AM. (9). xvii. 1926. 375.	Kuala Lumpur. Java., Sumatra.
X*	**julieni** Lesne Bull. Mus. Paris ii. 1896. 240; id. Miss. Pàv. 66. t. 8. f. 4.	Indo-China.
A*	**lewisi** Andr. AM. (9). xx. 1927. 268.	Japan.
T*	**lobata** Bon. Obs., Ent. ii., Mém. Acad. Sci. Turin. xx. 1813. 481; Dej. Sp. i. 1825. 414; Putz. Mon. 599; Redt. RN. 8 (?); Putz. Rév. 120; id. AB. xi. 1868. t. 1. f. 8. Andr. TES. 1919. 209; id. RIM. xxii. 1921. 340; id. AG. xlix. 1922. 392; id. TES. 1923 (1924). 462; id. MB. 11; id. AM. (9). xvii. 1926. 375; id. FBI. 355 and 375. f. 55.	India., Burma. Siam.
E*	*anceps* Putz. Postscr. 50; id. Rév. 124.	
O*	*hydropica* Putz. Rév. 121.	
E*	*angularis* Putz. Rév. 122; Bates AG. xxxii. 1892. 276.	
O*	*capitata* Putz. Rév. 122.	
O*	*divaricata* Putz. Rév. 122.	
G*	*laeviceps* Bates AG. xxxii. 1892. 277; Andr. MB. 12.	
X*	**mekongensis** Lesne Bull. Mus. Paris ii. 1896. 241. f. 4. ; id. Miss. Pav. 67. t. 8. f. 6 and test-fig. 4.	Cambodia.
O*	**memnonia** Dej. Sp.v 1831. 503; Putz. Mon. 588; id. Rév. 108; Bouch. AF. 1903. 169. Andr. AM. (9). xvii. 1926. 373; id. FBI. 354 and 362. f. 53.	India., Ceylon. Burma., Indo-China. Sumatra.
H*	*indica* Putz. Mon. 585; id. Postscr. 35; id. Rév. 108; id. CRB. 1875. 69; Bates AM. (5). xvii. 1886. 72; id.	Java. Borneo.

	AF. 1889. 262; id. CRB. 1891. 325; id. AG. xxxii. 1892. 274; Andr. TES. 1919. 206; id. MB. 115.	
I*	*rugosifrons* Nietn. JASB. xxv. 1856. 390; id. AM. (2) xix. 1857. 245; Andr. AM. (9) xix. 1927. 105.	
B*	*recta* Walk. AM. (3). ii. 1858. 203; Bates AM. (5). xvii. 1886. 212; Andr. TES. 1919. 187.	
E*	**moerens** Putz. Suppl. ii. 15; Andr. AM. (9). xvii. 1926. 375.	China.
A*	**mustela** Andr. Spol. Zeyl. 1923. 226; id. FBI. 355 and 377.	Ceylon: Anuradhapura.
B*	**niponensis** Bates TES. 1873. 239; Andr. AM. (9). xvii. 1926. 377; id. FBI 361.	China. Japan.
X*	**parallela** Lesne Bull. Mus. Paris. ii. 1896. 239. f. 2; id. Miss. Pav. 64 t. 8. f. 3 and text-fig. 2.	Laos.
G*	**pileolata** Bates AG. xxxii. 1892. 278; Andr. FBI. 354 and 372.	Mangaldai. Rangoon., Katha. Palon., Tharrawaddy. Toungoo.
B*	**sabulosa** Macl. Ann. Jav. 1825. 24; Putz. Mon. 600; id. Rév. 119 (note) and 124; Andr. TES. 1919. 161; id. AM. (9). xvii. 1926. 375.	Java. Sumatra.
E*	*humeralis* Putz. Postscr. 48; id. Rév. 125.	
O*	*advena* Putz. Rév. 123.	
G*	**sagittaria** Bates AG. xxxii. 1892. 275; Andr. FBI. 353 and 358.	Pusa., Rangoon. Kuala Lumpur. Singapore.
A*	**saundersi** Andr. AM. (9). xviii. 1926. 273.	Malay Peninsula. Sumatra.
G*	**sectifrons** Bates AG. xxxii. 1892. 276; Andr. FBI. 355 and 379.	Burma., Rangoon. Palon and Tikekee (Pegu). Shenmaga., Paungde. Tharrawaddy. Toungoo. Kondan (Insein). Seikpyu., Pakokku. Pyinmana.

S* **semicarinata** Putz. CRB. 1877. 44; Andr. AM. (9). xvii. 1926. 377; id. FBI. 353 and 361. — Garo Hills. Mangaldai., Calcutta.

G* *occipitalis* Bates AG. xxxii. 1892. 281. — Pusa., Haldwani. Tanakpur. Hoshangabad. Bhamo. Upper Chindwin. Toungoo., Senmigion. Sumatra.

G* **siamica** Putz. Rév. 124; Bates AF. 1889. 261; Andr. TES. 1919. 208; id. TES. 1921. 175; id. AM. (9). xvii. 1926. 375. — Siam. Cochin-China. Tonkin.

B* **spatulifera** Andr. FBI. 354 and 369. — Tovoy and Maungmagan (Tavoy Dist.).

H* **striata** Putz. Mon. 592; id. Rév. 110; Andr. TES. 1919. 206; id. MB. 115; id. AM. (9). xvii. 1926. 374; id. FBI. 353 and 357. — Kierpur (Purnea). Ranchi. India, south of the Central Provinces. Ceylon.

L **stricta** Putz. Postscr. 49; id. Rév. 125; Andr. AM. (9). xvii. 1926. 375 and 376. — Java.

A* **talpa** Andr. AM. (9). xx. 1927. 269. — Borneo.

T* **tranquebarica** Bon., Obs. Ent. ii. Mém. Acad. Sci. Turin. xx. 1813. 484; Andr. AG. xlix. 1922. 394; id. AM. (9) xvii. 1926. 377; id. FBI. 353 and 359. f. 52. — India., Ceylon. Burma., Indo-China. Malay Peninsula and Archipelago.

M* *fulvaster (Eupalamus ?)* Motch. BM. 1861. i. 101; Andr. TES. 1928. 7.

M* *brunnescens (Eupalamus ?)* Motch. BM. 1861. i. 101; Andr. TES. 1928. 7.

M* *rufipes (Eupalamus ?)* Motch. BM. 1861. i. 102; Putz. Rév. 134; Bates AM. (5) xvii. 1886. 72; Andr. AG. xlix. 1922. 394; id. AM. (9). xvii. 1926. 377; id. TES. 1928. 7.

M* *cordicollis (Eupalamus ?)* Motch. BM. 1861. i. 102; Andr. TES. 1928. 7.

E* *convexicollis* Putz. Postscr. 52; id. Rév 133; Bates AF. 1889. 262; id. AG. xxxii. 1892. 279.

E* *foveicollis* Putz. Postscr. 61; id. Rév. 133

E* *obesicollis* Putz. Postscr. 66; id. Rév. 132; id. AB. xi. 1868. t. 1. f. 10.

E* *mordax* Putz. Postscr. 67; id. Rév. 133; Bates AF. 1889. 263; id. AG. xxxii. 1892. 280; Andr. RIM. xxii. 1921. 340; id. MB. 12.

O* *marginicollis* Putz. Rév. 133.

O* *placida* Putz. Rév. 134.

E* *stigmatica* Putz. Rév. 134.

G* *scuticeps* Bates AG. xxxii. 1892. 280.

transversa Putz. Rév. 125; Bates AF. 1889. 262; Andr. TES. 1919. 208; id. AM. (9). xvii. 1926. 375. — Siam. Cochin-China.

L(E)* **truncata** Putz. CRB. 1877. 45; Andr. AM. (9). xvii. 1926. 373. — Engano I. Mentawei Is. Sumatra., Amboina.

L **unicolor** (*Scarites*) Herbst. Nat. Syst. Inst. Käf. x. 1806. 265. t. 176. f. 9; Putz. Mon. 626; Andr. AM. (9). xix. 1927. 98; id. FBI. 381. — "East Indies."

N **vulgivaga** Boh. ER. 9; Putz. Rév. 115; Bates TES. 1873. 238; Andr. Am. (9). xvii. 1926. 375. — Tonkin., China. Japan.

Q *humilis* Moraw. Beitr. 22. — Philippine Is.

L (E)* **wallacei** Putz. Rév. 127; Andr. AM. (9). xvii. 1926. 374 and 375; id. AM. (9). xix. 1927. 109. — Sumatra., Celebes. New Guinea.

E* *custanea* || Putz. (not Westw.). Postscr. 35. (part).

westwoodi × Putz. Rév. 109 (New Guinea).

wetwoodi × Putz. Rév. 109 (Ceylon); Andr. AM. (9). xvii 1926. 373; id. FBI. 351 and 365. — Dehra Dun. Belgaum., Nagpur. Colombo., Japan.

E* *castanca* || Putz. (not Wetw.) Postscr. 35 (part).

COLEOLISSUS* Bates.

Coleolissus Bates AG. xxxii. 1892. 338.

G* **bicoloripes** Bates AG. xxxii. 1892. 339. — Karen Hills., Palon. Indo-China. Sarawak.

X* **biseriatus** Lesne Bull. Mus. Paris ii.1896. 244; id. Miss. Pav. 74. t. 8. f. 9.	Cambodia. Siam.
G* **eulamprus** Bates. AG. xxxii. 1892. 340.	Palon. Kawkareik.
A* **iris** Andr. Spol. Zeyl. 1924. 135.	Dehra Dun., Mhow. Khandesh., Nagpur. Balaghat., Mysore. Nandidrug., Madras. Kodaikanal. Nilgiri Hills. Ceylon : Hambegamuwa. Kotte.
G* **lamprotus** Bates AG. xxxii. 1892. 338.	Burma. Teinzo. Karen Hills., Bhamo.
I* **periucens** *(Hypolithus)* Bates PZS. 1878. 715; id. SYM. 9. id. AG. xxxii. 1892. 339.	Jhelum Valley.
G* **viridellus** Bates AG. xxxii. 1892. 340.	Shwegu.

COLFAX Andrewes.

Colfax Andr. AM. (9). vi. 1920. 504; id. MB. 54.

A* **stevensi** Andr. AM. (9). vi. 1920. 505; id. MB. 55. t. 3. f. 6.	Sadiya. Sikkim : Gopaldhara, Pashok. Chapra., Mussoorie. Dehra Dun., Surada. Bassein Fort (Bombay). Mergui., Indo-China.

COLPODES Macleay.

Colpodes Macl. Ann. Jav. 1825. 17; Lac. 361; Chaud. Mon. du genre Colpodes, AF. 1859. 289; id., Rév. des genres Onychopterygia, Dicranoncus, et Colpodes, AF. 1878. 278; Bates Biol. Centr. Am. Ins. Col. I. 1. 1882. 101; Sloane PLS. NSW. 1910. 453; Alluaud. AF. lxxxv. 1916. 77 and 78; Andr. TES. 1919. 146 and 164; id. MB. 50.

Loxocrepis Eschsch. Zool. Atl. ii. 1829. 6; Lac. 362.

Dyscolus Dej. Sp. v. 1831. 437; Lac. 356.

Vulcanophilus, Heller Abh. Mus. Dresd. vi. 3. 1896. 2; Andr. AM. (9). xx. 1927. 271.

O* **abropoides** Chaud. Rév. 361. — Philippine Is.

G* **acroglyptus** Bates AG. xxxii. 1892. 374. — Naga Hills. Manipur. Karen Hills., Laos. Tonkin.

O* **aeneipennis** *(Dyscolus)* Dej. Sp. v. 1831. 441; Chaud. Mon. 359; id. Rév. 333. — Java.

O* **aenescens** Chaud. Rév. 368. — ? India.

Y* **andrewesi** Heller Ent. Mitt. 1926. 92. — Philippine Is.

rufitarsis † Heller (not Chaud.). DEZ. 1916. 276.

O* **baconi** Chaud. Rév. 311. — Bengal. Sikkim : Karponang.

O* **bengalensis** Chaud. Rév. 312. — Bengal., Kurseong. Simanabhasti, (Nepal-Sikkim frontier).

B* **bipars** *(Lebia)* Walk. AM. (3). ii. 1858. 203; Bates AM. (5). xvii. 1886. 148; Andr. TES. 1919. 185. — Ceylon : Hadley, Dikoya, Kandy, Patipola, Hakgalla.

O* *dohrni* † Chuad. (not Nietn.) Rév. 375; Andr. Spol. Zeyl. xiii. 1924. 137.

B* *lampriodes* Bates AM. (5). xvii. 1886. 147.

M* **bispinus** *(Euplynes)* Motch. Etud. Ent. 1859. 33; Andr. TES. 1928. 17. — ? Sumatra.

B* **brunneus** Macl. Ann. Jav. 1825. 17. t. 1. f. 3; Chaud. Mon. 359; Andr. TES. 1919. 146. — Java.

B* **buchanaui** Hope Zool. Misc. 1881. 21; Chaud. Mon. 359; id. Rév. 369; Andr. TES. 1919. 173; id. MB. 50. — Throughout South East Asia, including Southern China and Japan, Korea.

O* *amoenus* Chaud. Mon. 326; id. Rév. 367; Bates AM. (5). xvii. 1886. 147; id. Entom. 1890. 213; Heller DEZ. 1916. 276.

Q* *splendens (Dyscolus)* Moraw. Mel. Biol. Acad. Sci. Patr. iv. 1862. 241; id. BAP. v. 1863. 324; id. Beitr.

	40; Bates TES. 1873. 275; Putz. CRB. 1875. 50; Har. DEZ. 1878. 67 and 213.	
G*	**caelitis** Bates AG. xxxii. 1892. 372.	Karen Hills.
Y*	**calathoides** *(Vulcanophilus)* Heller Abh. Mus. Dresd. vi. 3. 1896. 2. t. f. 1.	Celebes.
G*	**castaniventris** Bates AG. xxxii. 1892. 374.	Karen Hills.
O*	**chloropterus** Chaud. Rév. 339.	Celebes., Tonda.
O*	**cruralis** Chaud. Rév. 376.	Garo Hills.
G*	*ischioxanthus* Bates AG. xxxii. 1892. 376; Andr. TES. 1921. 148.	Silonibari. Sikkim : Gopaldhara. Manipur. Karen Hills., Satara. Eddakara (Malabar). Nilgiri Hills. Cochin State: Param Bikulam. Annam.
G*	**eucnemis** Bates AG. xxxii. 1892.	Karen Hills.
O*	**eulabes** Bates PZS. 1889. 215.	Kashmir; Goorais Valley.
A*	**euparyphus** Andr. AM. (9). xii. 1923. 448.	Kodaikanal. Mysore. Chikkaballapura.
B*	**fletcheri** Andr. Spol. Zeyl. 1923. 240.	Ceylon : Madulsima.
B*	**fryi** Bates PZS. 1889. 384; Andr. TES. 1919. 197.	Sumatra., Barneo.
B*	**hardwicki** Hope Zool. Misc. 1831. 21; Chaud. Mon. 359; Andr. TES. 1919. 172.	Nepal. Sikkim : Kurseong, Gopaldhara, Darjeeling.
O*	**hirmocoelus** Chaud. Rév. 365.	Darjiling., Kumaon. Lebong., Nainital. West Almora. Sunderdhunga Valley. Swal River basin. Mussoorie.
A*	**impunctatus** Andr. AM. (9). xii. 1923. 685.	Sikkim., Tongu. Darjiling.

O*	**incertus** Chaud. Rév. 369.	"East Indies."
B*	**iteratus** Bates AM. (5). xvii. 1886. 149.	Ceylon., Dikoya. Nuwara Eliya.
G*	**janellus** Bates AG. xxxii. 1892. 373.	Karen Hills.
M	**japonicus** *(Tanystola)* Motch. Etud. Ent. 1869. 8; Moraw. Mel. Biol. Acad. Sci. Petr. iv. 1862. 242; id. BAP. v. 1863. 324; id. Beitr. 41; Bates TES. 1873. 277; Har. DEZ. 1878. 213.	Tonkin. China. Japan.
J*	**laetus** *(Anchomenus)* Erichs. Nov. Act. Leop. Carol. xvi. 1834. Suppl. 222. t. 37. f. 2.; Chaud. Mon. 326; Heyne-Tasch. EK. 25; Andr. AM. (9). xix. 1927. 101.	Philippine Is.
O*	*apicalis* Chaud. Rév. 367.	
O*	**luzonicus** Chaud. Rév. 366.	Philippine Is.
O*	**maculicollis** Chaud. Rév. 376.	Celebes.
O*	**melittus** Bates PZS. 1889. 215.	Kashmir : Goorais Valley.
G*	**muleyitus** Bates AG. xxxii. 1892. 372.	Mt. Mooleyit (Tenasserim).
M*	**nigriceps** *(Loxocrepis)* Motch. BM. 1864. ii. 310. Andr. TES. 1928. 10.	East Indies.
O*	**nilgherriensis** Chaud. Rév. 301.	Malabar. Ootacamund. Naduvatam.
O*	**obscuritarsis** Chaud. Rév. 375; Heller DEZ. 1916. 276.	Rangoon., Borneo. Philippine Is.
O*	**olivius** Bates TES. 1873. 330; id. AG. xxxii. 1892. 373.	Sikkim : Gopaldhara, Tonglu, Kurseong. Shillong., Bengla.
O*	*coelopterus* Chaud. Rèv. 368; Andr. AM. (9). xvi. 1924. 591.	Karen Hills., Tonkin. Cambodia., China.
I*	**ovaliceps** Bates PZS. 1878. 719; id. SYM. 12 t. 1. f. 7.	Kashmir. Pir Panjal., Murree.
O	**ovipennis** *(Dyscolus)* Fairm. AB. 1888. 14.	Yunnan.
O*	**parallelus** Chaud. Mon. 326; id. Rév. 368.	Sumatra.
O*	**phaeoderus** Chaud. Rév. 364.	Celebes.

Y	**placidus** Heller Abh. Mus. Dresd. ix. 5. 1900. 3. 5. t. f. 1.	Celebes.
O*	**plagioderus** Chaud. Rév. 374.	"East Indies."
O*	**pleuralis** Jord. NZ. 110.	Timor., Burn.
G*	**punctulicollis** Bates. AG. xxxii. 1892. 375.	Karen Hills.
B*	**repletus** Bates AM. (5). xvii. 1886. 148.	Ceylon: Bogawantalawa, Hakgalla.
B*	**retusus** Bates AM. (5). xvii. 1886. 148.	Kandy.
O*	**rotundatus** Chaud. Rév. 302.	Malabar. Nilgiri Hills.
B*	**ruficeps** *(Lamprias)* Macl. Ann. Jav. 1825; 25; *(Loxocrepis)* Eschsch. Zool. Atl. ii. 1829. 6. t. 8. f. 3.; Chaud. Mon. 348; id. Rév. 376; Andr. TES. 1919. 164.	Calcutta., Pusa. Chapra., Haldwani. Ranchi.
J*	*dohrni (Euplynes)* Nietn. AM. (3) ii. 1858. 429; Bates AM. (5). xvii. 1886. 147; Andr. Spol. Zeyl. 1924. 137; id. AM. (9). xix. 1927. 105.	Nagpur., Belgaum. North Kanara. "Coromandel." Anaimalai Hills.
O*	*cyanipennis* † Chaud. (not Schm. Goeb). Mon. 350 and 360; id. Rév. 376.	Ceylon: Colombo, Negombo, Pusilawa, Peradeniya.
	schmidti × Chaud. Mon. 350 and 360; id. Rév. 376.	Rangoon., Java. Sumatra. Philippine Is.
O*	**rufitarsis** *(Dyscolus)* Chaud. BM. 1850. i. 385; id. Mon. 351; id. Rév. 375; Bates AG. xxxii. 1892. 377; Bouch. AF. 1903. 172.	Bhamo. Malay Peninsula. Sumatra., Java. Borneo.
O*	**saphyrinus** Chaud. Rév. 366.	Penang., Laos. Borneo., Banguey. Tonda.
L	**saphyripennis** Chaud. Rév. 334.	East Indies.
B*	**sebosus** Andr. AM. (9). xvii. 1926. 257.	Ceylon : Pidurutalagalla.
O*	**semiaeneus** Fairm. AF. 1886. 315.	Yunnan.

O*	**semistriatus** Chaud. Rév. 365; Andr. EMM. 1926. 69.	Kumaon : Almora, Gori Valley, Pindar Valley, Sunderdhunga Valley, Dudhatoli Chakrata., Mundali. "Massouna."
A*	**sikkimensis** Andr. AM. (9). xii. 1923. 683.	Sikkim : Gopaldhara, Pedong, Maria, Basti Pahang.
O*	**smaragdipennis** Chaud. Mon. 351; id. Rév. 375; Andr. TES. 1928. 18.	Sumatra.
Y	**spinifer** *(Vulcanophilus)* Heller Abh. Mus. Dresd. vi. 3. 1896. 3. t. f. 6.	Celebes.
G*	**spinulifer** Bates AG. xxxii. 1892. 377.	Gopaldhara. Karen Hills.
O*	**stricticolis** Fairm. AF. 1886. 316.	Yunnan.
O	**subsericatus** Fairm. AF. 1886. 316.	Yunnan.
O	**superlita** Bates PZS. 1888. 383.	China.
O*	**sycophanta** Fairm. AF. 1886. 316.	Yunnan.
G*	**undulipennis** Bates AG. xxxii. 1892. 375.	Bhamo., Katha. Singapore.
A*	**vikara** Andr. AM. (9). xii. 1923. 684.	Sikkim : Tonglu. Kumaon : Pindar Valley.
O*	**violaceus** Chaud. Mon. 351; id. Rév. 375; Sloane DEZ. 1907. 179; Maindr. BF. 1908. 186; Sloane PLS. NSW. 1910. 458; id. PLS. NSW. 1920. 322.	Aru Is., Kei Is. New Guinea. Australia.
M'	*mucronatus* W. Mack. PLS. NSW. 1887. 218.	
A*	**vulpinus** Andr. AM. (9). xii. 1923. 685.	Darjiling., Kurseong.
B*	**xenos** Bates AM. (5). xvii. 1886. 146.	Ceylon : Bogawantallawa, Pidurutalgalla.

COPTODERA* Dejean.

Coptodera Dej. Sp. i. 1825. 273; Schm. Goeb. FCB. 53; Lac. 140; Bates EMM. vi 1869. 69; Chaud. Mém. sur les Coptodérides, AB. xii. 1869. 163.

Belonognatha Chaud. BM. 1843. ii. 383; Lac. 142.

O* **cyanella** Bates EMM. vi. 1869. 74; (id.) Putz. CRB. xiii. 1870. 17. — Borneo., Celebes. Batchian. New Guinea.

O* *discogutlata* Chaud. Mém. 195.

O* **eluta** × Andr. TES. 1923. 30. — Khasi Hills.

O* *interrupta* ‡ Chaud. (not Schm. Goeb.) Mém. 194; Bates AM. (5). xvii. 1886. 203. — Lakhimpur., Bengal. Shembaganur.

G* *elegantula* ‡ Bates (not Schm. Goeb.) AG. xxvii. 1889. 111; id. AG. xxxii. 1892. 411. — Travancore. Wallardi., Colombo. Burma : Bhamo, Teinzo, Karen Hills, Kawkareik, Mt. Mooleyit, Haungdaran Valley (Tenaserim). Andaman Is., Siam. Indo-China., Penang. Malay Archipelago. Formosa.

P* **flexuosa** Schm. Goeb. FCB. 55; Chaud. Mém. 196; Bates AF. 1889. 283; id. AG. xxxii. 1892. 412; Heller DEZ. 1916. 273; Andr. TES. 1923. 31. — Patkai Hills. Garo Hills., Cachar. Si sagar. Sikkim : Gopaldhara. Dehra Dun. Nilgiri Hills. Karen Hills., Bhamo. Kawkareik. Andaman Is. Malay Peninsula and Archipelago. Philippine Is. Indo-China. Formosa.

O* **impicta** Chaud. Mém. 196. — "Moluccas.", Ternate.

P*	**interrupta** Schm. Goeb. FCB. 53; Bates AG. xxxii. 1892. 411 (part); Andr. TES. 1923. 29.	Sikkim : Pashok, Namsoo, Nurbong, Birik.
P*	*elegantula* Schm. Goeb. FCB. 54; Andr. TES. 1923. 30.	Dehra Dun. Kali Valley (Almora). Nilgiri Hills. Ceylon Matale, Anuradhapura, Horawupotana. Burma : Teinzo, Shwegu, Palon. Pinwe., Tharrawaddy. Kawkareik. Malay Peninsula. Laos., Tonkin., Sarawak.
B*	**japonica** Bates TES. 1883. 281. t. 13. f. 4.	Formosa., Japan.
D*	*formosana* Dupuis AB. 1912. 328.	
O*	**lineolata** Bates EMM. vi. 1869. 74; (id.) Putz. CRB. xiii. 1870. 16.	Mysol. Celebes.
O*	*misella* Chaud. Mém. 197.	Aru Is., New Guinea.
O*	**ocellata** Chaud. Mém. 188.	"North India."
O*	**oxyptera** Chaud. Mém. 175; Maindr. Nov. Guin. v. ii. 1908. 299.	Celebes. New Guinea.
O*	**picta** Chaud. Mém. 191.	Amboina.
O*	**piligera** Chaud. in Oberthür Col. Nov. i. 1883. 20; Bates AG. xxxii. 1892. 412.	Sikkim., Garo Hills. Shillong., Karen Hills., Ruby Mines. Perak., Pahang. Laos., Tonkin., China.
L	**subapicalis** Putz. DEZ. 1877. 84; Bates TES. 1883. 281.	Japan.
O*	**tetrastigma** Chaud. Mém. 174.	Perak.
G'*	*tetracolon (Dolichoctis)* † Bouch. (not Chaud.) AF. 1903. 175.	Java., Sumatra. Borneo., Celebes. Great Sangi I. Philippine Is.
P*	**transversa** Schm. Goeb. FCB. 54; Chaud. Mém. 164; Andr. RIM. xxii. 1921. 346; id. TES. 1923. 31.	Bengal., Etawah. Dehra Dun. Barkuda and Gopkuda.

O* *bicincta* † Chaud. (not Hope) Mém. 187. — Is. (Lake Chilka)

G* *interrupta* † Bates (not Schm. Goeb,) AG. xxxii. 1892. 411 (part); Andr. TES. 1923. 29. — Nilgiri Hills.

chaudoiri × Andr. TES. 1919. 179. — Ceylon: Anuradhapura, Horawupotana, Colombo., Palon. Lashio (North Shan States)., Laos., Hongkong. Java., Sumatra.

COPTOLOBUS* Chaudoir.

Coptolobus Chaud. BM. 1857. ii. 59; id., Mon. des. Scaritides (i), AB. 1879. 159; Andr. FBI. 210.

O* **anodon** Chaud. Mon. 160; Andr. FBI. 212 and 215. — Ceylon : Colombo, Boralesgamuwa.

O* *taprobanae* Chaud. Mon. 161; Bates AM. (5). xvii. 1886. 72; Andr. AM. (9). xiv. 1924. 587.

O* **glabriculus** Chaud. BM. 1857. ii. 60; id. Mon. 162; Putz. CRB. 1875. 45; Bates AM. (5). xvii. 1886. 72; Andr. FBI. 212 and 214. t. 6. f. 4. — Ceylon : Colombo, Nuwara Eliya, Horton Plains, Bogawantalawa, Pattipola, Hakgalla, Perideniya, Pidurutalagalla.

B* *subsignans (Scarites)* Walk. AM. (3). ii. 1858. 203; Chaud. Mon. 162. (note); Andr. TES. 1919. 186.

B* *obliterans (Scarites)* Walk. AM. (3). ii. 1858. 203; Chaud. Mon. 162. (note); Andr. TES. 1919. 186.

B* **latus** Andr. Spol. Zeyl. 1923. 224; id. FBI. 212. — Ceylon : Bulutota, Bilihul-Oya.

O* **omodon** Chaud. Mon. 161; Bates AM. (5) xvii. 1886. 71; Andr. FBI. 212 and 213. f. 36. — Ceylon : Hadley, Dikoya, Bulutota.

CORYZA* Putzeys.

Coryza Putz., Rév. Gén. des Clivinides, AB. x. 1867. 194; Tchitch RRE. 1904. 258; Andr. AM. (9). xvii. 1926. 380; id. FBI. 344 and 346.

S* **maculata** *(Clivina)* Nietn. JASB. xxv. 1856. 391; id. AM. (2). xix. 1857. 246; Putz., Postscr. ad Cliv. Mon., Mém Liége xviii. 1863. 51; id. Rév. 196; Andr. AM. (9). xix. 1927. 105; id. AM. (9). xx. 1927. 265; id. FBI. 348 and 349. — Naga Hills. Eastern Duars., Pusa. Baxar (Garhwal). Dehra Dun., Almora. Haldwani., Nagpur.

L	*nietneri* × Putz. Rév. 196.	North Kanara.
O*	*cariniceps* Putz. Rév. 196.	Madras.
G*	*chlorizans (Clivina)* Bates AG. xxxii. 1892. 282; Andr. AM. (9). iii. 1919. 470.	Malabar. Nilgiri Hills. Murunkan (Ceylon). Teinzo., Shwebo.
B*	**semirubra** Andr. EMM. 1926. 67 and 71; id. AM. (9). xx. 1927. 265; id. FBI. 348 and 350. f. 50.	Sarju Valley, Upper Gumti Valley, Ranikhet, Swal River basin, Dwarahat and Kanarichina (all Almora). Tanakpur., Haldwani. Saiya and Binal Gad (Chakrata). Aglar River (Tehri Garhwal). Dehra Dun. Mussoorie. Seraj (Punjab).

COSMODISCUS* Sloane

Cosmodiscus Sloane PLS. NSW. 1907. 371; Andr. AM. (9). v. 1920. 445.

A*	**picturatus** Andr. AM. (9). v. 1920. 447; id. RIM. xxii. 1921. 345.	Fyzabad., Nagpur. Barkuda and Gopkuda Is. (Lake Chilka). Jammalamadugu. Rambha (Ganjam).
B*	**platynotus** *(Stomonaxus)* Bates TES. 1873. 283; Andr. AM. (9). v. 1920. 447; id. RIM. xxii. 1921. 345.	Sikkim : Gopaldhara. Palni Hills., Java. Sumatra., Japan.
A	**rubripictus** Sloane PLS. NSW. 1907. 372; id. PLS. NSW. 1920. 322; Andr. AM. (9). v. 1920. 447.	Kei Is. Australia.

CRASODACTYLUS* Guérin.

Crasodactylus Guér. Lefeb. Voy. Abyss. vi. 1849. 258; Lac. 279; Tchitch. HER. xxxiv. 1900; 369; id. HER. xxxv. 1901. 251.

O **punctatus** Guér. Lefeb. Voy. Abyss. vi. 1849. 262. t. 1. f. 5; Bedel Fn. med. 1899. 146; Tchitch. HER. xxxv. 1901. 137. — Bannu., Arabia. Abyssinia. Normth Africa.

CRASPEDONOTUS* Schaum.

Craspedonotus Schaum. BEZ. 1863. 86; Putz., Les Broscides, SEZ. 1868. 314; Sem. HER. xxxiv. 1900. 45; id. RRE. x. 1910. 216; Andr. MB. 21.

Pseudobroscus Sem. WEZ. 1888. 248; id. HER. xxxiv. 1900. 45.

Q **himalayanus** Sem. RRE. x. 1910. 217; Andr. MB. 21. t. 2. f. 4.m — Punjab : Baddia. Kangra. Kulu : Parbatti Valley., Dehra Dun.

D **margelanicus** Kraatz. DEZ. 1884. 229; SEM. WEZ. 1889. 60; id. HER. xxxiv. 1900. 45; id. RRE. x. 1910. 218. — Turkestan.

Q *leucocnemis* Sem. WEZ. 1888. 248.

J* **tibialis** Schaum BEZ. 1863. 87. t. 1. f. 5; Putz. SEZ. 1868. 314; Bates TES. 1873. 243; Heyd. DEZ. 1879. 331; Fairm. AF. 1889. 10; Sem. RRE. x. 1910. 216; Andr. AM. (9). xix. 1927. 108. — China. Japan.

CRASPEDOPHORU* Hope.

Craspedophorus Hope Col. Man. ii. 1838. 165; Lac. 210; Chaud., Mon. sur less Panagéides, AB. 1878. 90; Andr. TES. 1919. 126; id. MB. 22.

Eudema Cast. Hist. Nat. i. 1840. 137; Lac. 210. note (4); Chaud., Mon. sur les Panagèides, AB. 1878. 133.

Epicosmus Chaud. BM. 1846. ii. 512 (note); id., BM. 1854. i. 338; id. Rev. des espéces qui rentrent dans l'ancien genre Panagaeus, BM. 1861. ii. 335; id., Mon. sur les Panagéides, AB. 1878. 104.

B* **angulatus** *(Carabus)* F. Spec. Ins. i. 1781. 302; id. Mant. Ins. i. 1787. 197; id. Ent. Syst. i. (1). 1792. 148; Oliv. Ent. 35. 38. t. 7. f. 76; Hope Col. Man. ii. 1838. 92; Chaud. Rév. 336; Heyne-Tasch. EK. 12 t. 3. f. 35; Andr. TES. 1919 125; id. TES. 1921. 154; id. MB. 115. — India south of the Central Provinces, Ceylon.

C* *fasciatus (Pimelia)* F. Spec. Ins. i. 1781. 318; id. Mant. Ins. i. 1787. 209; id. Ent. Syst. i. (1). 1792. 104; Schaum SEZ. 1847. 42; Motch. Etud. Ent. 1855. 69.

C* *reflexus (Cychrus)* † F. Syst. Eleuth. i. 1801. 166 (not F. Spec. Ins. i. 1781. 302); Andr. TES. 1919. 127.

B* *tomentosus (Panagaeus)* Vigors Zool. Journ. i. 1824. 537. t. 20. f. 1; Dej. Sp. ii. 1826. 284; id. Sp. v. 1831. 598; Laf. AF. 1851. 220; Redt. RN. 9; Andr. TES. 1919. 134; id. TES. 1923 (1924). 462.

bifasciatus (Epicosmus) || Chaud. (not Cast.) Rév. 336; id. Mon. 133.

michardi (Eudema) Fairm., Le Naturaliste i. 1880. 307.

O* **basifasciatus** *(Epicosmus)* Chaud. RMZ. 1869. 115; id. Mon. 127. — Laos.

L **bifasciatus** *(Panagaeus)* Cast. Etud. Ent. 1835. 155; Chaud. Rév. 336; Heyne-Tasch. EK. 21; Andr. TES. 1919. 126; id. RIM. xxii. 1921. 341. — Barkuda I. (Lake Chilka)., Madras. Nilgiri Hills. Udayagiri. Horseley Konda. Trichinopoly. Pondicherry. Ceylon : Colombo, Trincomali, Suriya Ara, Tissa, Maha, Illupalama.

castelanvi (Epicosmus) × Chaud. Mon. 112; Bates AM. (5). xvii. 1886. 73.

G* **breviformis** *(Epicosmus)* Bates AG. xxxii. 1892. 299. — Karen Hills., Palon. Thagata (Tenasserim).

G **brevisternis** *(Epicosmus)* Bates AG. xxxii. 1892. 301. — Thagata (Tenasserim).

B* **cereus** *(Panagaeus)* Macl. Ann. Jav. 1825. 12; Chaud. Rév. 349; Andr. TES. 1919. 135. — Java.

O* **elegans** *(Panagaeus)* Dej. Sp. ii. 1826. 290; Laf. AF. 1851. 221; Andr. TES. 1921. 162; id. MB. 23. — Generally distributed throughout India and Ceylon.

notulatus (Epicosmus) † Chaud. (not F.) Rév. 347; id. Mon. 115.

Y **everetti** *(Epicosmus)* Heller Abh. Mus. Dresd. vii. 3. 1898. 2. t. 3. f. 1. — Celebes.

G* **feae** *(Epicosmus)* Bates AG. xxvii. 1889. 101; id. AG. xxxii. 1892. 301. — Burma : Bhamo, Teinzo. Thigyam, Prome, Magwa, Toungoo.

C* **geniculatus** *(Panagaeus)* Wied. Zool. Mag. ii. 1. 1823. 56; Chaud. Rév. 351; id. Mon. 112; Andr. TES. 1921 170 and 187. — Bengal. Pusa.

O* *? rufipalpis (Isotarsus)* Laf. in litt. AF. 1851. 221; Chaud. Mon. 110; Andr. AM. (9). xiv. 1924. 588.

G* **gracilipes** *(Epicosmus)* Bates AG. xxxii. 1892. 302. — Assam : Noa Dehing, Sadiya. Bhamo., Yunnan.

A* **halyi** Andr. Spol. Zeyl. 1923. 230. — Ceylon : Niroddumunai.

O* **hexagonus** *(Epicosmus)* Chaud. Rév. 338; id. Mon. 114. — Dehra Dun.

O* **hilaris** *(Epicosmus)* Chaud. Rév. 345; id. Mon. 110; Andr. TES. 1921. 170; id. AM. (9). xiv. 1924. 588. — Bengal.

O* **laticollis** *(Epicosmus)* Chaud. RMZ. 1869. 114; id. Mon. 125; Lesne Miss. Pav. 69. — Siam. Indo-China.

G* **latigenis** *(Epicosmus)* Bates AG. xxxii. 1892. 300. — Karen Hills.

X* **lesnei** × Andr. AM. (9). xvii. 1926. 253. — Siam.

hilaris (Epicosmus)† Lesne (not Chaud.) Miss. Pav. 69.

G* **mandarinellus** *(Epicosmus)* Bates AG. xxxii. 1892. 299; Andr. TES. 1921. 187. — Bhamo. North Kanara.

J* **mandarinus** *(Isotarsus)* Schaum AF. 1853. 436; Chaud. Rév. 337; id. Mon. 113; Redt. RN. 9; Gestro AG. 1882. 304; Fairm. AF. 1888. 336; Bates AF. 1889. 265; id. AG. xxxii. 1892. 299; Heyne-Tasch. EK. 21; Vuillet Ins. ii. 1912. 18; Andr. AM. (9). xiii. 1924. 468; id. TES. 1923 (1924). 462; id. AM. (9). xix. 1927. 107. — Mandalay. Minhla. Siam., Tonkin. Annam. China. Formosa.

H* *fabricii* Hope in litt. Col. Man. ii. 1838. 165; Andr. TES. 1919. 127.

A* **microspilotus** Andr. Spol. Zeyl. 1924. 131. — Ceylon.

O* **mouhoti** *(Epicosmus)* Chaud. RMZ. 1869. 69; id. Mon. 124. — Laos., Cambodia.

O* **pubiger** *(Epicosmus)* Chaud. Rév. 337; id. Mon. 122. — "East Indies".

O*	**saundersi** *(Epicosmus)* Chaud. RMZ. 1869. 114; id. Mon. 125; Lesne Miss. Pav. 69.	Laos., Cochin-China. Siam.
Z*	**sundaïcus** *(Eudema)* Oberth. NLM. v. 1883. 221.	Sumatra.
L	**transversalis** *(Panagaeus)* Cast. Etud. Ent. 1835. 154; Chaud. Rév. 349; id. Mon. 151.	Java.

CREAGRIS* Nietner.

Creagris Nietn. JASB. xxvi. 1857. 139; id. AM. (2), xx. 1857. 277; Gestro AG. 1875. 868; Sloane PLS. NSW. 1914. 570.

G*	**binoculus** Bates AG. xxxii. 1892. 390.	Jorhat., Khasi Hills. Bhamo., Laos.
C*	**distacta** *(Helluo)* Wied. Zool. Mag. ii. 1. 1823. 49; Andr. TES. 1919. 169; id. TES. 1921. 168 and 186.	Singapore. Java., Sumatra.
O*	**hamaticollis** Bates AG. xxxii. 1892. 390 (note).	Sarawak.
J*S*	**labrosa** Nietn. JASB. xxvi. 1857. 139; id. AM. (2). xx. 1857. 278; Chaud. RMZ. 1872. 213; Gestro AG. 1875. 871. fig.; Bates AM. (5). xvii. 1886. 200; Sloane PLS. NSW. 1914. 570; id. PLS. NSW. 1920. 322; Andr. AM. (9). 1927. 105. *picea (Acanthogenius)* × Schaum. BEZ. 1863. 80; id. BEZ. 1864. 116. t. 2. f. 6. (larva).	Mysore : Chikkaballapura. Madas : Nilambur, Iyerpadi. Ceylon : Colombo., Perideniya: Matale, Horawupotana. Tharrawaddy.
G*	*affinis* Gestro. AG. 1875. 870. fig; Andr. AM. (9). xvii. 1926. 258.	Siam., Java., Tonkin.
B*	**lineola** Andr. AM. (9). xvii. 1926. 281.	Selangor., Sumatra.
G*	**wilsoni** Gestro AG. 1875. 869. fig.; Sloane PLS. NSW. 1914. 570.	Java. Australia.

CREPIDOGASTER* Boheman.

Crepidogaster Boh. Ins. Cafir. i. 1848. 68; Lac. 101; Chaud., Mon. des Brachynides, AB. 1876. 92.

D*	**horni** Dupuis AB. 1914. 132.	Ceylon : Nalanda.
O*	**humerata** Chaud. Mon. 96.	Malabar. Shembaganur.

CRYPTOCEPHALOMORPHA* Ritsema.

Cryptocephalomorpha Rits. Tijds. Ent. 1875. Verslag 92.

Z*	**gaverei** Rits. Tijds. Ent. 1875. Verslag 93.	Java., Sumatra.
B*	*marginata (Adelotopus)* Waterh. TES. 1877. 2; Rits. Tijds. Ent. 1879. Verslag 87; Andr. TES. 1919. 197.	Singapore.

CYCHROPSIS* Boileau.

Cychropsis Boileau BF. 1901. 350; Roeschke, Mon. Cychrini, Ann. Mus. Hung. 1907. 201; Andr. FBI. 100.

O*	**sikkimensis** *(Cychrus)* Fairm. BF. 1901. 265; Roeschke Mon. 202; Andr. FBI. 101. f. 17.	Sikkim : Gnatong, "above Jalep".

CYCHRUS Fabricius.

Cychrus F. Ent. Syst. iv. 1794. 440; Dej. Sp. ii. 1826. 4; Lac. 62; Roeschke, Mon. Cychrini, Ann. Mus. Hung. 1907, 203.

Cychrys F. Skrivt at Naturhist. Selskab. iii. (2). 1794. 68.

O*	**davidis** Fairm. AF. 1886. 307; Roeschke Mon. 234.	Yunnan.
A	**seriatus** Roeschke Mon. 225.	Tibet.
O*	yunnanus Fairm. AB. 1887. 90; Bates Entom. 1891. Suppl. 70; Roeschke Mon. 233.	Yunnan. China.

CYCLOSOMUS* Latreille.

Cyclosomus Latr. in Cuvier's Règne Anim. Ed. 2 iv. 1829 (Aprd). 394 (note); Dej. Sp. iv. 1829 (Dec.). 23; Lac. 258; Chaud, Etude monographique des Masoriédes, des Tetragonodérides, et du genre Nematotarsus, BM. 1876. ii. 29; Andr. MB. 61.

Scolytus || F. (not Müll) Ent. Syst. i. (1). 1792. 180 (part).

C*	**flexuosus** *(Carabus)* F. Syst. Ent. 1775. 246; *(Scolytus)* id. Ent. Syst. i. (1). 1792. 180; Dej. Sp. iv. 1829. 24; Brullé HN. v. 1835. 140; Lac. Atl. t. 10. f. 4; Motch. Etud. Ent. 1855. 70; Chaud. Mon. 32; Heyne-Tasch EK. 23. t. 3. f. 53; Lesne Miss. Pav. 80; Andr. TES. 1921. 153; id. TES. 1923 (1924). 464.	Berhampur. Chandipur. Almora. Chota Nagpur. Konbir., Belgaum. North Kanara. Madras : Mahe, Coimbatore Palur Farm and
J*	*dytiscoides* Nietn. JASB xxvi. 1857. 132; id. AM. (2). xx. 1857. 272; Chaud. Mon. 31; Bates CRB. 1891. 338; Andr. AM. (9), xix. 1927. 105.	

C*	**flexuosus** —	Panicuppam (South Arcot), Madras. Ceylon : Colombo, Negombo., Laos. Cochin-China.
A*	**inustus** Andr. TES 1923 (1924). 464.	Chandipur.
V*	*flexuosus* † Redt (not. F.) RN. 12.	Laos., Hongkong.
Y*	**philippinus** Heller PJS. xxiii. 1923. 302. t. 1. f. 3.	Philippine Is.
	sumatrensis Bouch. AF. 1908. 174.	Sumatra.
C*	**suturalis** *(Scolytus)* Wied. Zool. Mag. i. 3. 1819. 169; Andr. TES. 1921. 166; id. TES. 1923 (1924). 464; id. MB. 62.	Gauhati. Sikkim : Gopaldhara. Malda., Kulu.
M	*marginatus* Motch. BM. 1864. ii. 200; Chaud. Mon. 32; Bates AF. 1889. 282; id. Entom 1891. Suppl. 9; id. CRB. 1891. 338; Andr. TES. 1928. 22.	Delhi., Laos. Annam.

CYMINDIS Latreille.

Cymindis Latr. Gen. Crust. et Ins. i. 1806. 190; Latr. et. Dej. HN. et Ic. 132; Dej. Sp. i. 1825. 202; Lac. 103; Schm. Goeb. FCB. 31; Chaud., Essai monographique sur le genre Cymindis proprement dit, BEZ. 1873. 57; Bedel Fn. med. 1905. 252 and note (3); Andr. MB. 63.

Tarus Schell. Helv. Ent. ii. 1806. 94-5.

Anomoeus Fisch. Ent. Russ. i. 1822. 125.

Psammastes Motch. BM. 1864. ii. 240.

Q	**attenuata** Jak. HER. xxi. 1887. 150.	Pamir.
A*	**babaulti** Andr. MB. 113. t. 4. f. 6.	Leh.
B*	**championi** Andr. AM. (10). ii. 1928. 589.	Kumaon : Leptel, Shelshel.
	alticola | | Andr. (not Casey) EMM. 1926. 70 and 79.	Tibet : Supi River.
I*	**glabrella** Bates PZ. 1878. 719; id. SYM. 20. t. 1. f. 17; Andr. MB. 64.	Kashmir : Baltal. Ladak.
B*	**longstaffi** Andr. AM. (9). xi. 1923. 277.	Mount Everest Expeditions base camp.

O*	**mannerheimi** Gebl. BAP. 1843. i. 36; Chaud. BM. 1850. ii. 81; id. Mon. 99; id. BM. 1875. ii. 59; Bates SYM. 26; Andr. MB. 64.	Kashmire: Pangong Valley, Baltal, Mulbeck, Kargil, "betwenn Dras and Leh.", Pamir Turkestan. Tarbagatai Mts.
I*	*altaïca* † Bates (not Gebl.) SYM. 20.	
Q	**obsoleta** Jak. HER. xixi. 1885. 288.	Pamir.

CYMINDOIDEA* Castelnau.

Cymindoidea Cast. AF. 1832. 390; Chaud., Genres aberrants du groupe des Cymindides, BM. 1875. ii. 9; Andr. MB. 63.

Philotecnus Mann. BM. 1837. i. 42.

A*	**aequa** Andr. AM. (9). xii. 1923. 451.	Kodaikanal., Madras. Mysore : Nandidrug.
O*	**distigma** Chaud. Mon. 15.	Bengal. "Hardwar and Roorkee."
P*	**indica** *(Cymindis)* Schm. Goeb. FCB. 31; Chaud. Mon. 16; Bates AG. xxxii. 1892. 417; Andr. TES. 1923. 11; id. MB. 63.	Saran., Pusa., Purnea. Orissa : Puri, Khandagiri. C. P. : Mukhi. North Kanara. Nilgiri Hills. Pondicherry. Madras. Trichinopoly. Bhamo, Mandalay. Shwegu., Naba.
O*	*geerini (Cymindis)* Chaud. BM. 1850 i. 49.	
A*	**munda** Andr. Spol. Zeyl. 1923. 247.	Nilgiri Hills. Ceylon : Bandarawela.
O*	**nigra** Chaud. Mon. 19.	"Coromandel". Naduvatam.
V	**quadrimaculata** *(Cymindis)* Redt. HK. 499. t. 23. f. 3.	Kashmir.

DESERA* Hope.

Desera Hope Col. Man. ii. 1838. 105; Lac. 80. note (2).

DESERA* **Hope.—**

Dendrocellus Schm. Goeb. FCB. 24; Lac. 80.

J* **coelestina** *(Drypta)* Klug Jahrb. Ins. 1834. 54; Andr. AM. (9). xix. 1927. 100.	Java.
D* **elegans** *(Dendrocellus)* Sloane DEZ. 1907. 473; Andr. AM. (9). xix. 1927. 110.	New Guinea.
J* **geniculata** *(Drypta)* Klug Jahrb. Ins. 1834. 52; Schm. Goeb. FCB. 25; Chaud. RMZ. 1872. 102; Bates TES. 1883. 279; id. AF. 1889. 280; id., CRB. 1891. 336; Bouch. AF. 1903. 173; Andr. TES. 1921. 173; Heller PJS. xix. 1921. 530; Andr. TES. 1923. 8; Heller PJS. xxiii. 1923. 303; Andr. AM. (9). xiii. 1924. 469.	Throughout South East Asia. including Japan.
var. **rugicolis** *(Dendrocellus)* × Chaud. BM. 1861. i. 546; id. RMZ. 1872. 102; Andr. TES. 1919. 170.	
P* *flacipes (Dendrocellus)* † Schm. Goeb (not Wied) FCB. 24; Andr. TES. 1921. 173 (note); id. TES. 1923. 8.	
G* **gestroi** *(Dendrocellus)* Bates AG. xxxii. 1892. 385; Heller PJS. xxiii. 1923. 364.	Karen Hills.
E **gilsoni** Dapuis AB. 1912. 319.	Formosa.
O* **longicollis** *(Drypta)* Macl. Ann. Jav. 1825. 28; Dej. Sp. i. 1825. 185; Bates AG. xxxii. 1892. 385; Andr. TES. 1919. 168; Heller PJS. xxiii. 1923. 363.	India., Karen Hills. Thagata. Tenasserium.
B* **nepalensis** Hope Zool. Misc. 1831. 21; Chaud. RMZ. 1872. 192; Dohrn SEZ. 1879. 457; Andr. TES. 1919. 170; Heller PJS. xxiii. 1923. 303.	Assam Valley. Khasi Hills. Cachar., Nepal. Sikkim : Darjiling Gopaldhara, Pedong. Chota Nagpur : Tetara, Konbir. Shembaganur Manipur., Martaban. Tonkin.
P* *discolor (Dendrocellus)* Schm. Goeb. FCB. 24; Bates CRB. 1891. 336; Andr. TES. 1923. 7; Heller PJS. xxiii. 1923. 304.	
O* **parallela** *(Dendrocellus)* Chaud. RMZ. 1872. 101; Heller PJS. xxiii. 1923. 303.	Sumatra.
Y* **schultzei** Heller PJS. xxiii. 1923. 303 and 304.	Philippine Is.
O* **ternatcusis** *(Dendrocellus)* Chaud. RMZ. 1872. 102; Heller PJS. xxiii 1923. 203.	Ternate.

B* **unidentata** *(Drypta)* Macl. Ann. Jav. 1825. 28; Andr. TS. 1919. 167; Heller PJS. xxiii. 1923. 303. — Java.

DHANYA* Andrewes.

Dhanya Andr. AM. (9). vi. 1919. 295 and 296; Bänn DEZ. 1927. 190 and 203.

B* **bioculata** Andr. AM. (9). iv. 1919 295 and 297. — Tonkin Philippine Is. Borneo., Sumatra. Sumbawa.

A* **parallela** Andr. AM. (9). iv. 1919. 295 and 298. — Sarawak.

DICAELINDUS* Macleay.

Dicaelindus Macl. Ann. Jav. 1825. 18; Schaum. BEZ. 1863. 86.

Dicoclindus Lac. 366; Andr. AM. (9). xiv. 1924, 589.

Melanodes Chaud., Mon. des Oodes (ii), AF. 1882. 545 (part).

Arsenoxenus Bates AG. xxxii. 1892. 365; Andr. TES. 1919. 149.

B* **felspaticus** Mad. Ann. Jav. 1825. 19. t. 1. f. 6; Chaud. BM. 1869. i. 356; Andr. TES. 1919. 148; id. AM. (9) xiv. 1924. 589. — Java. Borneo.

B* **impunctatus** *(? Lagarus)* Bates AM. (5). xvii. 1886. 145; Tchitch. HER. xxxiv. 1990. 476; Andr. AM. (9). xiv. 1924. 589. — Ceylon., Colombo. Calie., Kotte.

X **nitescens** *(Arsenoxenus)* Tchitch. HER. xxxiv. 1960. 475. — Berneo of Sumatra.

O* **pernitidus** *(Melanodes)* Chaud. Mon. 550; Andr. AM. (9). xiv. 1921. 589. — Jorbat., Khasi Hills. Cathuti.

O* *subelongatus (Melanodes)* Chaud. Mon. 550. — Dacca., Sahibganj.

G* *harpaloides (Arsenoxenus)* Bates AG. xxxii. 1892. 365; Tchitch. HER. xxxiv. 1900. 476. — Bhamo., Tharrawaddy. Palon., Rangoon.

DICEROMERUS* Chaudoir

Diceromerus Chaud., Essai monographique sur les Drimostomides, AB. xv. 1872. 15.

M **orientalis** *(Stomonaxus)* Motch. Etud. Ent. 1859. 36; Chaud. Mon. 15; Bates AM. (5). xvii. 1886. 143; id. CRB. 1891. 334; id., AG. xxxii. 1892. 351; Andr. TES. 1928. 22. — India., Ceylon. Burma. Kuala Lumpur. Singapore., Java.

X* *chaudoiri* Fleut. AF. 1887. 60; Bates AF. 1889. 277; Andr. TES. 1921. 179. — Sumatra., Timor. Laos., Cochin-China., Hongkong.

DICRANONCUS* Chaudoir.

Dicranoncus Chaud. BM. 1850. i. 392; Lac. 358; Chaud. Rév. des genres Onychopterygia, Dicranon-cus, Colpodes, AF. 1878. 277; Sloane PLS. NSW. 1910. 453.

Menera Motch. Etud. Ent. 1859. 32; Andr. TES. 1928. 17.

Monacanthonyx Bates AG. xxxii. 1892. 367.

O* **cinctipennis** Chaud. Rév. 278. — Kandy, Peradeniya. Hongkong (Chaudoir).

O* **femoralis** Chaud. BM. 1850. i. 393; id. Rév. 277; Bates TES. 1873. 278; id. CRB. 1891; 335; id. AG. xxxii. 1892. 378. — Himalayan tract from Simla to Manipur. Chota Nagpur: Tetara., Burma: Ruby Mines

M *coelestinus (Loxocrepis)* Motch. BM. 1864. ii. 310; Andr. TES. 1928. 22. — Take Pum Mt., Karen Hills. Perak., Laos. Tonkin., Japan.

G* **pocillator** *(Monacanthonyx)* Bates AG. xxxii. 1892. 367. — Karen Hills.

X **pallidicornis** Fairm. CRB. 1891. 188. — China.

M* **quadridens** *(Menera)* Motch. Etud. Ent. 1859. 32; Andr. TES. 1928. 17. — Assam : Rotung. Shillong.

X* *ruficeps (Loxocrepis)* † Brullé (not Macl). HN. iv. 1834. 325. t. 12. f. 2; *(Colpodes)* Bates TES. 1883. 263; id. AM. (5). xvii. 1886. 147; id. AG. xxxii. 1892. 376; Andr. TES. 1919. 164. — Sikkim: Mungphu, Gopaldhara. Almora., Bhimtal. Mussoorie. Nilgiri Hills.

O* *amabilis* Chaud. AF. 1859. 350 (note) and 359; id. Rév. 277; ?Schm. Goeb. FCB. t. 2. f. 7; Andr. TES. 1923. 57. — Anaimalai Hills. Coorg., Travancore. Burma : Gokteik, Karen Hills, Moulmein, Momeit, Tavoy., Andaman Is. Malay Peninsula and Archipelago. Philippine Is. Indo-China.

DICRASPEDA* Chaudoir.

Dicraspeda Chaud. BM. 1862. ii. 300; Sloane PLS. NSW. 1923. 30.

Eudalia Cast. Trans. Roy. Soc. Vict. viii. 1868. 102; Bates EMM. viii. 1871. 32; Sloane PLS. NSW. 1917. 415 and 417.

O* **brunnea** Chaud. BM. 1862. ii. 300. — Java., Celebes. Timor., Philippine Is.

DINOPELMA* Bates.

Dinopelma Bates PZS. 1889. 385; Andr. AM. (9). xiv. 1924. 591.

B* **plantigradum** Bates PZS. 1889. 385; Andr. TES. 1919. 197. — Bornco.

DIORYCHE* Macleay.

Dioryche Macl. Ann. Jav. 1825. 21; Lac. 300. note (1); Bates TES. 1873. 271; Alluaud BF. 1917. 321; Andr. TES. 1919. 156; id. MB. 32.

G* **cavernosa** *(Platymetopus)* Putz. AG. 1875. 737. — Java., Sumatra.

J* *aerea (Gnathaphanus)* Shauf. HER. xxi. 1887. 105; Andr. AB. 1920. 106; id. AM. (9). xix. 1927. 110. — Celebes.

A* **chinnada** Andr. RIM. xxii. 1921. 342. — Bengal., Pusa. Gorakhpur (U.P.). Throughout the Central Provinces. Bombay and Madras. Ceylon.

B* **clara** Andr. AM. (9). ix. 1922. 290 and 292; id. AM. (9). xiii. 1924. 468. — Tonkin., Annam. Hongkong., Macao. Pescadores Is.

L	**colombensis** *(Selenophorus)* Nietn. JASB. xxvi. 1857. 151; Id. AM. (2). xx. 1857. 373; Bates AM. (5). xvii. 1886. 76; id. AB. 1892. 231; Andr. RIM. xxii. 191. 342; id. MB. 32; id. AM. (9). xix. 1927. 105.	Throughout India and Ceylon. Maldive Is.
B*	*scita (Cardiaderus)* Walk AM. (3). ii. 1858. 203; Andr. TES. 1919. 188.	
E*	var. **braccata** *(Platymetopus)* Bates CRB. 1891. 330; id. AB. 1892. 231; Andr. EMM. 1926. 68.	U. P.: Almora. Dehra Dun. Chota Nagpur. Nagpur., Pachmarhi. Bombay. Nilgiri Hills.
A*	**convexa** Andr. MB. 90.	Majgaon (Jabbulpore). Belgaum.
X*	**indochinensis** *(Platymetopus)* Bates AF. 1889. 270; id. AG. xxxii. 1892. 330; Lesne Miss. Pav. 73; Andr. TES. 1921. 177; id. AM. (9). ix. 1922. 291 and 292.	Tharrawaddy. Shwegu., Rangoon. Siam., Indo-China.
G*	**longula** *(Platymetopus)* Bates AG. xxxii. 1892. 330.	Mandalay.
B*	**melanauges** Andr. AM. (9). ix. 1922. 288 and 292.	Siam., Tonkin.
E*	**nagpurensis** *(Platymetopus)* Bates CRB. 1891. 329; Andr. RIM. xxii. 1921. 342; id. MB. 116.	Dehra Dun. Almora., Mussoorie. Pusa., Chapra. Purnea., Calcutta. Cuttack., Saran. Sarda., Sunderbans. Berhampur. Murshidabad. Tinpahar (Rajmahal). Manbhum. Chota Nagpur : Konbir Barway. Puri., Bhubaneswar. Chandipur. Balugaon., Surada. Barkuda I (Lake, Chilka). Ceylon : Anuradhapura, Vavuniya.

B*	**sericea** Andr. AM. (9). ix. 1922. 289 and 292.	Ceylon: Halupahani (Haldummulle). Annam., Tonkin.
B*	**torta** Macl. Ann. Jav. 1825. 21; Hope Col. Man. ii. 1838. t. 2. f. 4; Bates TES. 1873. 271; Andr. TES. 1919. 154; id. EMM. 1926. 68.	India., Ceylon. Burma. Malay Peninsula.
O*	*amoena (Platymetopus)* Dej. Sp. iv. 1829. 73; Bates AG. xxxii. 1892. 330; id. AB. 1892. 231; Bouch. AF. 1903. 171; Andr. TES. 1919. 155; id. TES. 1921. 147; id. AM. (9). ix. 1922. 291 and 292; id. AM. (9). xiv. 1924. 590.	Java., Sumatra. Borneo., Indo-China. Formosa.
X*	*laetula (Platymetopus)* Bates AF. 1889. 270; Andr. TES. 1921. 177.	

DIPLOCHEILA* Brullé.

Diplocheila Brullé HN. iv. 1834. 407.

Diplochila Bedel Fn. med. 1897. 102; Andr. TES. 1919. 90.

Rembus || Dej. (not Germ.) Sp. ii. 1826. 380; Laf. AF. 1851. 278; Lac. 233.

Eccoptogenius Chuaud. BM. 1852. i. 72; id. BM. 1874. i. 2; Bates AF. 1889. 267; Andr. TES. 1921. 176; id. AM. (9). xiv. 1924. 590.

Symphyus Nietn. AM. (3). ii. 1858. 180.

Rhembus Mun. Cat. 238.

G*	**colossus** *(Rhembus)* Bates AG. xxxii. 1892. 326; Andr. AM. (9). ix. 1922. 285 and 286.	Palon., Rangoon. Siam.
V*	*impressa* † Redt. (not F.) RN. 10; Andr. TES. 1923 (1924). 464.	Indo-China., Java. Sumatra. Philippine Is.
O*	**cordicollis** *(Rhembus)* Laf. AF. 1851. 279.	Pusa., Chapra. Malda., Dacca. Jalpaiguri (Bengal). Delhi., Bombay.
O*	**distinguenda** *(Rhembus)* Laf. AF. 1851. 278; Andr. TES. 1921. 147; id. AM. (9). xiv. 1924. 590.	Pusa., Purnea. Dinapore.
O*	*moesta (Eccoptogenius)* Chaud. BM. 1852. 1. 74.	Balugaon.

B*	**elongata** *(Rhembus)* Bata TES. 1873. 256.	China., Japan.
C*	**impressa** *(Carabus)* F. Suppl. Ent. Syst. 1798. 57; Dej. Sp. ii. 1826. 383; Hope Col. Man. ii 1838. 81; Laf. AF. 1851. 279; Motch. Etud. Ent. 1855. 62; Bates AG. xxxii. 1892. 326; Andr. TES. 1921. 159; id. AM. (9). ix. 1922. 285 and 286.	Assam. Calcutta. Diamond Harbour. Berhampur. Birbhum. Muzaffarpur. Pusa., Chapra Madras. Trichinopoly. Kandy., Prome. Palon., Moulmein. Singapore. Cochin-China.
G*	**laevigata** *(Rhembus)* Bates AG. xxxii. 1892. 326; Andr. TES. 1921. 177; id. AM. (9). ix. 1922. 282 and 286.	Mandalay. Kawkareik. Malacca., Siam.
B*	*polita (Rhembus)* † Macl. (not F.) Ann. Jav. 1825. 16; Redt. RN. 10; Lesne Miss. Pav. 72; Andr. TES. 1919. 141; id. TES. 1923. (1924). 463.	Indo-China. Java., Sumatra. Berneo. Philippine Is. ? Japan.
X*	*moesta (Eccoptogenius)* † Bates (not Chaud.) AF. 1889. 267; Andr. TES. 1921. 176.	
X*	**laevis** *(Rhembus)* Lesne Bull. Mus. Paris. ii. 1896. 243. f. 6; id. Miss. Pav. 72. t. 8. f. 8. and text fig. 6; Bouch. AF. 1903. 171; Andr. AM. (9) ix. 1922. 284 and 286.	Ceylon., Rangoon. Andaman Is. Siam., Indo-China. Java., Sumatra. Borneo., China.
O*	**latifrons** *(Rhembus)* Dej. Sp. v. 1831. 679; Dej. et Boisd. Ic. et HN. ii. 204. t. 97. f. 4; Laf. AF. 1851. 279; Bates AG. xxxii. 1892. 327; Lesne Miss. Pav. 72; Andr. AM. (9). ix. 1922. 283 and 286.	Calcutta., Birbhum. Murshidabad. Rajmahal., Kashmir. Bangalore., Palon.
O*	*opaca (Rhembus)* Chaud. BM. 1852. i. 67; Bates TES. 1873. 255; id. AF. 1889. 267.	Siam., Indo-China. Java., Chusan., Japan.
	perscissa × Andr. TES. 1921. 156; id. AM. (9). ix. 1922. 283.	Jorhat. Sylhet.
L	*? indica (Carabus)* Herbst in Arch. Ins.-gesch. (Fuessly). v. 1784. 138. t. 29. f. 11; Chaud. BM. 1852. i. 67; Andr. TES. 1919. 144; id. AM. (9). xix. 1927. 98.	Sardah., Rajmahal. Sunderbans., Pusa. Chapra., Purnea.

O*	*polita (Rhembus)* † Dej. (not F.) Sp. ii. 1826. 381; Laf. AF. 1851. 279; Bates AG. xxxii. 1892. 325.	Bombay., Burma. Tharrawaddy. Rangoon., Bhamo. Teinzo., Palon., Java.
B*	**pinodes** Andr. AM. (9). ix. 1922. 281 and 286.	Annam., Tonkin.
C*	**polita** *(Carabus)* F. Ent. Syst. i. (1). 1792. 146; Brullé HN. iv. 1834. 408. t. 16. f. 6; Motch. Etud. Ent. 1855. 62; Andr. TES. 1921. 155; id. AM. (9). ix. 1922. 283.	Throughout India, Ceylon and Burma.
O*	*subpunctata (Rhembus)* Laf. AF. 1851. 278.	
J*	*unicolor (Symphyus)* Nietn. AM. (3). ii. 1858. 180; Motch. Etud. Ent. 1862. 55; Andr. AM. (9). xixi. 1927. 105.	
B*	*retinens (Platysma)* Walk. AM. (3). iii. 1859. 51; Bates AM. (5). xvii. 1886. 212; Andr. TES. 1919. 192.	
E*	*rectificata (Rhembus)* Bates CRB 1891. 329; id. AG. xxxii. 1892. 325; Andr. TES. 1919. 193.	
	distinguenda † Andr. (not Laf.) TES. 1919. 193; id. TES. 1921. 147.	
L	**quadricollis** *(Rhembus)* Laf. AF. 1851. 279.	? Dehra Dun (Forest Res. Inst. Coll.)
Q	**transcaspiea** Sem. HER. xxv. 1890. 278; Andr. RIM. xviii. 1919. 100; Sem. RRE. 1926. 42.	Seistan : Labi Baring. Trans-Caspian Province.
V*	**zeelandica** *(Rhembus)* Redt. RN. 10. t. 1. f. 5; Heyd. DEZ. 1879. 330; Andr. TES. 1923 (1924). 464.	Formosa., China. Japan.
B*	*gigas (Rhembus)* Bates TES. 1873. 256; Putz. CRB. 1875. 49.	

DIROTUS* Macleay.

Dirotus Macl. Ann. Jav. 1825. 16; Lac. 312.

Pirantillus Bates AG. xxvii. 1889. 108.

G*	**extensicollis** *(Pirantillus)* Bates AG. xxxii. 1892. 370.	Karen Hills., Laoss.
G*	**feae** *(Pirantillus)* Bates AG. xxvii. 1889. 109; id. AG. xxxii. 1892. 370.	Sikkim : Tarkhola (Tista Valley). Burma : Palon, Meetan, Victoria Point., Laos.

B*	**subiridescens** Macl. Ann. Jav. 1825. 17; Hope Col. Man. ii. 1838. t. 2. f. 1; Andr. TES. 1919. 145.	Pahang., Java. Sumatra., Borneo.

DISCHISSUS* Bates.

Dischissus Bates TES. 1873. 243; Chaud. RMZ. 1878. 86; id. Mon. sur less Panagéides, AB. 1878. 149.

G*	**alaticollis** Bates AG. xxxii. 1892. 302.	Thagata (Tenasserim). Andaman Is.
B	**borneensis** Friv. Trrmés. Füzetek. vi. 1882. 134.	Borneo.
	chaudoiri × Andr. TES. 1919. 135.	Java.
O*	*cereus* † Chaud. (not Macl.) RMZ. 1869. 116; id. Mon. 150.	
J*	**guttiferus** *(Isotarsus)* Schaum. AF. 1853. 437; Chaud. BM. 1861. ii. 349; id. Mon. 151; Andr. AM. (9). xix. 1927. 107.	Java.
B*	**mirandus** Bates TES. 1873. 244; id. TES. 1876. 5. t. 1. f. 2; Chaud. Mon. 152.	Japan.
C*	**notulatus** *(Carabus)* F. Syst. Eleuth. i. 1801. 201; Hope Col. Man. ii. 1838. 90; Schaum SEZ. 1847. 48; id. AF. 1853. 432; Motch. Etud. Ent. 1855. 70; Andr. TES. 1921. 162.	Sikkim : Namsoo. Calcutta., Almora. Nilgiri Hills. Insein., Toungoo.
J*	*longicornis (Craspedophorus)* Schaum BEZ. 1863. 84; Chaud. Mon. 153; Bates AG. xxxii. 1892. 303; Heller DEZ. 1916. 276; Andr. AM. (9). xix. 1927. 108.	Rangoon. Kawkareik. Kuala Lumpur. Singapore., Sumatra. Hongkong.
S*	*sumatranus (Panagaeus)* Dohrn SEZ. 1891. 253 and 254; Andr. TES. 1922. 246.	Shanghai.
M	**quadrinotatus** *(Peronomerus)* Motch. BM. 1864. ii. 33; Bates TES. 1873. 244; Chaud. Mon. 152; Andr. TES. 1928. 22.	India (Oxford Mus.) Annam. Japan.

DISPHAERICUS* G.R. Waterhouse.

Disphaericus G.R. Waterh. TES. 1842. 211; Lac. 249; Kolbe SEZ. 1895. 346; id. in Deutsch-Ost-Africa vi. 4. Käf. 1898. 64.

Dyschiridium Chaud. BEZ. 1861. 130.

Spanus Westw. PES. 1864. 3; Kolbe SEZ. 1895. 347; id. in Deutsch-Ost-Africa iv. 4. Käf. 1898. 65.

J*	**marginicollis** Shaum BEZ. 1864. 122. t. 2. f. 3; Andr. AM. (9). xix. 1927. 109.	Madras. Tranquebar. Horseley Konda.
B*	**myrmex** Andr. Spol. Zeyl. 1923. 228.	Anaimalai Hills. Ceylon : Koggala, Velverry.
B*	**ovicollis** Bates AM. (5). xvii. 1886. 73.	Ceylon : Dikoya.

DOLICHOCTIS* Schmidt-Goebel.

Dolichoctis Schm. Goeb. FCB. 62; Lac. 136; Chaud., Mém. sur les sur les Coptodérides. AB. xii. 1869. 245.

O*	**aculeata** Chaud. Mém. 251.	Batchian., Tenimber.
O*	*spinipennis* Chaud. Mém. 251.	Celebes. New Guinea.
O*	**angulicollis** Chaud. Mém. 250; Bates AG. xxxii. 1892. 413.	Burma : Karen Hills, Rangoon, Palon, Meetan. Penang., Java. Laos., Tonkin.
G*	**angusticollis** Bates AG. xxxii. 1892. 413.	Burma. : Bhamoo, Zigon, Meetan, Moulmein.
A*	**chitra** Andr. Spol. Zeyl. 1923. 246.	Ceylon : Nalanda, Negombo, Kandy.
A*	**convexa** Andr. AM. (9). xii. 1923. 450.	Kodaikanal. Shembaganur.
G	**expansicollis** Bates AG. xxxii. 1892. 415.	Bhamo., Karen Hills.
O*	**gilvipes** *(Coptodera)* Dej. Sp. v. 1831. 396; Chaud. Mém. 248.	Philippine Is.
B*	**goniodera** Bates AM. (5). xvii. 1886. 204.	Ceylon : Kitugalle.
V*	**immaculata** *(Mochtherus)* Redt. RN. 7; Andr. TES. 1923 (1924). 461.	Java.
G	**incerta** Bates AG. xxxii. 1892. 414.	Kawkareik.

G*	**iridea** Bates AG. xxxii. 1892. 414.	Karen Hills.
B*	**lunigera** Andr. AM. (9). xviii. 1926. 288.	Sumatra.
B*	**marginifer** *(Dromius)* Walk. AM. (3). ii. 1858. 202; Bates AM. (5). xvii. 1886. 210; Andr. TES. 1919. 184.	Ceylon : Nuwara Eliya, Anuradhapura, Kandy, Pattipola.
B*	*marginicollis (Colpodes)* Walk. AM. (3). iii. 1859. 51; Andr. TES. 1919. 192.	Chota Nagpur : Tetara (Bates).
M	*quadriplagiata (Cyrtopterus)* Motch. BM. 1861. i. 106. t. 9. f. 4; Chaud. Mém. 245; Putz. CRB. 1875. 45; Bates AM. (5). xvii. 1886. 204; id. CRB. 1891. 339; Andr. TES. 1928. 22.	
O*	**parvicollis** Chaud. Mém. 249.	Borneo.
G*	**rutilipennis** Bates AG. xxxii. 1892. 412.	Karen Hills.
P*	**striata** Schm. Goeb. FCB. 62; Chaud. Mém. 246; Andr. TES. 1923. 36 and 45.	Cochin State: Parambikalam.
P*	*rotundata (Mochtherus)* Schm. Goeb. FCB. 77; Chaud. Mém. 241 and 246; Bates AG. xxvii. 1889. 111; id. AG xxxii. 1892. 413; Andr. TES. 1923. 45.	Ceylon : Kandy, Balangoda, Alutumoara Nalanda.
O*	*tetrastigma* Chaud. Mém. 248; Sloane PLS. NSW. 1920. 322.	Henaratgoda, Horawupotana.
B*	*ornatella* Bates TES. 1883. 282.	Burma : Bhamo, Teinzo, Katha,
B*	*fasciola* Bates AM. (5). xvii. 1886. 205.	Shwegu, Martaban, Thagata.
A*	*macleayi (Mochtherus)* Sloane PLS. NSW. 1907. 377; id. PLS. NSW. 1917. 432.	Andaman Is. Malay Peninsula and Archipelago. Arus Is. New Guinea., Siam. Laos., Tonkin. Philippine Is. Japan., Australia.
Z*	**tenuilimbata** Oberth. NLM v. 1883. 219.	Sumatra.
O*	**tetracolon** Chaud. Mém. 248.	Palni Hills. Trichinopoly. Tenasserim., Annam. Malay Peninsula and Archipelago.

B* **vitticollis** Bates AM. (5). xvii. 1886. 204. Ceylon : Dikoya

DOLICHUS Dejean.

Dolichus Dej. Sp. iii. 1828. 36; Lac. 344; Andr. TES. 1919. 91.

X* **bicolor** Maindr. BF. 1910. 15. Yunnan.

X* **viduus** Maindr. BF. 1910. 16. Yunnan.

X* var. **phaeopus** Maindr. BF. 1910. 17. Yunnan.

DROMIUS Latreille et Dejean.

Dromius Latr. et Dej. HN. et Ic. 175; Dej. Sp. i. 1825. 233; Schm. Goeb. FCB. 39; Lac. 119; Motch. Etud. Ent. 1855. 50. note (1); Reitt. Rév. (palaearctic species) WEZ. 1905. 229; Andr. TES. 1919. 91.

B* **adoxus** Andr. AM. (9). xii. 1923. 687. Sikkim ; Kureseong, Gopaldhara. Chakrata. Kumaon : Almora, Nainital, Ranikhet, Bhatkot, Dudhatoli. Simla Hills : Kotagarh., Dharmsala. Lyallpur.

X* **alienus** Bates AF. 1889. 285; id. AG. xxxii. 1892. 120. Sadiya., Karen Hills. Annam.

A* **indicus** Andr. AM. (9). xii. 1923. 686. Buxar Duars. Sikkim : Gopaldhara, Kurseong, Gangtok, Mungphu. Kumaon : Almora, Lebong, Phoobsering

B* **orthegonioides** Bates AM. (5) xvii. 1886. 205. Nilgiri Hills. Ceylon : Dikoya.

B* **steno** Bates AM. (5). xvii 1886. 206. Nuwara Eliya.

DRYPTA* Latreille.

Drypta Latr., Précis des caractéres génériques des Insectes, 1796. 75; F. Syst. Eleuth. i. 1801. 230;

	Latr. et Dej. HN. et Ic. 117; Dej. Sp. i. 1825. 182; Schm. Goeb. FCB. 22; Lac. 79; Andr. MB. 51.	
G*	**aeneipennis** Bates AG. xxvii. 1889. 109; id. AG. xxxii. 1892. 384.	Bhamo.
G*	**aeneipennis** † Bouch. (not Bates) AF. 1903. 173.	Borneo.
C*	**aeneipes** Wied. Zool. Mag. ii. 1. 1823. 60; Andr. TES. 1921. 173; *(Desera)* Heller PJS. xxiii. 1923. 304.	Dacca.
G*	var. **fimbriata** Bates AG. xxxii. 1892. 384.	Burma: "between Prome and Minhla," Teinzo.
O*	**amabilis** Chaud. BM. 1852. i. 35.	Tibet.
A*	**argillacea** Andr. MB. 106. t. 3. f. 5.	Dacca., Bijnor. Chakrata. Dehra Dun., Nainital. Nagpur., Surada.
O*	**crassiuscula** Chaud. BM. 1861. i. 550; Andr. TES. 1923. 6.	Dehra Dun.
Z*	**dimidiata** Putz. NLM. ii. 1880. 191; id. Midden-Sumatra iv. 6. 3. 1887. t. 2. f. 2.	Sumatra.
G	**feae** Gestro AG. 1875. 857; Chaud. BM. 1877 i. 259.	Ternate.
C*	**flavipes** Wied. Zool. Mag. ii 1. 1823. 60; Dej. Sp. ii. 1826. 442; Schm. Goeb. FCB. 24; Andr. TES. 1921. 173.	Northern India from Simla to Assam.
O*	*pallipes* Chaud. BM. 1850. i. 33; id. BM. 1861. i. 548.	
O*	**formosana** Bates TES 1873. 333; Chaud. BM. 1877. i. 257.	Formosa.
B*	**fulveola** Bates TES. 1883. 279.	Japan.
G	**fumigata** Putz. AG. 1875. 721; Chaud. BM. 1877. i. 258.	New Guinea.
B*	**japonica** Bates TES. 1873. 303; Chaud. BM. 1877. i. 262; Bates TES. 1883. 279; Andr. TES. 1919. 167.	Japan.
B*	**lineola** Macl. Ann. Jav. 1825. 27; Dej. Sp. i. 1825. 184; Redt. RN. 4; Chaud. BM. 1877. i. 262; Bates TES. 1883. 279; id. CRB. 1891. 336; id. AG. xxxii. 1892. 383; Heyne-Tasch. EK. 13. t. 2. f. 25; Bouch. AF. 1903. 173; Andr. TES. 1919. 167; id. AM. (9). xiii. 1924. 469; id. TES. 1923 (1924). 460; id. MB. 52.	Throughout S. E. Asia. Java. Sumatra., Borneo.

O*	var. **virgata** Chaud. BM. 1850. i. 34; Bates AF. 1889. 280.	Burma. Indo-China., Yunnan. China., Formosa.
O*	var. **philippinensis** Chaud. BM. 1877. i. 263.	Philippine Is.
P*	**lugens** Schm. Goeb. FCB 23; Andr. TES. 1923. 7.	Burma.
L	**mandibularis** Cast. Etud. Ent. 1834. 43.	Borneo., Java.
O*	**mouhoti** Chaud. RMZ. 1872. 102.	Laos.
P*	**obscura** Schm. Goeb. FCB. 23; Andr. TES. 1923. 6.	Burma.
G*	**sidera** Bates AG. xxxii. 1892. 382; Andr. TES. 1923. 6.	Garo Hills. Dehra Dun. Karen Hills., Laos. China (Brit. Mus.).
G	**sulcicollis** Putz. AG. 1875. 721; Chaud. BM. 1877. i. 258.	New Guinea.
P*	**tristis** Schm. Goeb. FCB. 23; Andr. TES. 1923. 7.	Burma.
	tuberculata × Andr. AM. (9). xiv. 1924. 592.	Shwegu.
G*	*crassiuscula* † Bates (not Chaud.) AG. xxxii. 1892. 383; Andr. TES. 1923. 6.	

DYSCHIRIUS Panzer.

Dyschirius Panz. Index. Ent. 1813. 67; Steph. Ill. Brit. Ent. i. 37 and 40; Putz., Mon. des Clivina et genres voisins, Mém. Liége ii. 1846. 522 and 524; Lac. 202; Schioedte Nat. Tidsskr. 1866 161-3 (Danish); Putz. Rév. Gén. des Clivinides, AB. x. 1867. 32; id., Suppl. Rév. Gén., AB. xi. 1868. 9-10; id. Suppl. ii. Rév. Gén. AB. xvi. 1873. 11-15; Fleisch. Best. Tab. eur. Col. xxxix. Scaritini, 1899, 8; Tchitch. RRE. 1904. 259; Andr. TES. 1919. 91; Müller, Best. Tab. Dyschirius Arten Eur., Kol. Rundesch. x. 1922. 33; Andr. AM. (9). xvii. 1926. 377; id. FBI. 344 and 390.

Phreoryctes Schm. Goeb. in litt. FCB (cover).

B*	**aeger** Andr. FBI. 393 and 404.	Kumaon : West Almora and Upper Gumti Valley.
B*	**asper** Andr. FBI. 392 and 398.	Sardah.
B*	var. **intricatus** Andr. FBI. 392 adn 398.	Pusa.

B*	**bengalensis** Andr. FBI. 392 and 397.	Calcutta. Sunderbans.
B*	**bifrons** Andr. FBI. 394 and 411.	River Sarda Gorge (Kumaon).
L	**binodosus** Putz. CRB. 1878. 173; Andr. AM. (9). xvii. 1926. 378; id. FBI. 414.	Calcutta.
B*	**championi** Andr. FBI. 393 and 406.	Haldwani.
B*	**constrictus** Andr. FBI. 393 and 408.	Shillong., Sardah. Kierpur (Purnea district). West Bhatkot (Kumaon). Dehra Dun.
B*	**disjunctus** Andr. FBI. 394 and 411.	Sardah., Haldwani. West Almora.
L	**fusus** Putz. CRB. 1878. 172; Andr. AM. (9). xvii. 1926. 378; id. FBI. 414.	Calcutta.
B*	**hingstoni** Andr. FBI. 392 and 400.	Calcutta., Pusa. Sitapur (U. P.). Fyzabad.
O*	**indicus** Putz. Rév. 91; Andr. AM. (9). xvii. 1926. 378; id. FBI. 393 and 406.	Calcutta. "North India."
I*	**lacustris** Andr. FBI. 393 and 401.	Barkuda I. (Lake Chilka).
A*	**ladakensis** Andr. FBI. 394 and 410.	Ladak.
A*	**mahratta** Andr. FBI. 392 and 396.	Belgaum. Anuradhapura.
A*	var. **paucipunctus** Andr. FBI. 392 and 397.	Belgaum., Poona. Colombo. Anuradhapura.
B*	**mitodes** Andr. FBI. 394 adn 409.	Fyzabad.
L (E*)	**niteus** Putz. CRB. 1878. 173; Andr. AM. (9). xvii. 1926. 378; id. FBL. 392 and 399.	Calcutta.
I*	**ordinatus** † Bates (not Bates 1873) SYM. 4; Andr. AM. (9). xvii. 1926. 379; id. FBI. 415.	Pamir.
E*	**orientalis** Putz. Rév 92; Andr. AM. (9). xvii. 1926. 378.	China.

E*	**ovicollis** Putz. Suppl. ii. 14; Andr. AM. (9). xvii. 1926. 378.	China.
B*	**perfectus** Andr. FBI. 392 and 399.	Haldwani., Bombay.
S*	**porosus** Putz. CRB. 1877. 40; Andr. AM. (9). xvii. 1926. 378; id. FBI. 394 and 412. f. 61.	Burma.
L	**pusillus** *(phreoryctes)* Schm. Geob. in litt. FCB. t. 3. f. 6; Andr. TES. 1923. 59; id. AM. (9). xvii. 1926. 378; id. FBI. 415.	? Burma.
	debilis Putz. in litt. Rév. 97; id. CRB. 1878. 174.	
L	**rugifer** Putz. CRB. 1878. 173; Andr. AM. (9). xvii. 1926. 378; id. FBI. 415.	Calcutta.
I*	**seistanus** Andr. FBI. 391 and 394. f. 59.	Seistan.
B*	**singhalensis** Andr. AM. (9). xvii. 1926. 379 and 380; id. FBI. 393 and 407.	Kandy.
	ordinatus † Bates (not Bates 1873) AM. (5). xvii. 1886. 73.	
B*	**speculifer** Andr. FBI. 393 and 403. f. 60.	Sardah., Calcutta. Haldwani.
E*	**stenoderus** Putz. Suppl. ii. 13; Andr. AM. (9). xvii. 1926. 378.	China.
B*	**tamil** Andr. FBI. 394 and 408.	Barkuda I. (Lake Chilka)., Madras. Chipurupalle.
B*	**tricuspis** Andr. FBI. 391 and 395.	West Almora. Ranikhet. Dehra Dun.
B*	**variabilis** Andr. FBI. 393 and 405.	Calcutta., Sardah. West Almora. Ranikhet., Haldwani. Sitapur (U. P.).
L(E*)	**verticalis** Putz. CRB. 1878. 172; Andr. AM. (9). xvii. 1926. 378; id. FBI. 393 and 402.	Calcutta., Burma. Senmigion.
	aeneobrachialis Bates AG. xxxii. 1892. 282.	
B*	var. **bispicifer** Andr. FBI. 393 and 403.	Haldwani.
B*	var. **tennescens** Andr. FBI. 393 and 403.	Sardah., Pusa.

ECNOMALAUS* Bates.

Ecnomalaus Bates AG. xxxii. 1892. 357.

G* **clivinoides** Bates AG. xxxii. 1892. 358. — Bhamo., Katha. Rangoon.

EGAPLOA* Alluaud.

Egaploa Alluaud AF. lxxxv. 1916. 73.

O* **crenulata** *(Stenolophus)* Dej. Sp. iv. 1829. 432; Fairm. AF. 1868. 772; Bates AB. 1892. 232; All. AF. lxxxv. 1916. 73; Andr. MB. 39; All. AG. lii. 1928. 80. — Throughout India. Seychelles Is. Amirante Is. Comoro I. Madagascar. Mauritius.

J. *fulvipes (Stenolophus)* Erichs. in Arch. Naturg. ix. 1843. 216; Gerst. in Peter's Reise Moss. Zool. v. 1862. 172; Kolbe Mitt. Zool. Mus. Berl. 1910. 18; Scott TLS (2). xv. 1912. 249; Andr. AM. (9). xix. 1927. 102. — Africa : Obock. Kilimandjaro (foot), Mozambique, Angola, Portuguese Guinea.

O *basicollis (Anisodactylus)* Fairm. BF. 1892. 151; id. AB. 1893. 522; Kolbe Mitt. Zool. Mus. Berl. 1910. 17. — Cape Verde Is.

ELAPHRUS* Fabricius.

Elaphrus F. Syst. Ent. 1775. 227'; Dej. Sp. ii. 1826. 268; Lac. 44.

Q **tibetanus** Sem. RRE. iv. 1904. 21. — Tibet.

ENDYNOMENA* Chaudoir.

Endynomena Chaud., Mon. des Callidides, AB. xv. 1872. 186.

Saronychium Blackb. MEM. xiv. 1877. 142; Andr. AM. (9). iii. 1919. 483.

G **discoidalis** Bates AG. xxxii. 1892. 423. — Karen Hills.

O* **pradieri** *(Lebia)* Fairm. RMZ. 1849. 34; Chaud. Mon. 186; Bates AF. 1889. 283; Andr. Ins. Samoa iv. Col. 1927. 12. f. 9. — Nagpur., Bombay. Pondicherry. Trichinopoly.

B* *inconspicua (Saronychium)* Blackb. EMM. xiv. 1877. 142; Andr. AM. (9). iii. 1919. 483. — Kandy. Kuala Lumpur. Tonkin.

Z* *huebneri* Fairm. Pet. Nouv. Ent. ii. 1878. 286; id. AF. 1881. 245. — Cochin-China. Sumatra.

V* *paroeca (Thyreopterus)* Csiki Denkschr. Akad. Wiss. Wien, Mathnat. Kl., xci. 1915. 164. — Philippine Is. New Caledonia. Samoa. Cocos-Keeling Is. Ellice Is. Marquesas Is. Tahiti. Sandwich Is. Tonga Is., Hawaii. Fiji.

EULEPTUS* Klug.

Euleptus Klug. Ins. Madag. Phys. Abh. K. Akad. Wiss. Berlin. 1832 (1833). 131; Lac. 353; Alluaud AF. lxxxv. 1916. 77.

O* **ooderus** Chaud. BM. 1850. i. 365; Andr. EMM. 1926. 69. — Kumaon : Almora, Upper Gumti Valley, Gori Valley, East Ramganga Valley. Mussoorie. Dehra Dun. Lansdowne. Simla Hills; Theog, Matiana, Fagu. Kangra Valley.

EUPLYNES* Schmidt-Goebel.

Euplynes Schm. Goeb. FCB. 52; Lac. 131; Chaud. AF. 1859. 350; Bates TES. 1883. 264; id. AM. (5). xvii. 1886. 147; Andr. TES. 1923. 28.

B* **aurocinctus** Bates PZS. 1889. 384; Andr. TES. 1919. 197. — Sumatra. Borneo.

J* **batesi** Har. DEZ. 1877. 381; Bates TES. 1883. 264; Andr. AM. (9). xix. 1927. 109. — Japan.

P* **cyanipennis** Schm. Goeb. FCB. 52; Bates AM. (5). xvii. 1886. 147; Andr. TES. 1923. 28. — India (Brit. Mus.). Moulmein. Andaman Is. Celebes., Batchian. Gilolo, Morty. Philippine Is.

X* **limbipennis** Bates AF. 1889. 279. — Annam.

A* **marginatus** Andr. Spol. Zeyl. 1923. 241. — Ceylon : Kekirawa.

EURYAPTUS* Bates.

Euryaptus Bates AG. xxxii. 1892. 355.

O* **assamensis** Bates AG. xxxii. 1892. 356. (note). — Assam. U. P. : Sitapur, Dehra Dun. Almora.

G* **nigellus** Bates AG. xxxii. 1892. 355. — Karen Hills. Bhamo., Toungoo. Palon., Tharrawaddy. Rangoon.

O* **rufipes** Bates AG. xxxii. 1892. 356 (note). — Andaman Is.

EUSCHIZOMERUS* Chaudoir.

Euschizomerus Chaud. BM. 1850. i. 413; Lac. 212; Chaud. Mon. sur les Panagéides, AB. 1878. 157.

O* **aeneipennis** Chaud. RMZ. 1869. 118; id. Mon. 159; Bates AG. xxxii. 1892. 303; Vuillet Ins. ii. 1912. 18. — Bhamo., Rangoon. Malacca., Cambodia.

O* **aeneus** Chaud. RMZ. 1869. 118; id. Mon. 160; Bates AG. xxxii. 1892. 303. — Dacca. Tikekee (Pegu).

V* **denticollis** *(Panagaeus)* Koll. Ann. Wien. Mus. I. 1835. 334. t. 31. f. 2; Chaud. Mon. 161. — Calcutta. Ceylon.

J* **metallicus** Harold SEZ. 1879. 331; Vuillet Ins. ii. 1912. 19; Andr. AM. (9). xix. 1927. 109. — Karen Hills. Rangoon. Tenasserim. Cambodia.

Y* **rufipes** Heller PJS. xix. 1921. 526. — Philippine Is.

Y* var. **pilosulus** Heller PJS. xxiii. 1923. 301. — Philippine Is.

O* **vitalisi** Vuillet Ins. ii. 1912. 18. fig. — Penang., Borneo. Cochin-China. Cambodia.

EUSTRA* Schmidt-Goebel.

Eustra Schm. Goeb. FCB. 65; Lac. 161; Chaud. BM. 1854. i. 309; id., Révision du groupe des Ozénides, AB. xi. 1868. 71; Bates TES. 1873. 237; Lewis EMM. xviii. 1881. 39; Andr. AM. (9). iv. 1919. 295; Bänn. DEZ. 1927. 189 and 206; Andr. FBI. 167.

A* **bryanti** Andr. AM. (9). iv. 1919. 295, 299, and 300. Sarawak.

japonica × Bates AG. xxxii. 1892. 270; Dupis AB. 1913. 418; Andr. AM. (9). iv. 1919. 295 and 299. Japan.

B* *plagiata* † Bates (not Schm. Goeb.) TES. 1873. 237; id. TES. 1876. 5. t. 1. f. 1.

batesi × Lewis AM. (6). xvii. 1896. 330.

A* **matanga** Andr. AM. (9). iv. 1919. 295 and 299. Sarawak.

P* **plagiata** Schm. Goeb. FCB. 66. t. 3. f. 1; Chaud. BM. 1854. i. 310; id. Rév. 71; Bates AG. xxxii. 1892. 269; Lewis AM. (6). xvii. 1896. 330; Dupuis AB. 1913. 418; Andr. AM. (9). iv. 1919. 295 and 299; id. TES. 1923. 37; id. FBI. 169. f. 30. Kobo., Nilgiri Hills. Burma Teinzo, Palon, Martaban. Tonkin., Sumbawa. Formosa.

FOUQUETIUS* Maindron.

Fouquetius Maindr. BF. 1906. 252.

Holconotus [Schm. Goeb. in litt. FCB. (cover);] || Chaud. (not Forst.) RMZ. 1876. 352; Tchitch., Note sure le genre Holconotus Chaud., AB. 1898. 451.

Q* **crassmargo** *(Holconotus)* × Tchitch. Note 453. Cambodia.

ferrugineus (Holconotus) † Bates (not Chaud.) AF. 1889. 277; Tchitch. Note 454; Andr. TES. 1921. 179.

O* **ferrugineus** [*(Holconotus)* Schm. Goeb. FCB. t. 2. f. 6;] (*Abacetus)* Chaud. BM. 1869. i. 399; Bates AG xxxii. 1892. 364; Tchitch. Note 453; Andr. TES. 1923. 57. Sunderbans. Nagpur., Belgaum Madras., Mandalay. Bhamo., Katha. Palon., Tharrawaddy. Rangoon. Tenasserim. Vcitoria. Point. Siam., Cambodia.

2. var **lioderus** *(Holconotus)* Tchitch. Note 453. Cochin-China. Cochin-China. Cambodia., Siam.

GALERITA* Fabricius.

Galerita F. Syst. Eleuth. i. 1801. 214; Latr. et Dej. HN. et Ic. 117; Dej. Sp. i. 1825. 186; Shm. FCB. 26; Lac 82; Andr. MB. 52.

A* **batesi** Andr. TES 1923. 9. — Sylhet.

orientalis † Bates (not Schm. Goeb.) AG. xxvii. 1889. 109 (part); id. AG. xxxii. 1892. 385 (part). — Burma : Karen Hills, Teinzo, Palon, Thagata (Tenasserim). Annam.

J* **carinifrons** Schauf. HER. xxi. 1887. 103; Andr. AM. (9). xix. 1927. 109. — Celebes.

G* **feae** Bates AG. xxxii. 1892. 386. — Karen Hills., Laos.

O* **indica** Chaud. BM. 1861. i. 557; id. BM. 1877. i. 255. — "North India."

B* **japonica** Bates TES. 1873. 304; Putz. CRB. 1875. 52; Chaud. BM. 1877. i. 255; Heyd. DEZ. 1879. 329. — Japan.

O* **nigripennis** Chaud. BM. 1861. i. 557; id. BM. 1877. i. 255; Andr. MB. 52. — Dacca., Mhow. Nagpur., Surada.

P* **orientalis** Schm. Goeb. FCB. 26; Bates AG. xxvii. 1889. 109 (part); id. AG. xxxii. 1892. 385 (part); Andr. TES. 1923. 8. — Khasi Hills. Patkai Hills. Bhamo.

S* **peregrina** Dohrn SEZ. 1880. 291; Andr. TES. 1922. 246. — Bhamo., Tonkin. Annam. Hongkong.

ruficeps † Bates (not Chaud.) AG. xxvii. 1889. 109.

G* *birmanica* Bates AG. xxxii. 1892. 385; Andr. AM. (9). iii. 1919. 480.

O* **ruficeps** Chaud. BM. 1861. i. 556. — U. P. : Harki Dun, Nainital, Almora.

GIDDA* Andrewes.

Gidda Andr. AB. 1920. 26.

B* **barthelemyi** Andr. AB. 1920. 27. — Laos.

GLYCLA* Chaudoir.

Glycia Chaud. BM. 1842. 805; Lac. 110; Chaud. Mon. des Callidides, AB. xv. 1872. 98; Bedel Fn. med. 1907. 265.

Neotarus Reitt. DEZ. 1884. 36.

J **ornata** *(Cymindis)* Klug in Symb. Phys. Decas iii. 1832. t. 22. f. 3; Maindr. AF. 1899. 379; id. AF. 1905. 333; Bedel Fn. med. 1907. 267; Andr. RIM. — Karachi., Seistan. Nasratabad. Central Asia.

xviii. 1919. 101; id. AM. (9) xix. 1927. 100. *karelini* Motch. Ins. Sib. 1844. 41; Chaud. BM. 1850. ii. 92. — Mesopotamia. Mediterranean and Red Sea basins.

B' *krueperi* Reitt. DEZ. 1844. 36. t. 1. f. 1.

X* **rufolimbata** Maindr. AF. 1905. 332; Bedel Fn. med. 1907. 266. — Karachi., Arabia. Erythrea., Somaliland.

GNAPHON* Andrewes.

Gnaphon Andr. AM. (9). v. 1920. 454; id. FBI. 210 and 216.

Anomoderus || † Fairm. (not Chaud.) BF. 1883. 55.

Scaritoderus × Fairm. BF. 1883. 55.

I* **costatus** Andr. FBI. 217 and 218. t. 6. f 5. — Tinnevelly.

O* **humeralis** *(Haplogaster)* Putz. in Chaudoir's Mon. des Scaritides (i), AB. 1879. 151; Andr. AM. (10). ii. 1928. 586; id. FBI. 217 and 219. — Madras.

O* **layolae** *(Scaritoderus)* Fairm. BF. 1883. 55; Andr. FBI. 217. t. 6. f. 6. — Ramnad., Madura. Shembaganur.

X* *favrei (Crepidopterus)* Maindr. BF. 1904. 264. fig. Andr. AM. (9). iii. 1919. 469. — Kodaikanal. Ootacamund.

GNATHAPHANUS* Macleay.

Gnathaphanus Macl. Ann. Jav. 1825. 20; Lac. 299; Chaud. AG. xii. 1878. 503; Sloane PLS. NSW. 1898. 456; id. PLS. NSW. 1920. 131; Andr. AB. 1920. 106; id. MB. 20.

Pachauchenius W. Macl. TES. NSW. i. 1864. 117.

G* **exaratus** Bates AG. xxxii. 1892. 329; Andr. AB. 1920. 108. — Dafla Hills. Naga Hills "Between Sorbhog and Golakganj." Karen Hills., Laos.

B* **festivus** Andr. AB. 1920. 107 and 109; id. TES. 1921. 177. — Cambodia. Cochin China.

G* **impressipennis** *(Harpalus)* Cast. Trans. Roy. Soc. Vict. viii. 1868. 186; Chaud. AG. xii. 1878. 510; Sloane DEZ. 1907. 468; id. PLS. NSW. 1920. 320; Andr. AB. 1920 107; id. Ins. Samoa iv. Col. 1927. 3. f. 2. — Kuala Lumpur. Singapore., Java. Sumatra., Sumbawa. New Guinea. Australia.

V*	*upolensis (Dioryche)* Csiki Denkschr. Akad. Wiss. Wien. Mathnat. Kl. xci. 1915. 163.	New Caledonia. Samoa.
J*	**laevistriatus** *(Harpalus)* Sturm Deutschl. Faun. Ins. iv. 1818. 80. t. 91. f. B; Andr. AM. (9). xix. 1927. 99.	Throughout India and Burma.
O*	*orientalis (Selenophorus)* Dej. Sp. iv. 1829. 128; Andr. AB. 1920. 107.	
G*	*acutipennis* Bates AG. xxxii. 1892. 328; Andr. TES. 1919. 151; id. TES. 1921. 147.	
H*	**licinoides** Hope AM. ix. 1842. 427; Sloane DEZ. 1907. 468; id. PLS. NSW. 1920. 320; Andr. TES. 1919. 202; id. AB. 1920. 106.	New Guinea. New Caledonia. Australia.
	impressus (? Catadromus) Montr. AF. 1860. 240.	
L	*alternans (Harpalus)* Cast. Trans. Roy. Soc. Vict. viii. 1868. 185.	
H*	**philippensis** *(Amblygnathus)* Chevr. RZ. 1841. 221; Desmarest, Voy. la Bonite, 1841. 299. t. 2. ff. 5-10; Bates AG. xxxii. 1892. 327; Andr. TES. 1919. 207; id. AB. 1920. 107; Sloane PLS. NSW. 1920. 320.	Purnea., Nagpur. Belgaum. North Kanara. Rangoon. Tenasserim.
M'	*laeviceps (Pachauchenius)* W. Macl. TES. NSW. i. 1864. 117; Chaud. AG. xii. 1878. 511; Sloane DEZ. 1907. 468; id. PLS. NSW. 1917. 411.	Sumatra., Celebes. New Guinea, Philippine Is. Australia. New Britain.
B*	**punctilabris** *(Harpalus)* Macl. Ann. Jav. 1825. 20; Andr. TES. 1919. 150; id. AB. 1920. 107; id. MB. 30.	Throughout South East Asia (Except N. W. India).
O*	*javanus (Anisodactylus)* Dej. Sp. iv. 1829. 146; Baer. AF. 1886. 102.	Java., Sumatra. Celebes. Philippine Is. Formosa.
B*	*dispellens (Harpalus)* Walk. AM. (3). iii. 1859. 51; Bates AM. (5). xvii. 1886. 75; id. AF. 1889. 268; id. AG. xxxii. 1892. 328; Andr. TES. 1919. 193.	
O*	*gnathaphanoides (? Platymetopus)* Bates AG. xxxii. 1892. 332 (note); Andr. AM. (9). iii. 1919. 473; id. TES. 1921. 147.	
G*	**rufitactor** Bates AG. xxxii. 1892. 329 Andr. AB. 1920. 108.	Bhamo.

B*	**rusticus** Andr. AB. 1920. 107 and 110; id. MB. 30.	Pusa., Chapra. Muzaffarpur. Janakpur., Purnea. Patna., Dehra Dun. Samstipur. "Hurdwar and Roorkee.", Lucknow. Nagpur., Chikalda. Surat., Belgaum. Dharwar. North Kanara Surada., Kandy.
B*	**subolivaceus** *(Amara)* Macl. Ann. Jav. 1825. 21; Andr. TES. 1919. 153; id. AB. 1920. 107.	Java. Sumatra.
B*	*subaeneus (Amara)* Macl. Ann. Jav. 1825. 21; Andr. TES. 1919. 153.	
B*	**vulneripennis** Macl. Ann. Jav. 1825. 20; Hope Col. Man. ii. 1838. t. 2. f. 2; Chaud. AG. xii. 1878. 508; Bates AF. 1889. 267; id. AG. 1892. 327; Andr. TES. 1919. 149; id. AB. 1920. 107; id. TES. 1921. 180.	Sylhet., Dacca. Manbhum., Purnea. Pusa., Almora. Kali Valley (Kumaon).
O*	*subcostatus (Harpalus)* Dej. Sp. iv. 1829. 261.	Chanda., Balaghat.
N*	*melanarius (Platymetopus)* Boh. ER. 10.	Ceylon., Bhamo. Karen Hills. Tenasserim. Kuala Lumpur. Indo-China. Hongkong., Java. Sumatra., Borneo. Celebes. Philippine Is.

GNOPHODES* Andrewes.

'*Gnophodes* Andr. Spol. Zeyl. 1923. 244.

A*	**pearsoni** Andr. Spol. Zeyl. 1923. 245.	Ceylon : Hakgalla, Horton Plains, Pidurutalagalla.

GRANIGER* Motchulsky.

Graniger Motch. BM. 1864. ii (Oct.) 197.

Coscinia || Dej. (not Huebn.) Sp. v. 1831. 478; Lac. 167; Chaud., Mon. des Siagonides, BM. 1876. i. 115.

Cymbionotum Baudi BEZ. 1864. (Dec.) 211.

O* **fasciger** *(Coscinia)* Chaud. BM. 1852. i. 92; id. Mon. 121; Bates AF. 1889. 264.	Calcutta., Fyzabad. Kheri. Barkuda I. (Lake Chilka)., Madras. Ceylon : Hambegamuwa, Horawupotana. Cochin-China. Annam.
C* **helferi** *(Coscinia)* Chaud. BM. 1850. i. 441; id. Mon. 122; Bates AG. xxxii. 1892. 286.	Bhamo., Mandalay. Tharrawaddy. Toungoo., Prome. Insein., Rangoon. Martaban. Victoria Point. Siam., Indo-China.
O* **semelederi** *(Coscinia)* Chaud. BM. 1861. i. 6; id. Mon. 124; Mars. Ab. 1867. xxxiii; Fairm. AF. 1868. 472; Sem. HER. xxv. 1891. 288; Bedel Fn. med. 1897. 109.	Baluchistan. South Western Asia. Southern Russia. North Africa. Cyprus.

M *algirinus* Motch. BM. 1864. ii. 198; Andr. TES. 1928. 22.

T *collaris (Cymbionotum)* Baudi BEZ. 1864. 213.

HAPLOGASTER* Chaudoir.

Haplogaster Chaud., Mon. des Scaritides (i), AB. 1879. 149; Andr. FBI. 210 and 220.

G* **ampliata** Bates AG. xxxii. 1892. 270; Andr. FBI. 221. and 224.	Kachin.
G* **mollita** Bates AG. xxxii. 1892. 270; Andr. FBI. 221 and 224.	Karen Hills.
O* **ovata** Chaud. Mon. 150; Andr. FBI. 221 and 222. f. 37.	Sylhet., Shillong. Khasi Hills. Khatmandu. Sikkim : Gopaldhara, Darjiling, Pashok, Kurseong, Pedong Allahabad., Siliguri.
B* **wardi** Andr. FBI. 221 and 223.	Burma : Seinghku.

HARPAGLOSSUS* Motchulsky

Harpaglossus × Motchulsky Etud. Ent. 1858. 173.

Ceroglossus || Chaud. (not Sol.) BM. 1857. ii. 6.

Penthimus × Mun. Cat. 229; Chaud., Mon. des Chléniens, AG. 1876. 291.

O* **opacus** *(Ceroglossus)* Chaud. BM. 1857. ii. 8; id. Mon. 291. — Bengal. Sitapur (U. P.).

L *indicus (Sogines)* Motch. BM. 1865. ii. 257; Chaud., Mon. du genre Poecilus, AB. xiv. 1876. 2 and 6; Andr. TES. 1928. 22. — Ajmere., Kathiawar. Gogo., Trichinopoly Ceylon.

HARPALISCUS* Bates.

Harpaliscus Bates AG. xxxii. 1892. 340.

O* **birmanicus** Bates AG. xxxii. 1892. 341; Andr. EMM. 1926. 68. — Sadiya., Shillong. Margherita., Patkai Hills. Khasi Hills., Manipur., Sikkim: Gopaldhara., Almora., Sarju Valley. South Garhwal, Burma : Karen Hills, Teinzo, Shwegu, Prome, Ruby Mines, Rangoon. Laos. Tonkin.

A* **punctulatus** Andr. AM. (9). xii. 1923. 447. — Nilgiri Hills. Palni Hills. Shevaroy Hills. Madura. Shembaganur.

HARPALUS Latreille.

Harpalus Latr. Hist. Nat. Crust. et. Ins. iii. 1802. 92; Dej. Sp. iv. 1829. 190; Lac. 295; Motch. BM. 1864. ii. 209; Tchitch. HER. xxxiv. 1900. 360; id. HER. xxxv. 1901. 233; Andr. MB. 36.

S* **advolans** Nietn. JASB. xxv. 1856; 526; id. AM. (2). xix. 1857. 377; Andr. RIM. xxii. 1921. 344; id. AM. (9). xix. 1927. 106. — Barkuda I. (Lake Chilka). Pondicherry. Trichinopoly. Manaparai. Ramnad., Pamban.

	HARPALUS Latreille.—	Chatrapur., Mahe. Madras., Dindigul. Ceylon : Colombo, Kotte, Trincomali.
Q	**agonus** Tchitch. HER. xxviii. 1894. 259.	Pamir
L	**amarellus** Bates Entom. 1891. Suppl. 10; Andr. AM. (9). xix. 1927. 111.	Kulu.
E'	**anxius** Dufts Faun. Austr. ii. 1812. 101; Dej. Sp. iv. 1829. 375; Andr. AM. (9). xix. 1927. 111.	Kashmir : Goorais Valley, Sonemarg, Pir Panjal.
I*	*quadricallis,* † Bates (not Redt.) SYM. 30 ; Andr. MB. 37.	Chamba., Jhelum.
A*	*amarellus* Andr. MB. 37.	Kalka., Kangra. Mandi. Lahoul: Cimur, Keiling, Sumdeo, Sisu. Bashahr: Poo, Nichar.
I*	**indicola** Bates PZS. 1878. 714; id. SYM. 7. t. 1. f. 3; id. Entom. 1891. Suppl. 10; Andr. EMM. 1926. 68.	Himalayan tract from Murree to Darjiling. Chota Nagpur : Konbir (Bates).
E*	*praticola* Bates CRB. 1891. 332.	
G*	**karennius** Bates AG. xxxii. 1892. 341.	Assam : Shillong, Rotung, Yambung, Khasi Hills. Sikkim : Gopaldhara. Manipur. Karen Hills. Ruby Mines. Simla (Pusa Coll). Tonkin., Cambodia.
O*	**kashmirensis** Bates PZS. 1889. 213.	Kashmir : Goorais Valley, Pir Panjal. Seistan.
I*	**masorecides** Bates PZS. 1878. 715; id. SYM. 8. t. 1. f. 5; Tchitch. HER. xxxv. 1901. 147 and 240 (note 54).	Pamir. Turkestan.

I* **melaneus** Bates PZS. 1878. 714; id. SYM. 6. t. 1. f. 4; id Entom. 1891. Suppl. 10; Andr. MB. 36; id. EMM. 1926. 68.

Kashmir : Sind Valley, Leh, Gond, Baltal, Sonemarg, Lidder, Kazi, Nag range, Goorais Valley, Skardo, Yusimarg, Gulmarg, Pir Panjal, Kishtwar Panjal, Uri.
Chamba. Murree., Dalhousie.
Simla Hills : Phagu, Baghi, Kufri, Narkanda.
Spiti : Pulga.
Lahoul: Zingzinbar, Sumdeo, Patseo, Sisu, Keiling, Gandla, Cimur.
Kulu : Kotv. Rotang Valley. Bashahr : Chim, Taranda.
Mundali (Dehra Dun).
Burphu and Milam, (Almora).

V* **quadricollis** *(Selenophorus)* Redt. HK. 562; Andr. MB. 37.

O* *idiotus* Bates PZS. 1889. 213; Andr. AM. (9). xix. 1927. 111.

Kashmir : Goorais Valley, Skardo, Sonemarg, Baltal.
Rawalpindi.
Lahoul : Sisu, Zingzinbar, Patseo, Sumdeo, Keiling.
Spiti : Pulga.
Kumaon : Sunder-dhunga Valley, Kali Valley, Ranikhet.
Bashahr : Poo.

HELLUODES* Westwood.

Helluodes Westw. TES. 1817. 279; Lac. 92; Chaud. BM. 1862. ii. 302.

R **taprobanae** Westw. TES. 1847. 279. t. 21. f. B; Lac. Atl. t. 7. f. 1. (as *Physocrotaphus ceylonicus* Parry); Chaud. BM. 1862. ii. 303; Putz. CRB. 1875. 45; Bates AM. (5). xvii. 1886. 200; Andr. TES. 1919. 180. — Ceylon : Kitugalle, Galgamuwa, Kotte, Kandy, Peradeniya, Matale.

O* **westwoodi** Chaud. RMZ. 1869. 203. — Dacca.

HEMIAULAX* Bates.

Hemiaulax Bates AG. xxxii. 1892. 347.

G* **dentipennis** Bates AG. xxxii, 1892. 347. — Rangoon., Palon.

HEMICHLAENIUS* Bates.

Hemichlaenius Bates AG. xxxii. 1892. 307.

G* **microspilus** Bates AG. xxxii. 1892. 308. — Karen Hills.

HEXACHAETUS* Chaudoir.

Hexachaetus Chaud. Essai monographique sur les Orthogoniens, AB. xiv. 1871-124.

P* **angulatus** *(Orthogonius)* Schm. Goeb. FCB. 58; Chaud. Mon. 110; Bates AG. xxxii. 1892. 401; Andr. TES. 1923. 82. — Karen Hills. Maymyo. Tenasserim., Perak. Penang.

P* var. **plicatus** *(Orthogonius)* Schm. Goeb. FCB. 59; Chaud. Mon. 116; Andr. TES. 1923. 33. — Maymyo. Tenasserim., Laos. Tonkin.

O* **laevissimus** Chaud. BM. 1878. iii. 6. — Tenasserim., Mergui. Malacca., Borneo.

O **lateralis** *(Orthogonius)* Guér. Voy. Deless. ii. 1843. 35; Chaud. Mon. 125. — Penang., Sumatra. Borneo.

HEXAGONIA* Kirby.

Hexagonia Kirby TLS. xiv. 1825. 563; Cast. Etud. Ent. 1834. 51; Schm. Goeb. FCB. 49; Lac. 69; Schaum. BEZ. 1863. 72 and 433; van de Poll. NLM. xi. 1889. 247-50; Dupuis Gen. Ins. Fasc. 147, Hexagoniinae, 1913. 2; Andr. TES. 1923. 26.

Trigonodactyla Dej. Sp. v. 1831. 288; Lac. 70.

P* **apicalis** Schm. Goeb. FCB. 51; Dupuis Gen. Ins. Fasc. 147. t. f. 11; Andr. TES. 1923. 26. — Cossipore or Darjiling.

O* **bencoulensis** Pllde. Ins. iv. 1914. 167. ff. 6-7. — Sumatra.

O* **bouchardi** Pllde. Ins. iv. 1914. 164. ff. 1-3. — Java., Sumatra.

J* **bowringl** Schaum BEZ. 1863. 73 and 433. t. 3. f. 3; Andr. AM. (9). xix. 1927. 108. — Penang.

D* **elongata** Dupuis AB. 1913. 263. — Formosa.

X* **fleutiauxi** Dupuis AB. 1913. 325. — Tonkin.

O* **gracilis** Pllde. Ins. iv. 1914. 108. ff. 8-9. — Sumatra.

B* **insignis** (*Trigonodactyla)* Bates TES. 1883. 277. t. 13. f. 6; Dupuis AB. 1913. 263. — Formosa., Japan. China.

C* **longithorax** *(Lebia)* Wied. Zool. Mag. ii. 1. 1823. 58; Bates AG. xxxii. 1892. 366; Andr. TES. 1921. 172. — Assam: Rotung, Upper Rotung, below Damda.

O* *brunnea* Chaud. BM. 1861. i. 531; Schaum. BEZ. 1863. 433. — Bengal., Dehra Dun. Gauniarau (Haldwani). Burma: Malewoon, Palon.

lucasseni van de Poll. NLM. xi. 1889. 249. — Java.

nigrita van de Poll. NLM. xi. 1889. 247; Bates AG. xxxii. 1892. 367; Dupuis AB. 1913. 263. — Karen Hills., Java. Sumatra., Laos. Formosa.

D* **sauteri** Dupuis AB. 1912. 308; id. AB. 1913. 263. — Formosa.

B* **terminata** Kirby TLS. xiv. 1825. 564; Brullé HN. iv. 1834. 227; Schm Goeb. FCB. 50; Andr. TES. 1919. 132. (Lacordaire, Atl. t. 3. f. 1., gives a figure alleged to be *Trigonodactyla terminata* Kirby, but in the "Explication des planches" it appears correctly as *T. terminata* Dej. (=*terminalis* Mun. Cat.) an African species.) — Darjiling (Schmidt Goebel). Munshiganj. North Kanara. Palon., Tharrawaddy. Zigon., Andaman Is. Java., Laos.

P* *kirbyi* Schm. Goeb. FCB. 51. t. 2. f. 2; van de Poll. NLM. xi. 1889. 250; Bates AG. xxxii. 1892. 366; Andr. TES. 1923. 25.

O* *andamanensis* Pllde. Ins. iv. 1914. 165. ff. 4-5; Andr. AM. (9). xiv. 1924. 591.

HILETUS* Schioedte.

Hiletus Schioedte in Kroeyer Naturh. Tidsskr. (2). ii. 1847. 346; Lac. 48; Alluand PF. 1914. 439.

Camaragnathus Bertrand—Bocandé Mag. Zool. (2). vii. 1845 (1849). tt. 163-4. and text pp. 1-3; id. RMZ. 1849. 460. t. 12. ff. 2-7.

Z* **sumatrensis** Oberth. NLM. v. 1883. 215. — Sumatra.

HOLCOCOLEUS* Chaudoir.

Holcocoleus Chaud. Mon. des Oodides (ii), AF. 1882. 521.

O* **latus** *(Oodes)* Laf. AF. 1851. 269. — "Coromandel."

O* *sulcatulus* (*Oodes*) Chaud. BM. 1857. ii. 38; id. Mon. 522. — Hulical., Parur. Vriddachalam. Pondicherry. Madras.

Holcoderus* Chaudoir

Holcoderus Chaud., Mém. sur les Thyréopterides, AB. xii. 1869. 153.

O* **auripennis** Chaud. BM. 1877. i. 198. — Penang., Sumatra.

J* **elongatus** *(Catascopus)* Saund. TES. 1863. 466. t. 18. f. 5; Chaud. RMZ. 1872. 249; Andr. AM. (9). iii. 1919. 482; id. AM. (9) xvii. 1926. 259; id. AM. (9). xix. 1927. 107. — New Guinea.

Z* **gracilis** *(Catascopus)* Oberth. NLM. v. 1883. 220; Andr. AM. (9). iii. 1919. 482. — Kuala Lumpur. Sumatra.

O* **limbipennis** Chaud. BM. 1877. i. 199. — Penang.

O* **praemorsus** Chaud. Mém. 153; Bates AM. (5). xvii. 1886. 203. — Ceylon : Dikoya, Bogawantalawa.

A* **smaragdinus** Andr. AM. (9). xviii. 1926. 287. — Tonkin., Sumatra. Borneo.

HOLOLEIUS* Laferté

Hololeius Laf. AF. 1851. 214 and 274; Lac. 227; Sloane PLS. NSW. 1910. 436.

Hololius Chaud. BM. 1857. ii. 9; id. Mon. des Chléniens, AG. viii. 1876. 290; id. in Oberthür Col. Nov. i 1883. 37.

O* **nitidulus** *(Chlaenius)* Dej. Sp. ii. 1826. 341; Laf. AF. 1851. 274; Chaud. BM. 1857. ii. 10; Cast. Trans. Roy. Soc. Vict. viii. 1868. 147; Chaud. Mon. 290; — Throughout South East Asia, including Malay Archipelago,

Bates AM. (5). xvii. 1886. 75; id. AG. xxxii. 1892. 321; Sloane PLS. NSW. 1920. 322. — excluding Japan. Australia.

J*S* *ceylanicus (Chlaenius)* Nietn. JASB. xxv. 1856. 385; id. AM. (2). xix. 1857. 241; Andr. AM. (9). xix. 1927. 106.

O* *punctulatus* Chaud. BM. 1857. ii. 10; Vuillet. Ins. ii. 1912. 19.

? ornatus (Poecilus) Tryon Rep. Adm. Br. New Guinea (ii). 1890. 109; Sloane PLS. NSW. 1910. 437.

HYPHAEREON* Macleay.

Hyphaereon Macl. Ann. Jav. 1825. 22; Lac. 284; Andr. TES. 1919. 156.

B* **reflexus** Macl. Ann. Jav. 1825. 22; Hope Col. Man. ii. 1838. t. 2. f. 5; Andr. TES. 1991. 156. — Java.

HYPHARPAX* Macleay.

Hypharpax Macl. Ann. Jav. 1825. 22; Lac 282; Chaud. AG. xii. 1878. 496; Sloane PLS. NSW. 1898. 456 and 458; id. PLS. NSW. 1920. 131.

Sagraemerus Redt. RN. 13.

O* **celebensis** Chaud. AG. xii. 1878. 502. — Java., Sumatra. Celebes.

C* **dentipes** *(Harpalus)* Wied. Zool. Mag. ii. 1. 1823. 54; Chaud. AG. xii. 1878. 500; Veth Tijdschr. Ent. 1910. 305; Andr. TES. 1921. 170. — Java.

B* *tricolor (Amara)* Macl. Ann. Jav. 1825. 21; Andr. TES. 1919. 152.

B* *lateralis* Macl. Ann. Jav. 1825. 22; Hope Col. Man. ii. 1838. t. 2. f. 3; Andr. TES. 1919. 158.

V* *javanus (Sagraemerus)* Redt. RN. 14. t. 1. f. 7; Andr. TES. 1923. (1924). 467.

O* **simplicipes** Chaud. AG. xii. 1878. 502; Bouch. AF. 1903. 172; Veth Tijdschr. Ent. 1910. 307. — Java. Sumatra.

HYPOLITHUS* Dejean.

Hypolithus Dej. Sp. iv. 1829, 166; Lac 297.

E*	**acutangulus** Bates CRB. 1891. 331.	Sardah., Pusa. Allahabad. Sitapur (U. P.). Dehra Dun. Siwalik Hills. Delhi. Chota Nagpur Tetara Mhow., Surat. Bombay., Tranquebar. Ceylon., Java. Sumatra. Andonare I.
X*	**cyanellus** Bates AF. 1889. 269; Andr. AM. (9). ix. 1922. 287 and 288.	Federated Malay States Indo-China.
X*	**cyaneotinctus** Bates AF. 1889. 269; id. CRB. 1891. 331; Andr. AM. (9). ix. 1922. 288.	Dehra Dun. Chota Nagpur : Konbir, Tetara, Barwa. Ranchi. Mysore : Chikkaballa-pura, Nandidrug, Bangalore. Ceylon., Annam.
L	**javanus** Gory AF. 1833. 241.	Java.
A*	**lividus** Andr. Spol. Zeyl. 1923. 234.	Throughout the plains of India. Colombo.
G*	**subtilis** Bates AG. xxxii. 1892. 338.	Bhamo.
B*	**vitalisi** Andr. AM. (9). ix. 1922. 287. and 288.	Siam., Laos.

HYPSINEPHUS* Bates.

Hypsinephus Bates PZ. 1878. 715; id. SYM. 9.

I*	**ellipticus** Bates PZS. 1878. 716; id. SYM. 10. t. 1. f. 9; Andr. MB. 36; id. EMM. 1926. 68.	Kashmir : Baltal, Pangong Valley. Tibet : Supi River.

IDIOMORPHUS* Chaudoir.

Idiomorphus Chaud. BM. 1846. ii. 515; Lac. 254.

O*	**guerini** Chaud. BM. 1846. ii. 518; Lac. Atl. t. 12. f. 1; Bates CRB. 1891. 333.	Chota Nagpur : Nowatoli, Konbir, Tetara.

IDIOMORPHUS* Chaudoir.— Nilgiri Hills. Coimbatore. Koilpatti (Tinnevelly). Pondicherry. Trichinopoly.

IRIDESSUS* Bates.

Iridessus Bates TES. 1883. 240; Tchitch. Ab. xxix. 1897. 60; id. HER. xxxiv. 1900. 363; id. HER. xxxv. 1901 245; id. HER. xxxvii. 1906. 284.

Q **lucidus** *(Harpalus)* Moraw. Beitr. 72. Bates TES. 1873. 265; id. TES. 1883. 240; Tchitch. HER. xxxvii. 1906. 285 (note 9). — Japan.

B* **orientalis** *(Amara)* Hope TES. iv. 1845. 14; Andr. TES. 1919. 177. — China. Japan.

B* *relucens (Harpalus)* Bates TES. 1873. 264; id. TES. 1883. 240; Putz. CRB. 1875. 49; Tchitch. Ab. xxix 1897. 60; id. HER. xxxvii. 1906. 285 and note 9.

ITAMUS* Schmidt-Goebel.

Itamus Schum. Goeb. FCB. 67; Lac. 160; Chaud. BM. 1854. i. 297; Andr. AM. (9). iv. 1919. 295; Bänn. DEZ. 1927. 191 and 202; Andr. FBI. 163.

P* **castaneus** Schm. Goeb. FCB. 67; Chaud. BM. 1854. i. 297; id. AB. xi. 1868. 51; Bates CRB. 1891. 326; id. AG. xxxii. 1892. 269; Andr. AM. (9). iv. 1919. 295 and 296; id. TES. 1923. 38; Bänn. DEZ. 1927. 202; Andr. FBI. 164. f. 28. — Sadiya. Chota Nagpur. Madras., Bhamo. Katha., Prome. Mandalay. Pyinmana. Tenasserim. Andaman Is., Laos.

B* **dentatus** Andr. AM. (9). iv. 1919. 295; Bänn. DEZ. 1927. 202. — Tonkin.

KAREYA* Andrewes.

Kareya Andr. AM. (9). iii. 1919. 473; id. MB. 32.

G* **edentata** *(? Platymetopus)* Bates AG. xxxii. 1892.331; Andr. AM. (9). iii. 1919. 473; id. MB. 32. — Khasi Hills. Shillong. Sikkim : Gopaldhara,

KAREYA* Andrewes.— Kalimpong Manipur., Pusa. Chapra., Sahibganj. Rajmahal., Almora. Kangra., Mandla. Koyna River (Satara). Burma : Shan Hills, Bhamo, Karen Hills, Pyaung-gyaung, Palon, Tikekee, Rangoon. Tavoy., Laos. Tonkin.

G* **erebia** *(? Platymetopus)* Bates AG. xxxii. 1892. 331; Andr. AM. (9). iii. 1919. 473. — Sardah., Rajmahal. Puri., Mhow. Jubbulpore., Surada. Burma : Prome, Minhla, Palon, Tharrawaddy, Thyetmyo, Katha, Pyinmana, Rangoon, Tenasserim. Hongkong.

G* **sublaevis** *(? Platymetopus)* Bates CRB. 1891. 330; id. AG. xxxii. 1892. 331; Andr. AM. (9). iii. 1919. 473. — Assam : Rotung. Purnea. Chota Nagpur : Konbir., Lucknow. Kumaon., Almora. Shamket., Nagpur. Mandla., Baihar. Castle Rock. Vizagapatam. Burma : Shan Hills, Palon, Pyinmana, Tharrawaddy, Rangoon, Tenasserim. Siam., Indo-China. Hongkong.

G* var. **grandiceps** *(? Platymetopus)* Bates AG. xxxii. 1892. 332; Andr. AM. (9). iii. 1919. 473. — Karen Hills.

E var. **major** *(? Platymetopus)* Bates CRB. 1891. 330; Andr. AM. (9). iii. 1919. 473. — Chota Nagpur : Konbir, Tetara. Lucknow., Balghat.

LACHNOCREPIS* Leconte.

Lachnocrepis Lec. Trans. Am. Phil. Soc. x. 1853. 391; Lac. 394 (note 3); Chaud., Mon. des Oodides (i), AF. 1882. 377.

B* **japonicus** Bates TES. 1873. 255; Har. DEZ. 1877. 338; Heyd. DEZ. 1879. 330; Chaud. Mon. 378. — Tonkin., China. Japan., Korea.

O* *piceolus* (Oodes) Fairm. AB. 1887. 93; id. CRB. 1891. 190.

B* **prolixus** *(Oodes)* Bates TES. 1873. 254; Chaud. Mon. 378. — Japan.

LACHNODERMA* W. Macleay.

Lachnoderma W. Macleay TES. NSW. ii. 1873. 321; Gestro AG. 1875. 860.

B* **asperum** Bates TES. 1883. 285. t. 13. f. 2. — Japan

G* **biguttatum** Bates AG. xxxii. 1892. 424. — Patkai Hills. Sikkim. Kumaon : Almora, Kousanie. Dehra Dun. Shwegu.

O **hirsutum** *(Singilis)* Bates TES. 1873. 333; id. TES. 1883. 285. — Hongkong. Japan.

A* **tricolor** Andr. AM. (9). xviii. 1926. 289. — Pusa., Singapore.

LACHNOTHORAX* Motchulsky.

Lachnothorax Motch. Etud. Ent. 1862. 48; Sloane PLS. NSW. 1917. 414; Andr. AM. (9). iii. 1919. 476; Sloane PLS. NSW. 1923. 33.

M **biguttata** Motch. Etud. Ent. 1862. 50; Andr. TES. 1928. 22. — Calcutta. Dehra Dun.

O* *oculata (Casnonia)* Chaud. BM. 1862. ii. 291; id. BM. 1872. i. 407. — Mormugao. Madras.

LACHNOTHORAX* Motchulsky.— Tranquebar. Madura. Ceylon : Elephant Pass, Horawupotana.

G **tokkia** *(Casnonia)* Gestro. AG. 1875. 856; Chaud. BM. 1877. i. 268. — Celebes.

LAMPROPHONUS* Bates.

Lamprophonus Bates AG. xxvii. 1889. 101; Andr. TES 1919. 194.

G* **lucens** Bates AG. xxvii. 1889. 102; id. AG. xxxii. 1892. 329. — Saidpur (Bengal). Madhupur., Fyzabad. Chapra., Pusa. Patna. Chota Nagpur : Barwa. Barkuda I. (Lake Chilka). Khandesh., Berar. Chhindwara. Madras., Kunavaram. Ceylon : Kurunagala, Colombo, Kotte, Vavuniya. Burma : Teinzo, Meetan. Java., Indo China.

B* **marginalis** *(? Drimostoma)* Walk. AM. (3). iii. 1859. 51, Chaud. AB. xv. 1872. 13; Bates AM. (5). xvii. 1886. 212; Tchitch. HER. xxxiv. 1900. 264; Andr. TES. 1919. 193. — Nilgiri Hills. Ceylon : Peradeniya, Yatiyantota.

LASIOCERA* Dejean.

Lasiocera Dej Sp. v. 1831. 283; Lac. 376; Bates EMM. viii. 1871. 31; Maindr. BF. 1906. 75.

X* **coromandelica** Maindr. BF. 1906. 75. — Barkuda I. (Lake Chilka). Madras : Vizagapatam, Madras, Madura,

LASIOCERA* Dejean.— Genji. Ceylon : Elephant Pass, Anuradhapura, Jaffna, Vavuniya.

X* **malabrica** Maindr. BF. 1906. 76. — Matheran. Koyna River Valley (Satara)., Mahe.

O* **orientalis** Chaud. BM. 1850. i. 403. — Pusa., Nainital. Delhi., Ajmere. Nagpur., Bandra.

LASIOTRECHUS* Ganglbauer.

Lasiotrechus Ganglb. Käf. Mitteleur. i. 1892. 187; Jeann. AF. xc. 1922. 168 and 299; id. Mon. des Trechinae (iii), Ab. xxxv. 1928. 94.

K **discus** *(Carabus)* F. Ent. Syst. i. (1). 1792. 164; Bates TES 1883. 266; Jeann. AF. xc. 1922. 299; Andr. TES 1921. 157; Jeann. Mon. (iii). 96. ff. 1332. 1333 and 1336. — Europe. Japan.

A var. **orientalis** Jeann. Mon. (iii). 96. ff. 1334 and 1335. — Yunnan.

LEBIA Latreille.

Lebia Latr. Hist. Nat. Crust. et Ins. iii. 1802. 85; Dej. Sp. i. 1825. 253; Latr. et. Dej. HN. et Ic. 155; Schm. Goeb. FCB. 43; Lac. 127; Chaud. Mon. des Lebiides, (i) BM. 1870. ii. 111-255, (ii) BM. 1871. i. 1-87; Silvetri Redia ii. 1904. 68-84. tt. 3-7 (life history).

Echimuthus Leach in Brewster's Edinburgh Encyclopedia ix. i. 1815. 81.

Lebida Motch. Etud. Ent. 1862. 51.

Nematopeza Chaud. Mon. des Lebiides (i) BM. 1870. ii. 146.

Eulebia W. Macl. TES. NSW. ii. 1871. 86; Sloane PLS. NSW. 1917. 424.

O* **acutangula** Jord. NZ. 106. — Sumbawa.

O* **baconi** *(Nematopeza)* Chaud. Mon. (i). 150. — Chapra. Hoshangabad. Khandesh.

O*	**basalis** Chaud. BM. 1852. i. 43; id. Mon. (i). 149.	Simla., Dinapore. Simla.
O*	**boysi** Chaud. BM. 1850. i. 70; id. Mon. (i). 222; Andr. EMM. 1926. 70.	Mandi : Kataula. Kumaon : Almora, Gori River Gorge, Dhauli Ganga Valley.
P*	**calycophora** Schm. Goeb. FCB. 44; Bates AG. xxxii. 1892. 427; Andr. TES. 1923. 21. **cardoni** Bates AB. 1892. 233.	Patkai Hills. Naga Hills. Khasi Hills. Burma. : Thagata. Momeit. Malacca., Siam. Sumatra. Chota Nagpur : Konbir.
P*	**circumdata** Schm. Goeb. FCB. 44; Chaud. Mon. (i) 224; Andr. TES. 1923. 21.	Shwegyin. Tenasserim., Tonkin.
M*	*infuscata* Motch. BM. 1864. ii. 227 (table); Andr. TES. 1928. 9.	Cambodia.
O*	**decora** *(Nematopeza)* Chaud. Mon. (i). 150.	"North India."
A*	**dichroma** Andr. Spol. Zeyl. 1923. 249.	Dehra Dun. Chapra., Khandesh. Bombay. Secunderabad. Madras. Nilgiri Hills. Ceylon : Hambegamuwa, Anuradhapura.
B*	**exsanguis** Bates AM. (5) xvii. 1886. 209.	Ceylon : Dikoya.
O*	**gressoria** Chaud. Mon. (i). 223.	Sadiya. Chittagong., Pusa. Kali Gad (Dehra Dun). Khandesh. Rangoon.
G*	**karenia** Bates AG. xxxii. 1892. 426.	Karen Hills. Great Sangir. Batchian. Halmahera.

O*	**melanonota** Chaud. Mon. (i). 226; Sloane PLS. NSW. 1917. 424.	Adonare I. Australia.
M'	*plagiata (Eulebia)* || W. Macl. (not Germ.) TES. NSW. ii. 1871. 87; Sloane PLS. NSW. 1907. 376.	
A*	**monostigma** Andr. Spol. Zeyl. 1923. 250.	Kodaikanal., Ceylon. Sarawak.
P*	**sellata** Schm. Goeb. FCB. 45; Andr. TES. 1923. 22.	Momeit.
P*	**tau** Schm. Goeb. FCB. 45; Andr. TES. 1923. 22.	Bhamo., Momeit. Moulmein.
G*	*maharani* Bates AG. xxxii. 1892. 427.	

LEBIDIA* Morawitz.

Lebidia Moraw. Mel. Biol. Acad. Sci. Petr. iv. 1862. 239; id. BAP. v. 1863. 322; id. Beitr. 27; Chaud. BM. 1872. i. 417; Andr. AM. (9). xiv. 1924. 241.

Q	**bioculata** Moraw. Beitr. 29; Andr. AM. (9). xiv. 1924. 243.	Japan.
O*	var. **biomaculata** *(Sarothrocrepis)* Jord. NZ. 106; Andr. TES. 1921. 150; id. AM. (9). xiv. 1924. 243.	Sumbawa. Alor. I.
B*	var. **pupillata** Andr. AM. (9). xiv. 1924. 243.	Japan.
	bioculata † Bates TES. 1873. 318; id. TES. 1876. t. 5. 1. f. 7.	
A*	var. **amurensis** Andr. AM. (9). xiv. 1924. 243.	Amur.
B*	var. **himalayica** Andr. AM. (9). xiv. 1924. 243.	Kumaon : Almora, Bhatkot, Upper Gumti Valley, Nainital, Ranikhet. Karen Hills.
A*	**octocelis** Andr. AM. (9). xiv. 1924. 243 and 244.	Sikkim : Gopaldhara.
Q	**octoguttata** Moraw. Mel. Biol. Acad. Sci. Petr. iv. 1862. 239; id. BAP. v. 1863. 323; id. Beitr. 28; Bates TES. 1873. 318; Fairm AF. 1889. 6; Andr. AM. (9). xiv. 1924. 243.	China. Japan.

LEISTUS Froelich.

Leïstus Froelich Naturf. xxviii. 1799. 1; Dej. Sp. ii. 1826. 212; Lac. 52; Reitt. Rév. WEZ. 1885. 213 (European species); Bänn. Ent. Mitt. 1925. 329; Andr. FBI. 106.

Pogonophorus Latr. Hist. Nat. Crust. et Ins. iii. 1802. 88.

Leïstophorus Reitt. WEZ. 1905. 213; Bänn. Ent. Mitt. 1925. 332 and 333.

O **angulicollis** Fairm. AF. 1886. 307; Tchitch. HER. xxxvi. 1903. 32. — Yunnan.

B* **championi** Andr. AM. (9). v. 1920. 448; Bänn. Ent. Mitt. 1925. 331. and 333; Andr. FBI. 108 and 109. — Kumaon : Almora, Upper Gumti Valley.

Q **indus** Tchitch. RRE. 1903. 109; Bänn. Ent. Mitt. 1925. 333; Andr. FBI. 108 and 111. t. 3. f. 1. — Sikkim : Ghoom, Tonglu.

A* **kashmirensis** Andr. Eos ii. 1927. 65 f. 1; id. FBI. 108 and 110. f. 19. — Kashmir : Sintan Pass.

A* **nivium** Andr. AM. (9). xv. 1925. 312; Bänn. Ent. Mitt. 1925. 331 and 333; Andr. FBI. 108. — Kashmir : Dusu, Sintan Pass.

A **yunnanus** Bänn. Ent. Mitt. 1925. 334 and 335. — Yunnan.

LESTICUS* Dejean.

L. *Lesticus* Dej. Sp. iii. 1828. 189; Lac. 312; Chaud., Rév. des Trigonotomides, AB. xi. 1868. 152; Tchitch. HER. xxxiv. 1900. 186; id. HER. xxxv. 1902. 506; Kuntz. Trig. 43 et seqq.

T. *Triplogenius* Chaud. BM. 1852. i. 71; id. AB. xi. 1868. 151; Tchitch. HER. xxxiv. 1900. 169 (note 16); Kuntz. Trig. 47 et seqq.

Trigonomima Motch. BM. 1864. ii. 349; Andr. TES 1928. 14.

C. *Celistus* Tchitch. HER. xxxiv. 1900. 186 (note 45); Kuntz. Trig. 58.

S.L **amabilis** Chaud. Rév. 155; Tchitch. HER. xxxiv. 1900. 190; Kuntz. Ent. Rundsch. 1911. 164; id. Trig. 43 et seqq. — Java.

Q L var. **semenowi** Tchitch. HER. xxxii. 1898. 31; id. HER. xxxiv. 1900. 169 and 190; Kuntz. Ent. Rundsch. 1911. 164; id. Trig. 43 et seqq. — Java.

O* C **andamanensis** *(Triplogenius)* Chaud. BM. 1878. ii. 32; Tchitch. HER. xxxiv. 1900. 186. (note 45); Kuntz. Ent. Rundsch. 1911. 182; id. Trig. 58 et seqq. — Andaman Is.

J* T. **assamicus** Kuntz. Ent. Rundsch. 1911. 165; id. Trig. 58; Andr. AM. (9). xix. 1927. 110.	Assam : Lamin, Margherita.
X* T. **auricollis** Tchitch. HER. xxxiv. 1900. 174 and 176; Kuntz. Ent. Rundsch. 1911. 165; id. Trig. 56 et seqq.	Siam., Tonkin. China.
D* T. **bennigseni** Sloane DEZ. 1907. 471 and 472; Kuntz. Trig. 48 et seqq; Andr. AM. (9). xix. 1927. 110.	New Guinea. New Britain.
L T. **buqueti** *(Trigonotoma)* Cast. Etud. Entom. 1834. 77; Chaud. Rév. 162; id. BM. 1878. ii. 31; Tchitch. HER. xxxiv. 1900. 192; Kuntz. Trig. 52 et seqq.	Java.
X* *viridicollis (Trigonotoma)* † Brullé (not Macl.) H. N. iv. 1834. 333. t. 12. f. 5; (Triplogenius) Chaud. Rév. 154.	
Y* T. **busuangae** Heller PJS. xxiii. 1923. 299 and 300.	Philippine Is.
O* T. **chalcothorax** *(Triplogenius)* Chaud. Rév. 153; Bates AF. 1889. 276; id. AG. xxvii. 1889. 105; id. Entom. 1891. Suppl. 71; id. AG. xxxii, 1892. 357; Kuntz. Ent. Rundsch. 1911. 165; id. Trig. 52 et seqq.	Bhamo. Siam. Indo-China. China.
X* *buqueti (Triplogenius)* † Bates (not Cast.) AF. 1889. 276; Tchitch. HER xxxiv. 1900. 177. (Obs.); Andr. TES. 1921. 178.	
O* T. **chloronotus** Chaud. Rév. 156; Sloane DEZ. 1907. 472; Kuntz. Trig. 45 et seqq.	New Guinea. Australia.
J var. **dahli** Kuntz. Ent. Rundsch. 1913. 12; id. Trig. 48 et seqq.	New Guinea. Bismark Archipelago. New Britain.
Q T. **coelestis** Tchitch HER. xxx. 1897. 265; id. HER. xxxiv. 1900. 194; Kuntz. Trig. 52 et seqq.	Java.
J* T. **crenicollis** Schauf. HER. xxi. 1887. 107; Tchitch. HER. xxxiv. 1900. 187; Kuntz. Ent. Rundsch. 1911. 165; id. Trig. 43 et seqq.; Andr. AM. (9). xix. 1927. 110.	Celebes.
Y* T. **cupreatus** Heller PJS. 1923. xxiii. 299 and 300.	Philippine Is.
O* L. **cupricollis** Pllde. Ins. vi. 1914. 215. ff. 1 and 2.	Sikkim : Pedong.
Q T. **desgodinsi** Techitch. HER. xxviii. 1894. 446; Kuntz. Trig. 56.	Sikkim : Pedong Gangtok, Mungphu, Gopaldhara.
Q T. **dichrous** Tchitch. HER. xxx. 1897. 267; id. HER. xxxiv. 1900. 195; Kuntz. Trig. 52 et seqq.	Java.

G* T. **feanus** *(Triplogenius)* Bates AG. xxxii. 1892. 356; Tchitch. HER. xxxiv. 1900. 173 and 175; Kuntz. Trig. 56.	Karen Hills.
Q T. **fortis** *(Triplogenius)* Tchitch. HER. xxx. 1897. 266; id. HER. xxxiv. 1900. 176 and 192; Kuntz. Trig. 52.	Java.
V T. **ganglbaueri** Tchitch. HER. xxxii. 1898. 33; Kuntz. Trig. 56.	"East Indies." ? Bombay (Brit Mus.).
J* T. **gregori** *(Triplogenius)* Kuntz. Ent. Rundsch. 1911. 175; id. Trig. 48 et seqq.; Andr. AM. (9). xix. 1927. 110. *mac-gregori* Heller PJS. xxiii. 1923. 300.	Philippine Is.
X* L. **harmandi** Tchitch. HER. xxxiv. 1900. 167; Kuntz. Trig. 43 et seqq.	Maria Basti.
V T. **himaleyicus** *(Omaseus)* Redt. HK. 501; Chaud. BM. 1878. ii. 33; Tchitch. HER. xxxii. 1898. 29; Kuntz. Trig. 56 et seqq.	Mussoorie.
X* T. **indus** *(Triplogenius)* Tchitch. HER. xxxiv. 1900. 169; Kuntz. Trig. 56.	Gangtok.
G* T. **insignis** *(Triplogenius)* Gestro AG. 1882. 310; Tchitch. HER. xxxiv. 1900. 194; Kuntz., Ent. Rundsch. 1911. 175; id. Trig. 48 et seqq.	Borneo. Banguey.
J* var. **philippinicus** Kuntz., Ent. Rundsch. 1911. 175; id. Trig. 48 et seqq; Heller PJS. xxiii. 1923. 300; Andr. AM. (9). xix. 1927. 110.	Philippine Is.
O* L. **janthinus** Dej. Sp. iii. 1828. 190; Dej. et Boisd. Ic. et HN. ii 399. t. 124. f. 3; Chaud. Rév 155; Tchitch. HER. xxxiv. 1900. 169 and 188; Kuntz. Ent. Rundsh. 1911. 164; id. Trig. 43 et seqq.	Java.
X* T. **laeticolor** Tchitch HER. xxxiv. 1900. 178 and 196; Kuntz. Trig. 52 et seqq.	Java.
X* T. **lakhonus** Tchitch. HER. xxxiv. 1900. 171; Kuntz. Trig. 56.	Siam.
J T. **lemoulti** Kuntz. Ent. Rundsch. 1913. 11; id. Trig. 48 et seqq.	Yule I.
M T. **magnus** *(Omaseus)* Motch. Etud. Ent. 1860. 5. Har. Abh. nat. Ver. Bremen iv. 1875. 286; Heyd. DEZ. 1879. 332.	Japan. China.

ingens (Omaseus) Moraw. Beitr. 54; Chaud. Rév. 154; Bates TES. 1873. 284; Putz. CRB. 1875. 50; Heyne-Tasch. EK. 24; Kuntz. Ent. Rundsch. 1911. 175; id. Trig. 56 et seqq.

O* T. **mouhoti** *(Triplogenius)* Chaud. Rév. 152; Bates AG. xxvii. 1889. 105; id. Ag. xxxii. 1892. 356; Tchitch. HER. xxxiv. 1900. 173 and 175; Kuntz., Trig. 56; Andr. AM. (9). xvii. 1926. 255. — Bhamo. Cambodia.

D* T. **nitescens** Sloane DEZ. 1907. 470 and 472; Kuntz. Ent. Rundsch. 1911. 165; id. Trig. 48; Andr. AM. (9). xix. 1927. 110. — New Guinea.

X* T. **nubilus** Tchitch. HER. xxxiv. 1900. 173 and 175; Kuntz. Trig. 56; Andr. AM. (9). xvii. 1926. 255. — Cochin China.

?G* T. **obscurus** *(Trigonotoma)* Cast. Etud. Ent. 1834. 76; Tchitch. HER. xxxiv. 1900. 183. — Java.

G*T. **peguensis** *(Triplogenius)* Bates AG. xxxii. 1892. 357; Tchitch. HER. xxxiv. 1900. 175; Kuntz. Trig. 56. — Palon.

O* T. **planicollis** *(Trigonotoma)* Dej. Sp. iii. 1828. 185; Dej. et. Boisd. Ic. et HN. ii. 395. t. 124. f. 1; Chaud. Rév. 154; Heyne-Tasch. EK. 24; Kuntz. Trig. 56 et seqq. — Bengal. Dehra Dun.

M* T. **politocollis** *(Trigonomima)* Motch. BM. 1864. ii. 349; Andr. TES. 1928. 14. — Pusa., Dehra Dun. Almora.

O* *semiviolaceus (Triplogenius)* Chaud. Rév. 152; Kuntz. Trig. 56 et seqq. — Lansdowne.

O T. **politus** Chaud. Rév. 156; W. Macl. PLS. NSW. 1886. 141; Tchitch. HER. xxxiv. 1900. 180; Sloane DEZ. 1907. 472; Kuntz. Ent. Rundsch. 1911. 165; id. Trig. 48. — New Guinea.

B* **politus** † Arrow (not Chaud.) TZS. xx. 1915. 499. — New Guinea.

O* T. **praestans** *(Triplogenius)* Chaud. Rév. 154; Kuntz. Trig. 68. — Tonkin. Hongkong.

X* T. **prasinus** Tchitch. HER. xxxiv. 1900. 178 and 193; Kuntz. Trig. 52 et seqq.; Heller PJS. xxiii. 1923. 300. — Sulu Is.

O* T. **putzeysi** *(Triplogenius)* Chaud. BM. 1878. ii. 31; Tchitch. HER. xxxii. 1898. 31 (note 7); id. HER. xxxiv. 1900. 196; Kuntz. Trig. 52 et seqq. — Sumatra. Java. Krakatau.

O* T. **rectangulus** *(Triplogenius)* Chaud. Rév. 153; Bates AG. xxxii. 1892. 357; Kuntz. Trig. 56 et seqq. — Dacca. Bhamo (Bates).

J* T. **sauteri** Kuntz. Ent. Rundsch. 1911. 175; id. Trig. 56 et seqq; Andr. AM. (9). xix. 1927. 110. — Formosa.

O* T. **serraticollis** *(Triplogenius)* Chaud. Rév. 153; Kuntz. Trig. 56 et seqq. — Murshidabad. Dacca., Pusa.

V L. **strictus** Tchitch. HER. xxx. 1897. 269; id. HER. xxxii. 1898. 33; id. HER. xxxiv. 1900. 190; Kuntz. Trig. 43 et seqq. — Java.

A L. **suavis** Tchitch. HER. xxx. 1897. 268; id. HER. xxxiv. 1900. 189; Kuntz. Trig. 43 et seqq. — Borneo.

O* L. **tricostatus** Chaud. Rév. 157; Kuntz. Trig. 43 et seqq. — Sikkim : Darjiling, Kurseong, Gopaldhara.

X* *costiger* Tchitch. HER. xxxiv. 1900. 165; Kuntz. Trig. 43 et seqq.

B* T. **viridicollis** *(Omaseus)* Macl. Ann. Jav. 1825. 17; Chaud. BM. 1878. ii. 33; Tchitch. HER. xxxiv. 1900. 176; Kuntz. Trig. 52 et seqq; Andr. TES. 1919. 148. — Java. Sumatra.

L *bicolor (Trigonotoma)* Cast. Etud. Ent. 1834. 75; Chaud. Rév. 152; Heyne-Tasch. EK. 24. t. 3. f. 23.

X var. **concolor** Tchitch. HER. xxxiv. 1900. 177. — Java.

O* T. **waterhousei** *(Triplogenius)* Chaud. RMZ. 1862. 489; id. Rév. 155; Kuntz. Trig. 56 et seqq; Andr. AM. (9). xiv. 1924. 590. — Chandkhira.

LICINUS* Latreille.

Licinus Latr. Hist. Nat. Crust. et Ins. ii. 1802. 92; Dej. Sp. ii. 1826. 392; Dej. et Boisd. Ic et HN. ii. 207; Lac 233; Gangl. Käf. Mitteleur. i. 1892. 379; Reitt. Best. Tab. Eur. Col. Harp. u. Lic. 1900. 147.

Allolicinus Sem. RRE. xxi. 1927. 233.

B* **jaloricus** Andr. Eos iii. 1927. 73. f. 7; Sem. RRE. xxi. 1927. 233. — Punjab : Seraj, Jalori Pass.

LIODAPTUS* Bates.

Liodaptus Bates AG. xxvii. 1889. 102.

G* **birmanus** Bates AG. xxvii. 1889. 102; id. AG. xxxii. 1892. 344; Lesne Miss Pav. 75 t. 8. f. 10; *(Daptidius discipennis)* Chaud. in litt. Vuillet Ins. ii. 1912. 19. — Assam Valley. Sylhet., Calcutta. Rajmahal.

LIODAPTUS* Bates.—

Sunderbans.
Diamond Harbour.
Dacca., Damukhdia.
Sara., "Ganges."
Pusa.
Burma : Bhamo, Mandalay, Katha, Teinzo. Siam.
Indo-China.

X* **longicornis** Lesne Bull. Mus. Paris ii. 1896. 244. f. 7; id. Miss. Pav. 75. t. 8. f. 11 and text fig. 7. — Laos.

LIONYCHUS* Wissmann.

Lionychus Wissm. SEZ. 1846. 25; Schm. Goeb. SEZ. 1846. 389; id. FCB. 36; Lac. 123; Motch. Etud. Ent. 1855. 50 (note 1); Fleisch. WEZ 1900. 233 (European species); Andr. TES. 1923. 15.

B* **albivittis** Bates AM. (5) xvii. 1886. 207; Dupuis AB. 1913. 6. — Ceylon : Peradeniya, Nalanda, Rekirawa.

D* **horni** Dupuis AB. 1913. 6. — Ceylon : Trincomali, Negombo.

P* **marginellus** Schm. Goeb. FCB. 37. t. 3. f. 3; Andr. TES. 1923. 15. — Burma. China (Brit. Mus.)

LIOPTERA* Chaudoir.

Lioptera Chaud., Mém sur les Coptodérides, AB. xii. 1869. 208; Heller AB. 1903. 241.

O **brevicornis** Heller AB. 1903. 244. — Sikkim : Pedong, Maria Basti.

B* **erotyloides** Bates TES. 1883. 280. t. 13. f. 5; Heller AB. 1903. 245. — Tonkin. Japan.

O **malayana** Heller AB. 1903. 243. — Malacca.

O **oberthüri** Heller AB. 1903. 243. — Sikkim : Maria Basti.

O* **plato** Bates TES. 1883. 281 (note); Heller AB. 1903. 245. — Borneo.

O* **pseuda** Heller AB. 1903. 244; Andr. TES. 1921. 150. — Laos.

O* **quadriguttata** Chaud. Mém. 208; Heller AB. 1903. 242. — Philippine Is.

O **tetraspila** Heller AB. 1903. 245. — Celebes.

LISSOPOGONUS* Andrewes.

Lissopogonus Andr. AM. (9). xii. 1923, 213.

A* **glabellus** Andr. AM. (9). xii. 1923. 214; id. EMM. 1926. 68. — Assam : Rotung. Sikkim : Gopaldhara, Namsoo, Birik, Gielle Khola (Tista Valley). Haldwani. Almora. Sarju Valley. Tanakpur. River Sarda Gorge. Dehra Dun. Mussoorie. Siwalik Hills. Laos.

LOMASA* Andrewes.

Lomasa Andr. AM. (9). iii. 1919. 479.

C* **xanthacra** *(Chlaenius)* Wied. Zool. Mag. ii. 1. 1823. 51; Hope Col. Man. ii. 1838. 77; Laf. AF. 1851. 245; Chaud. AG. viii. 1876. 8; Andr. AM. (9). iii. 1919. 480; id. TES. 1921. 170. — Calcutta. Karachi.

V* *kügeli (Chlaenius)* Redt. RN. 9; Andr. TES. 1923 (1924). 463.

LORICERA* Latreille.

Loricera Latr. Hist. Nat. Crust. et Ins. iii. 1802. 88; Dej. Sp. ii. 1826. 292; Lac. 214; Andr. FBI. 170.

Lorocera Bedel AF. 1878. 248; Rousseau Gen. Ins. Fasc. 86, Lorocerinae, 1908. 2.

B* **aparupa** Andr. AM. (9). v. 1920. 450; id. FBI. 171 and 173. — Kumaon : Sunderdhunga Valley, Pindar Valley, Kali Valley.

A* **stevensi** Andr. AM. (9). v. 1920. 449; id. FBI. 171. f. 31. — Sikkim : Tonglu.

LOROSTEMA* Motchulsky.

Lorostema Motch. BM. 1864. ii. 329; Andr. TES. 1928. 11.

Feanus Bates AG. xxvii. 1889. 107; Andr. MB. 48.

M* **alutacea** Motch. BM. 1864. ii. 330; Andr. TES. 1928. 11. — Sadiya. Garo Hills.

A* *nigripes (Feanus)* Andr. AM. (5). xvii. 1926. 257. — Karimganj. Sunderbans. Parbatipur. Calcutta., Pusa. Chapra., Purnea. Simla Hills: Theog. Fagu., Kalka. Barkuda I. (Lake Chilka)., Tranquebar. Ceylon: Colombo, Kotte., Sumatra.

G* var. **spinipennis** *(Feanus)* Bates AG. xxvii. 1889. 108; id. AF. 1889. 279; id. AG. xxxii. 1892. 370; Andr. MB. 48; id. TES. 1928. 11. — Chapra., Pusa. Ceylon: Kandy, Weligama, Colombo, Kollonawa, Kotte. Burma: Bhamo, Teinzo, Tharrawaddy, Insein. Federated Malay States. Sumatra. Indo-China.

LOXANDRUS* Leconte.

Loxandrus Lec. Journ. Acad. Nat. Sci. Phil. (ii). ii. 3. 1853. 250; Lac. 401; Tchitch. HER xxxv. 1902. 507; Sloane PLS. NSW. 1920. 152.

O* **celebensis** Bates EMM. viii. 1871. 133. — Celebes.

LUPERCA* Astelnau.

Luperca Cast. Hist. Nat. Ins. Col. i. 1840. 63; Lac. 163; Andr. MB. 13; id. FBI. 174 and 205.

Holoscelis Chaud. BM. 1850. i. 1. 38; id., Mon. des Siagonides, BM. 1876. i. 71.

B* **laevigata** *(Carabus)* F. Spec. Ins. i. 1781. 304; *(Scarites)* Oliv. Ent. 36. 7. t. 2. f. 18; *(Enceladus)* Dej. sp. v. 1831. 474; Hope Col. Man. ii. 1838. 99; Lac. Atl. t. 6. f. 1; Chaud. Mon. 74; Dohrn. SEZ. — Sultanpur. Husainabad., Pusa. Orissa: Satpara, Balugaon., Nagpur.

1881. 309; Heyne-Tasch. EK 16. t. 3. f. 10; Andr. TES. 1919. 122; id. TES. 1921. 154; id. MB. 13; id. FBI. 206. t. 35. — Hoshangabad. Belgaum., Surada. Madras., Coimbatore.

L *herculeana (Siagona)* Cast. Etud. Ent. 1835 151; id. Hist. Nat. Ins. Col. i. 1840. 63.

LYMNASTIS* Motchulsky.

Lymnastis Motch. Etud. Ent. 1862. 27; Reitt. WEZ. 1884. 124.

Limnastus La Brûl. AF. 1875. 436; Bedel Fn. med. 1896. 79.

Limnastis Ganglb. Käf. Mitteleur. i. 1892. 181; Andr., Revision of the oriental species of the genus *Tachys,* AG. li. 1925. 493.

Limnastes Bedel Fn. med. 1896. 54.

Zuphiolum Fairm. BF. 1896. 344.

G* **atricapillus** Bates AG. xxxii. 1892. 297; Andr. Rev. 494 and 496. — Katha. Borneo.

M* **indicus** *(Lympaeum)* Motch. BM. 1851. ii. 507; id. Etud. Ent. 1862. 27; Andr. Rév. 401; id. TES. 1928. 3. — Calcutta. Victoria Point.

G* **pilosus** Bates AG. xxxii. 1892. 296; Sloane PLS. NSW. 1921. 208; Andr. Rev. 494 and 495. — Bengal: Sardah, Calcutta, "between Sorbhog and Golakganj"

A* *setiger (Tachys)* Sloane PLS. NSW. 1903. 582; id. PLS. NSW. 1920. 321. — Chota Nagpur: Barwa. Haldwani. Bhamo., Katha. Shwegu., Palon. Tenasserim. Kuala Lumpur. Singapore., Tonkin. Java., Sumatra. Borneo., Celebes. New Guinea. Philippine Is. Formosa., Australia.

M* **pullulus** Motch. Etud. Ent. 1862. 31; Andr. Rev. 498; id. TES. 1928. 19. — Calcutta.

I* *levatus* Andr. Rev. 494 and 497.

MACROCHEILUS* Hope.

Macrocheilus Hope Col. Man. ii. 1838. 166; Schm. Goeb. FCB. 64; Lac. 93; Sloane PLS. NSW. 1914. 570.

Macrochilus Andr. AM. (9). vi. 1920. 494; id. MB. 54.

Helluo † Dej. (not Bon.) Sp. i. 1825. 283 (part).

Acanthogenius Rreiche AF. 1842. 334; Lac. 93; Schaum. BEZ. 1863. 80; Chaud. BM. 1877. i. 247.

B* **astericus** *(Acanthogenius)* White AM. xiv. 1844. 422; Chaud. RMZ. 1872. 172; Bates AG. xxxii. 1892. 389; Andr. TES. 1919. 180; id. AM. (9). vi. 1920. 497 and 500; id. AM. (9). xiii. 1924. 470. — Assam (Bates). Bhamo (Bates). Macao. Hongkong.

V* *crucifer (Planetes)* Redt. RN. 4. t. 1. f .1; Andr. TES. 1923 (1924). 460.

A* **bicolor** Andr. AM. (9). vi. 1920. 498 and 503. — Belgaum. Patkai Hills.

chaudoiri × Andr. TES. 1919. 130; id. AM. (9). vi. 1920. 497 and 502; id. AM. (9). xiii. 1924. 470. — Dacca., Pusa. Nagpur.

O* *trimaculatus (Acanthogenius)* || Chaud. (not Oliv.) RMZ. 1872. 171; Bates AF. 1889. 280; id. AG. xxxii. 1892. 390. — Bhamo., Sumatra. Philippine Is. Annam.

Y* *ruficollis* Heller PJS. xxiii. 1923. 296. t. 1. f. 1; Andr. AM. (9). xvii. 1926. 258. — Laos., Cambodia. Macao.

A* **immanis** Andr. AM. (9). vi. 1920. 496 and 498. — Toungoo.

C* **impictus** *(Helluo)* Wied. Zool. Mag. ii. 1. 1823. 49; Dej. Sp. i. 1825. 287; Andr. AM. (9). vi. 1920. 497 and 503; id. TES. 1921. 168. — Calcutta. Dehra Dun., Mhow. Nagpur. Yenna Valley (Satara). Belgaum. Mysore : Chikaballapura, Bangalore. Saidapet Coimbatore., Madras. Malabar., Mangalore. Java (Dejean).

A* **niger** Andr. AM. (9). vi. 1920. 497 and 499. — Giridih., Pusa. Jhajra., Saharanpur. Mhow., Nagpur.

A* **niger** Andr.— — Castle Rock. Nilgiri Hills. Coimbatore. Vizagapatam. Ceylon : Diyantalana.

Y **nigrotibialis** Heller Abh. Mus. Dresd. ix. 5. 1900. 3; Andr. AM. (9). vi. 1920. 497. — Sumatra.

B* **trimaculatus** *(Carabus)* Oliv. Enc. Méth. 1790. 347. t. 179. f. 11; id. Ent. 35. 88. t. 7. f. 85; Andr. TES. 1919. 129; id. AM. (9). vi. 1920. 497 and 502; id. AM (9). xiii. 1924. 469; id. MB. 54. — India., Ceylon. Burma., Penang. Selangor. Laos.

B* *bensoni* Hope Col. Man. ii. 1838. 166. t. 1. f. 5; Chaud. RMZ. 1872. 212; Bates AG. xxxii. 1892. 389; Andr. TES. 1919. 176 and 202. — Tonkin. Philippine Is. Hongkong., Macao.

O* *quadrimaculatus (Helluo)* Guér RZ. 1840. 38; id. Voy. Deless. ii. 1843. 34; Chaud. RMZ. 1872. 212.

O* *tripustulatus (Helluo)* † Guér. (not Dej.) Voy. Deless. ii. 1843. 34; Schm. Goeb. FCB. 65; Andr. TES. 1923. 37.

V* *tripustulatus* † Redt. (not F.) RN. 4; Heller DEZ. 1916. 269; Andr. TES. 1923 (1924). 460.

G* var. **infuscatus** *(Acanthogenius)* Bates AG. xxxii. 1892. 389; Andr. AM. (9). vi. 1920. 493. — Assam. Bhamo.

O* **tripustulatus** (Helluo) † Dej. (not F.) Sp. i. 1825. 286; Chaud. RMZ. 1872. 212; Andr. TES. 1919. 124; id. AM. (9). vi. 1920. 497 and 501. — Java.

B* **vitalisi** Andr. AM. (9). vi. 1920. 497 and 500. — Laos., Tonkin. China., Borneo.

MASOREUS* Dejean.

Masoreus Dej. Sp. iii 1828. 536; Zim., Einige Worte über die Gattung Masoreus, GF. 119; Lac. 134; Schaum BEZ. 1863. 76; Chaud. Etude monographique des Masoréides, Tetraponodérides, et du genre Nematotarsus, BM. 1876. ii. 12; Andr. MB. 59.

O* **orientalis** Dej. Sp. iii. 1828. 539; Zimm. GF. 121; Schaum BEZ. 1863. 77; Motch. BM. 1864. ii. 234; Chaud. Mon. 14; Bedel Fn. med. 1904. 228; Andr. MB. 60; Alluaud. AG. lii. 1928. 84. — Sitamarhi. Muthwapur. Madhupur. Janakpur., Pusa. Chapra.

O*	*laticollis* Chaud. BM. 1843. ii. 778.	"Hurdwar and Roorkee."
B*	*nobilis* Woll. Cat. Col. Canar. 1864. 22.	Siwalik Hills.
L	*striatus* Motch. BM. 1864. ii. 234.	Dehra Dun.
B*	*ascendens* Woll. Col. Hesp. 1867. 17; Alluaud. AG. lii. 1928. 66 and 84.	Punjab : Nalagarh, Baddia., Helvak. Ceylon., Arabia. Palestine. North Africa. Canary Is. Cape Verde Is.

MASTAX* Fischer.

Mastax Fisch. Ent. Russ. iii. 1825. 111; Schm; Goeb. FCB. 68; Lac. 101; Schaum BEZ. 1863. 82; Chaud. Mon des Brachynides, AB. 1876. 97; Andr. MB. 57.

A*	**annulata** Andr. MB. 111. t. 4. f. 4; id. EMM. 1926. 70.	U. P : Almora Sarju Valley, Malwa Tal, Khairna, Haldwani, Kanarichina. Lansdowne. Someswar. Spiti : Mankaran. Sutlej Valley near Simla.
G	**carissima** Bates AG. xxxii. 1892. 398.	Burma : Teinzo.
P*	**elegantula** Schm. Goeb. FCB. 69. t. 2. f. 1; Chaud. Mon. 99; Andr. TES. 1923. 39.	Sikkim : Gopaldhara, Kalimpong. Burma (Schmidt-Goebel).
B*	**euanthes** Andr. Spol. Zeyl. 1924. 138.	Ceylon : Hambantota.
A*	**florida** Andr. MB. 112. t. 4. f. 3.	C. P. : Rajadhar. Ratnagiri, Belgaum. Nilgiri Hills. Mysore.
D*	**formosana** Dupuis. AB. 1912. 282.	Formosa.
G*	**gestroi** Bates AG. xxxii. 1892. 399.	Karen Hills.
C*	**histrio** *(Brachinus)* F. Syst. Eleuth. i. 1801. 219; Motch. Etud. Ent. 1855. 55; Chaud. Mon. 101; Andr. TES. 1921. 164.	Calcutta., Nagpur. Raipur., Bombay. Kalyan.

C*	**histrio** *(Brachinus)*—	Mysore : Teppukadu, Chikkaballapura, Nandidrug, Bangalore. Pondicherry. Madras. Ceylon : Hambegamuwa, Horawupotana.
E*	**laeviceps** Bates CRB. 1891. 337; Andr. MB. 58; id. EMM. 1926. 70.	Kurseong. Eastern Duars. Dehra Dun. Almora : Sarju Valley. Haldwani. Kanarichina. Spiti : Mankaran. Palon.
G*	*alveolatus* Bates AG. xxxii. 1892. 398.	
P*	**moesta** Schm. Goeb. FCB. 70. t. 2. f. 3; Chaud. Mon. 100; Bates AG. xxxii. 1892. 399; Andr. TES. 1923. 39.	Palon.
P*	**ornata** Schm. Goeb. FCB. 70; Chaud. Mon. 100; Andr. TES. 1923. 40.	Tavoy.
J*	**poecila** Schaum. BEZ. 1863. 82; Chaud. Mon. 101; Bates AF. 1889. 282; Andr. AM. (9). xix. 1927. 108.	Singapore. Cambodia. Hongkong.
O*	**pulchella** *(Brachinus)* Dej. Sp. v. 1831. 433; Chaud. Mon. 99.	"East Indies." Hongkong.
G*	**rugiceps** Bates AG. xxxii. 1892. 397.	Prome., Palon.
O*	**striaticeps** Chaud. Mon. 99.	Dacca.
A*	**vegeta** Andr. MB. 110. t. 4. f. 5.	Sikkim: Great Ranjit Valley, Tista Bridge, Birik. Eastern Duars. U. P. : Almora, Nainital, Phanduwala, Khairna, Tanakpur, Haldwani, River Sarda Gorge, Kangra.

MEGALONYCHUS* Chaudoir.

Megalonychus Chaud. BM. 1843. ii. 418; Lac. 352.

G* **birmanicus** Bates AG. xxxii. 1892. 369. — Assam: Noa Dehing Sadiya, Khasi Hills, Naga Hills. Sikkim : Gopaldhara. Pusa. Saharanpur. Dehra Dun. Burma: Bhamo, Karen Hills, Teinzo, Pyaunggyaung. Laos.

O* **cyanipennis** Bates AG. xxxii. 1892. 370 (note). — Assam : Noa Dehing, Khasi Hills. Sikkim : Tista Bridge.

G* **dilaticollis** Bates AG. xxxii. 1892. 369. — Karen Hills.

MEGALOODES* Lesne.

Megaloodes Lesne Bull. Mus. Paris ii. 1896. 242; id. Miss. Pav. 71.

X* **politus** Lesne Bull. Mus. Paris ii. 1896. 242. t. × 5; f. 5; id. Miss. Pav. 71. t. 8. f. 7 and text-fig. 5. — Siam. Cochin-China.

MELAENUS* Dejean.

Melaenus Dej. Sp. v. 1831. 481; Brullé HN. v. 1835. 85; Lac. 166; Andr. TES. 1921. 184.

C* **piger** *(Brachinus)* F. Syst. Eleuth. i. 1801. 219; Schaum SEZ. 1847. 49; Erichs. SEZ. 1847. 142; Motch. Etud. Ent. 1855. 42; Andr. TES. 1921. 164 and 185. — Bengal., Mhow. Nagpur., Chanda. Hoshangabad. Bilaspur., Balaghat. Bombay : Igatpuri, Belgaum, Bandra. Tinnevelly. Kunavaram., Kurnool. Madras., Chatrapur. Ootacamund. Ceylon : Horawupotana.

METABACETUS* Bates.

Metabacetus Bates AG. xxxii. 1892. 364.

G* **immarginatus** Bates AG. xxxii. 1892. 364. — Karen Hills.

METABLETUS* Schmidt-Goebel.

Metabletus Schm. Goeb. FCB. 38; id. SEZ. 1846. 390; Lac. 122; Motch. Etud. Ent. 1855. 50. (note 1); Zimm. Trans. Am. Ent. Soc. ii. 1869. 243; Bedel Fn. med. 1913. 281; Andr. MB. 64.

Dromoceryx Schm. Goeb. FCB. 40; Andr. TES. 1923. 18.

Apristomorphus Motch. BM. 1861. i. 104; Andr. TES. 1928. 7.

P* **angularis** *(Dromoceryx)* Schm. Goeb. FCB. 41; Bates AG. xxxii. 1892. 420; Andr. TES. 1923. 18. — Burma : Plapoo (Tenasserim).

G* **cymindulus** Bates AG. xxxii. 1892. 418. — Shillong. Sikkim : Gopaldhara, Darjiling, Kurseong. Calcutta., Pusa. Bajaura. U. P. : Almora, Dudbatoli, Tanakpur, Dehra Dun. Lebong., Karen Hills. Katha., Yunnan.

P* **dorsalis** *(Dromoceryx)* Schm. Goeb. FCB. 40; Andr. TES. 1923. 18. — "Burma". Pegu.

M **fuscomaculatus** *(Dromius)* Motch. Ins. Sib. 59; Bedel En. med. 1913. 285; Andr. RIM. xviii. 1919. 101; id. MB. 65; Andr. TES. 1928. 23. — Kashmir: Pir Panjal. Almora., Nainital. Punjab : Seraj (Jalori Pass), Mandi, Nalagarh, Baddia, Dharmsala. Chamba. Simla Hills : Fagu, Theog, Matiana.

O* *patruelis* *(Dromius)* Chaud. Enum. Carab. Caue. 1846. 60.

?Q *exclamationis* *(Dromius)* Mém. MAP. (6). vi. 1848. 22. t. 1. f. 4; Solsky HER. viii. 1871. 177.

O* *virgatus* *(Dromius)* Reiche AF. 1855. 575. — Seistan., Persia.

B* *arenicola* *(Dromius)* Woll. Ins. Mad. 1857. 4. — South Russia. Asia Minor.

O* ***vittula*** *(Dromius)* Fairm. RMZ. 1859. 59.	North Africa. Madeira.
L(Q) **lateralis** *(Dromius)* Motch. Etud. Ent. 1855. 82; Bedel Fn. med. 1913. 284; Andr. TES. 1928. 21.	Karachi., Peshawar. Multan.,
O* *mutabilis (Dromius)* Reiche AF. 1855. 574; Motch. Etud. Ent. 1862. 55.	Baghdad., Arabia. Palestine. North Africa.
A* **monticola** Andr. AM. (9). xii. 1923. 453.	Kodaikanal. Nilgiri Hills.
A* var. **curtus** Andr. AM. (9). xii. 1923. 454.	Nilgiri Hills.
E' **obscuroguttatus** *(Lebia)* Dufts. Fn. Austr. ii. 1812. 249; Schm. Goeb. FCB. 38; Schaum Nat. Ins. Deutschl. i. 1857. 279; Bedel Fn. med. 1913. 285 and 285 (note 2); Andr. TES. 1923. 16.	India (Schmidt-Goebel). Persia. Mediterranean basin. Europe., Madeira.
O* *spilotus (Dromius)* Dej. Sp. i. 1825. 246.	
P* **quadripunctatus** Schm. Goeb. FCB. 39; Bates TES. 1883. 284; id. AG. xxxii. 1892. 418; Andr. TES. 1923. 17.	Cossipur. Nilgiri Hills. Ceylon Burma : Prome, Toungoo, Karen Hills, Thagata, Mt. Mooleyit, Ruby Mines, Tavoy. Pahang., Siam. Java., Borneo., Japan.
M* *expunctatus (Apristomorphus)* Motch. BM. 1861. i. 105. t. 9. f. 2; Andr. TEZ. 1928. 7.	
G **subvittatus** Bates AG. xxxii. 1892. 417.	Karen Hills., Bhamo.
I* **tartarus** Bates PZS. 1878. 719; id. SYM. 19; Andr. MB. 64.	Kashmir : Sind Valley. Lahoul : Patseo. "Between Yangihissar and Sirikol."

MICROCOSMUS* Chaudoir.

Microcosmus Chaud., Mon. sur les Panagéides, AB. 1878. 139.

O* **flavopilosus** *(Isotarsus)* Laf. AF. 1851. 222; Chaud. BM. 1861. ii. 348; id. Mon. 142; Bates TES. 1873. 243.	Nepal (Motchulsky). Jorhat., Dacca. Sardah., Calcutta. Pusa., Chapra. Dehra Dun. Gopkuda I. (Lake Chilka).
M* *transversus (Epicosmus)* Motch. BM. 1864. ii. 332; Andr. TES. 1928. 11.	

O* **flavopilosus**— Nagpur., Bandra. Bombay., Nasik. Coimbatore. Madras., Minikoi. Paungde., Sumatra. Tonkin., Formosa. Japan., China.

MICROLESTES Schmidt-Goebel.

Microlestes Schm. Goeb. FCB. 41; *(Dromius)* Sloane PLS. NSW. 1898. 494; Holdh., Mon. der pal. Arten der Gattung Microlestes, Denkschr. Akad. Wiss. Wien. Math-nat. Kl., lxxxviii. 1913. 477-536; Bedel Fn. med. 1913. 287; Sloane PLS. NSW. 1920. 170.

Blechrus Motch. BM. 1847. ii. 219; id. BM. 1848. i. 543; Lac. 122 (note 1).

X* **annamensis** *(Blechrus)* Bates AF. 1889. 285; id. AG. xxxii. 1892. 419; Andr. TES. 1923. 19. — Annam. Cambodia.

corticalis *(Lebia)* Dufour. Ann. Gen. Sci. Phys. vi. 1820. 322; *(Dromius)* Latr. et. Dej. HN. et. Ic. 191. t. 15. f. 4; Dej. Sp. i. 1825. 245; Motch. Ins. Sib. 1844. 59; *(Blechrus)* Reitt. DEZ. 1900. 374; Holdh. Mon. 517; Bedel. Fn. med. 1913. 289; Andr. RIM. xviii. 1919. 101; id. TES. 1923. 18. — Cossipore (Schmidt-Goebel). Seistan : Labi Baring. South West Asia. Mediterranean basin.

L *suturellus (Dromius)* Motch. Ins. Sib. 1844. 58; Reitt. DEZ. 1900. 374; Andr. TES. 1928. 23.

P* *? plagiatus (Dramius)* † Schm. Goeb. (not Dufts.) FCB. 39; Holdh. Mon. 517; Andr. TES. 1923. 17.

B' *binotatus (Blechrus)* Reitt. DEZ. 1900. 373.

A* **demessus** Andr. Spol. Zeyl. 1923. 248. — Madras. Ceylon : Kanthalay, Trincomali, Anuradhapura.

P* **exilis** Schm. Goeb. FCB. 42; Bates AG. xxxii. 1892. 420; Andr. TES. 1923. 19. — Pusa., Tharrawaddy. Bhamo.

M* *fuscipennis* Motch. Etud. Ent. 1859. 122; Andr. TES. 1928. 19. — Rangoon. Siam.

V **glabratus** *(Metabletus)* † Redt. (not Dufts.) RN. 7 (note 1); Andr. TES. 1923 (1924). 461. — Philippine Is.

E **glabratus** *(Blechrus)* † Bates (not Dufts) CRB. 1891. 339. — Chota Nagpur : Tetara.

P* **inconspicuus** Schm. Goeb. FCB. 41; Bates AG. xxxii. 1892. 419; id. AB. 1892. 233; Andr. TES. 1923. 19. — Chota Nagpur (Bates)., ? Bhamo. Rangoon (Bates).

B* **xanthopus** *(Blechrus)* Bates AM. (5). xvii. 1886. 206. — Colombo.

MILTODES* Andrewes.

Miltodes Andr. AM. (9). x. 1922. 162.

A* **granum** Andr. AM. (9). x. 1922. 163. — Nagpur.

MISCELUS* Klug.

Miscelus Klug Jahrb. Ins. 1834. 82; Lac. 146; Chaud. BEZ. 1861. 125; Bates EMM. vi 1869. 72; Chaud. Mém. sur les Thyreopterides, AB. xii. 1869. 152; Sloane DEZ. 1907. 473; Andr. AM. (9). ix. 1922. 293.

Leptodactyla Brullé HN. iv. 1834. 130

B* **carinatus** Andr. AM. (9). ix. 1922. 292 and 295. — Assam : Rotung, Kobo, Garo Hills. Laos.

J* **javanus** Klug Jahrb. Ins. 1834. 82. t. 1. f. 9; Chaud. BEZ. 1861. 125; Putz. AG. 1875. 723; Bates AG. xxxii. 1892. 408; Lesne Miss. Pav. 80; Heller DEZ. 1916. 273; Andr. AM. (9). ix. 1922. 293 and 295; id. AM. (9). xix. 1927. 101. — Chittagong. Assam : Rotung, Kobo, Sadiya. Manipur. Chandkhira. Sikkim : Darjiling, Birik.

X* *apicalis (Leptodactyla)* Brullé HN. iv. 1834. 130. t. 4. f. 1.

E* *paradoxus* Putz. AG. 1875. 724; Andr. A.M. (9). ix. 1922. 294.

L (E*) *convexicollis* Putz. AG. 1875. 724; Andr. AM. (9). ix. 1922. 294.

E* *vulneratus* Putz. AG. 1875. 725; Andr. AM. (9). ix. 1922. 294.

J* *planatus* Schauf. HER. xix. 1885. 183; Andr. AM. (9). ix. 1922. 294; id. AM. (9). xix. 1927. 109.

— Ceylon : Niroddumunai, Naramulla. Bhamo. Myitkyina. Momeit. Tavoy. Kawkareik. Andaman Is. Nicobar Is., Siam. Malay Peninsula and Archipelago. Indo-China. Formosa.

E* var. **unicolor** Putz., Mém. Soc. Liége ii. 1845. 375; Chaud. BEZ. 1861. 125; Putz. AG. 1875. 725; Bates AG. xxxii. 1892. 408; Andr. AM. (9). ix. 1922. 294. — Pusa., Saharanpur. ? Kashmir., Raipur. North Kanara.

B* *rufiventris (Cymindis)* Walk. AM. (3). ii. 1858. 202; Andr. TES. 1919. 183. — Rewah: Bersingh-pur, Dhanpuri., Sidapur.

O* *ceylonicus* Chaud. BEZ. 1861. 125; Bates AM. (5). xvii 1886. 202; id. AF. 1889. 283. — Nilgiri Hills. Ceylon., Colombo. Kandy., Karen Hills.

G* *stygicus* Putz. AG. 1875. 726; Sloane DEZ. 1907. 474. — Siam., Indo-China. Hongkong (Oxford Mus.)

M' *morioformis* W. Macl. PLS. NSW. 1876. 168; id. PLS. NSW. 1886. 138; Sloane DEZ. 1907. 474 (note 2); id. PLS. NSW. 1920. 322. — Malay Peninsula and Archipelago. Australia.

J* *celebensis* Schauf. HER. xix. 1885. 184; Andr. AM. (9). xix. 1927. 109.

G* **luctuosus** Putz. AG. 1875. 725; Maindr. Nov. Guin. v. ii. 1908. 298; Andr. AM. (9). ix. 1922. 204. — New Guinea.

MNUPHORUS* Chaudoir.

Mnuphorus Chaud. BEZ. 1873. 55; id., Etude monographique des Masoréides, Tetragonodérides, et du genre Nematotarsus, BM. 1876. ii. 69; Glaz., Rév. der Mnuphorus-Arten, RRE. xiii. 1913. 231.

Hypercosmeton Reitt. WEZ. 1889. 291.

G* **discophorus** Chaud. Mon. 69. — Saran., Pusa. Chapra., Bhogaon. Dehra Dun. Katihar., Allahabad. Simla.

MOCTHEROIDES* Andrews.

Moctheroides Andr. TES. 1923.

P* **sericans** *(Masoreus)* Schm. Goeb. FCB. 87; Schaum BEZ. 1863. 78; Chaud. BM. 1876. ii. 28; Andr. TES. 1923. 50 and 51. — Burma. Singapore. Sumatra.

MOCHTHERUS* Schmidt-Goebel.

Mochtherus Schm. Goeb. FCB. 76; Lac. 137; Bates EMM. vi. 1869. 71; Chaud. Mém. sur les Coptoderides, AB. xii. 1869. 240.

Cyrtopterus Motch. BM. 1861. i. 106.

asemus × Andr. TES. 1923 (1924). 462. — Moluccas.

O* *immaculatus* † Chaud. (not Redt.) Mém. 243.

E* **luctuosus** Putz. CRB. 1875. 52; Bates TES. 1876. 5; id. TES. 1883. 281. — Japan.

D **obscurus** *(? Sinurus)* Slaone DEZ. 1907. 183; Andr. AM. (9). xixi 1927. 110. — New Guinea.

? immaculatus † Maindr. (not Redt.) Nov. Guin. v. ii. 1908. 299.

B* **tetraspilotus** *(Dromius)* Macl. Ann. Jav. 1825. 25; Schaum. BEZ. 1860. 187; Chaud. Mém. 241; Bates AM. (5). xvii. 1886. 203; id. AG. xxxii. 1892. 412; Heller DEZ. 273; Andr. TES. 1919. 163; id. EMM. 1926. 70; id. Ins. Samoa iv. 1927. 12. f. 8. — Throughout India, Ceylon and Burma. Andaman Is. Malay Peninsula and Archipelago. Philippine Is. Tonkin., Loas. Formosa.

O* *tetrasemus (Thyreopterus)* Dej. Sp. v. 1831. 448.

P* *angulatus* Schm. Goeb. FCB. 76; Redt. RN. 7; Andr. TES. 1923. 45; id. TES. 1923 (1924). 46. — Christmas I. Samoa.

B* *retractus (Panagaeus)* Walk. A1. (3). ii. 1858. 203; Andr. TM. 1919. 189. ES.

M *quadrinotatus (Cyrtopterus)* Motch. BM. 1861. i. 106; Andr. TES. 1928. 23.

G'* **tetraspilotus** † Bouch. (not Macl.) AF. 1903. 174. — Java., Sumatra.

MORION Latreille.

Morion Latr. Consid. gén. 1810. Tabl. méth. 159.

Morio Dej. Sp. i. 1825. 429; Lac. 183; Chaud. BM. 1854. i. 325; Putz. SEZ. 1879. 285; Chaud., Essai monographique sur les Morionides, BM. 1880. i. 327.

O* **angustum** Chaud. Mon. 346. — Philippine Is.

G* **brevior** Putz., AG. 1873. 217; id. AG. 1875. 727; Chaud. Mon 340. — Borneo. Sumatra.

B* **cucujoides** Walk. AM. (3). ii. 1858. 203; Bates AM. (5). xvii. 1886. 211; Andr. TES. 1919. 188. — Throughout India, Ceylon and Burma. Andaman Is.

B* *trogositoides* || Walk. (not Chaud.) AM. (3). ii. 1858. 203. — Malay Peninsula and Archipelago.

O* *cordicollis* Chaud. Mon. 343; Bates AM. (5). xvii. 1886. 143. — Siam. Indo-China.

G* **doriae** Putz. AG. 1873. 217; id. AG. 1875. 727; Chaud. Mon. 345. — Borneo.

jordani × Csiki Col. Cat. 104. Carab. 1929. 481. — Tenimber.

O* *gracile* || Jord. (not Fairm.) NZ. 107.

G* **longipennis** Putz. AG. 1875. 727; Chaud. Mon. 337; Sloane DEZ. 1907. 470; Maindr. Nov. Guin. v. ii. 1908. 295; Sloane PLS. NSW. 1920. 321. — Aru Is. New Guinea. Australia.

O* **luzonicum** Chaud. BM. 1852. i. 81; id. BM. 1854. i. 330; Putz. AG. 1875. 726; Chaud. Mon. 344; Bates AG. xxxii. 1892. 352. — Palon., Andaman Is. Siam., Laos. Tonkin.

O* *intermedium* Chaud. Mon. 344; Andr. AM. (9). xiv. 1924. 591. — Malay Peninsula and Archipelago.

O* **orientale** Dej. Sp. i. 1825. 432; id. Sp. v. 1831. 511; Chaud. BM. 1854. i. 330; Putz. AG. 1873. 216; id. CRB. 1875. 69; Chaud. Mon. 338; Schauf. HER. xix. 1885. 184; Bates AF. 1889. 278; id. AG. xxvii. 1889. 106; id. AG. xxxii. 1892. 352; Bouch. AF. 1903. 172; Andr. RIM. xxii. 1921. 345; Andr. AM. (9). xiv. 1924. 591. — Throughout South East Asia.

E* *walkeri* Putz. AG. 1873. 216; Chaud. Mon. 341; Bates AM. (5). xvii. 1886. 143.

O* *subconvexum* Chaud. Mon. 340.

O* *submarginatum* Chaud. Mon. 342.

O* *cucujoides* † Chaud. (not Walk.) Mon. 342.

O* **simulatum** Jord. NZ. 107 — Timor

MORIONIDIUS* Chaudoir.

Morionidius Chaud., Essai monographique sur les Morionides, B.M. 1880. i. 380; Andr. AB. 1921. 203.

B* **charon** Andr. AB. 1921. 204. — Laos., Tonkin.

G* **doriae** Chaud. Mon. 383. — Borneo.

MORMOLYCE* Hagenbach.

Mormolyce Hagenb., Mormolyce Nov. Col. Gen. 1825; id. Ann. Sci. Nat. vi. 1825. 500; Dej. Sp. v. 1831. 712; Mann. BM. 1837. i. 25; Hope Col. Man. ii. 1838. 112; Cast. Hist. Nat i. 1840. 119; Chaud. BM. 1848. i. 123; Lac. 144; Mulder Tijds. Ent. iii. 1860. 131 (Dutch.); Thoms., Mon du genre Mormolyce, Musée Scientifique, 1862. 93; Westw. AM. (3). x. 1862. 96; Chaud. Mém. sur les Thyreoptérides, AB. xii. 1869. 133; Girard Traité d'Ent. i. 1873. 263; Gestro. AG. 1875. 886; G. H. Horn Tr. Amer. Ent. Soc. ix. 1881. 150. t. 3. f. 3. and t. 8. f. 77; Rousseau Gen. Ins. Fasc. 40; Mormolycinae, 1906. 3.

castelnaudi Deyr. AF. 1862. 314. t. 11. f. 3; Thoms. Mon. 94; Gestro AG. vii. 1875. 886-9; Heyne-Tasch. EK. 15; Donck. BF. 1899. 272. fig. ; Rouss. Gen. Ins. Fasc. 40. t. f. 3. — Malacca.

H* **hagebachi** Westw. AM. (3). x. 1862. 97; Deyr. AF. 1862. 313. t. 11. f. 2; Gestro AG. vii. 1875. 886-9; Heyne-Tasch. EK. 15; Rouss. Gen. Ins. Fasc. 40. t. f. 5. *blattoides* Thoms. Mon. 94. — Malay Peninsula. ? Sumatra (Rousseau).

L **phyllodes** Hagenb. Nov. Col. Gen. 1825. fig.; id. Ann. Sci. Nat. vi. 1825. 503. f. 21. ff. 1-10; Dej. Sp. v. 1831. 714; Dej. et Boisd. Ic. et HN. ii. 324. t. 113. fig.; Gray in Griffiths Anim. Kingd. xiv. 1832. 220. t. 25. f. 7; Brullé HN. iv. 1834. 313; t. 11. f. 2; Mann. Bull. Mosc. 1837. i. No. 2. 25; Cast. Hist. Nat. i. 1840. 119. t. 7. f. 3; Guér (in Cuvier's Régne Animal) Icon. Ins. 23. 1844. t. 6. f. 7; Ver Huell, Sur les métamorphoses du *M. phyllodes,* Ann. Sci. Nat. (3) vii. 1847. 344. t. 7. ff. 1-6 (Biology); Overdijk, Briefe over *M. phyllodes,* Tijdschr. Ent. i. 1857. 41 (Dutch); Mulder Tijdschr. Ent. iii. 1860. 131. tt. 10 and 11. ff. 1-17. (Dutch); Bowr. AM. (3). vii. 1861. 423 (Biology); Deyr. AF. 1862. 315. t. 11. f. 1; Thoms. Mon. 94; Gestro AG. vii. 1875. 886-9. fig.; Giard Compt. rend. Soc. Biol. (9). iv. 1892. 156. (Biology); id. BF. 1892. 61 (Biology); Heyne-Tasch. EK. 15. t. 2. f. 36; Donck. BF. 1889. 271. fig.; Shelf. Rep. Br. Ass. Adv. Sci. lxxi. 1901. 690. (Biology); id. PES. 1905. 73 (Biology); Rouss. Gen Ins. Fasc. 40. t. tf. 6-8, 9. (larva), 10-11 (Pupa); Sharp and Muir TES. 1912. 489. t. 47. ff. 33 and 33*a* (Genitalia). — Malay Peninsula. Siam. Java. Sumatra. Borneo. ? New Guinea (Rousseau).

G var. **borneensis** Gestro AG. 1875. 886. fig.; Heyne-Tasch. EK. 15; Donck. BF. 1899. 272. fig.; Rouss. Gen. Ins. Fasc. 40. t. f. 2. — Borneo.

A **quadraticollis** Donck. BF. 1899. 271. fig.; Rouss. Gen. Ins. Fasc. 40. t. f. 4. — Borneo.

MOUHOTIA* Castelnau.

Mouhotia Cast. RMZ. 1862. 305; Pllde. Ins. vi. 1914. 327; Andr. FBI. 325.

Scaritarchus Schaum PES. 1862. 94.

O* **batesi** Lewis EMM. xv. 1879. 186; Waterh. Aid. xvi. 1883. t. 125. f. 3; Pllde. Ins. vi. 1914. 329; Andr. FBI. 326. t. 1. fig. — Tenasserim : Downdani. Valley, Thaungyin Valley Moulmein. Rangoon. Dawna Hills., Siam.

B* **convexa** Lewis EMM. xix. 1883. 193; Waterh. Aid. xvii. 1883. t. 129. f. 1; Pllde. Ins. iv. 1914. 329; Andr. TES. 1919. 197. — Laos.

G* **gloriosa** Cast. RMZ. 1862. 306 (August); Lucas BF. 1877. 173; Heyne-Tasch. EK. 19; Pllde. Ins. iv. 1914. 328. — Laos.

J* *inidas (Scaritarchus)* Schaun PES. 1862. 94 (Sept.); Andr. AM. (9). xix. 1927. 108.

O* var. **planipennis** Pllde. Ins. iv. 1914. 330. fig. (327). — Laos., Siam.

NEBRIA* Latreille.

N. *Nebria* Latr. Hist. Nat. Crust. et Ins. iii. 1802. 89; Dej. Sp. ii. 1826. 221; Lac. 50; Bänn. Kol. Rundsch. x. 1923. 129; Andr. MB. 5; Bänn. Ent. Mitt. 1925. 187; Andr. FBI. 112.

H. *Helobia* Curt. Brit. Ent. iii. 1826. t. 103; Dan. Münch. Kol. Z. i. 1903. 166; Bänn. Ent. Mitt. 1925. 191.

Neb. *Nebriola* Dan. Münch. Kol. Z. i. 1903. 164; Bänn. Kol. Rundsch. x. 1923. 135; id. Ent. Mitt. 1925. 192.

Ps. *Psilonebria* Andr. AM. (9). xi. 1923. 275; Bänn. Ent. Mitt. 1925. 191.

P. *Patrobonebria* Bänn. Kol. Rundsch. x. 1923. 131; id. Ent. Mitt. 1925. 192.	
B*N. **aborana** Andr. AM. (9). xv. 1925. 310; Bänn. Ent. Mitt. 1925. 194; Andr. FBI. 114 and 119. f. 21.	Assam : Naga Hills. Janakmukh.
B* **barbata** Andr. FBI. 115 and 130.	Kashmir.
B*H. **cameroni** Andr. AM. (9). xv. 1925. 309; Bänn. Ent. Mitt. 1925. 273 and 341; Andr. FBI. 114 and 116. f. 20.	Arni Gad (Mussoorie). Dehra Dun. Mohan Rau (Siwalik Hills). Almora.
B*N. **cinctella** Andr. AM. (9). xv. 1925. 311; Bänn. Ent. Mitt. 1925. 195; Andr. FBI. 115 and 124.	Mothronwala (Dehra Dun).
O* P **desgodinsi** Oberth. Col. Nov. i. 1883. 47; Bänn. Kol. Rundsch. x. 1923. 131; id. Ent. Mitt. 1925. 193; Andr. FBI. 115 and 127. f. 23.	Nepal : Soondrijal. Sikkim : Darjiling, Gopaldhara, Maria Basti., Tanakpur. River Sarda Gorge. Haldwani. Simla Hills : Gahan, Kotkhai.
B*P. **elegans** Andr. AM. (9). xv. 1925. 306; Bänn. Ent. Mitt. 1925. 193; Andr. FBI. 115 and 124. *himalayica* † Bänn. (not Bates) Kol. Rundsch. x. 1923. 131.	Sunderdhunga Valley (Almora).
O*P. **himalayica** Bates PZS. 1889. 212; Bänn. Kol. Rundsch. x. 1923. 131 and 138; id. Ent. Mitt. 1925. 193; Andr. FBI. 115 and 125.	Kashmir : Skardo, Goorais Valley, Ramsu (Ramban Div.). Bashahr : Poo. Kulu : Parbatti Valley.
A* *macrocephala* || Breit (not Motch.) Kol. Rundsch. 1914. 158; Bänn. Kol. Rundsch. x. 1923. 131; id. Ent. Mitt. 1925. 193.	
thibetica × Csiki Co. Cat. 91-2. Carab. 1927. 356.	
A* var. **pallidipes** Breit Kol. Rundsch. 1914. 158; Bänn. Kol. Rundsch. × 1923. 131; id. Ent. Mitt. 1925. 193; Andr. EMM. 1926; 66; id. FBI. 115 and 126.	Bashahr : Poo. Khedar Khud (Chakrata). Milam (Almora)

M N. **limbigera** Solsky FT. 13; Fairm. AF. 1889. 6; Bates SYM. 3; Glas. HER. xxxv. 1902. 481. t. 2. f. 6; Andr. MB. 5; Bänn. Ent. Mitt. 1925. 256; Andr. EMM. 1926. 66; id. TES. 1928. 21; id. FBI. 114 and 120.	Kashmir : Dras, Kargil, Leh, Mulbeck. Rukshu: Nima-Mud, Tsho-Morari. Lahoul: Zingzinbar. Kumaon: Shelshel. Western China. Turkestan.
A* var. **babaulti** Andr. MB. 67. t. 1. f. 1; Bänn. Ent. Mitt. 1925. 257 and 281; Andr. FBI. 114 and 121.	Kashmir: Baltal. Ladak: Mulbeck.
A* Neb. **masrina** Andr. MB. 68; Bänn. Ent. Mitt. 1925. 264; Andr. FBI. 115 and 128. f. 24.	Kashmir: Baltal.
B* **pharina** Andr. FBI. 116 and 132.	Tibet : Papri.
B* Neb. **pindarica** Andr. AM. (9). xv. 1925. 307; Bänn. Ent. Mitt. 1925. 264; Andr. FBI. 115 and 131.	Kumaon : Pindar Valley.
M N. **psammophila** Solsky FT. 12; Bates SYM. 2; Glas. HER. xxxv. 1902. 478. t. 2. f. 1; Bänn. Kol. Rundsch. x. 1923. 139; id. Ent. Mitt. 1925. 195; Andr. TES. 1928. 21; id. FBI. 114 and 118.	Kashmir : Dras, Kargil, Leh, Sonemarg, Ganderbal. Achhebal, Jhelum Valley. ? Nepal : Chitlong. Turkestan.
B*N. **pulcherrima** Bates TES. 1873. 236; Fairm. AB. 1887. 90; Bates Entom. 1890. 209; Maindr. BF. 1906. 226; Bänn. Kol. Rundsch. x. 1923. 140; id. Ent. Mitt. 1925. 194.	Yunnan. China. Japan.
X'* var. **pulchrior** Maindr. BF. 1906. 226; Bänn. Kol. Rundsch. x. 1923. 140; id. Ent. Mitt. 1925. 194.	Yunnan.
X'* var. **bourderyi** Maindr. FB. 1906. 250; Bänn. Kol. Rundsch. x. 1923. 140; id. Ent. Mitt. 1925. 194.	Yunnan.
B*Ps. **superna** Andr. AM. (9). xi. 1923. 273; Bänn. Kol. Rundsch. x. 1923. 142; id. Ent. Mitt. 1925. 266; Andr. FBI. 116 and 133 f. 25.	Mount Everest Expeditions base Camp.
O*N. **xanthacra** Chaud. BM. 1850. i. 423; Bänn. Kol. Rundsch. x. 1923. 140; id. Ent. Mitt. 1925. 195; Andr. FBI. 115 and 122. f. 22.	Kashmir: Ramsu. Almora., Ranikhet. Nainital., Dehra Dun. Bhowali. Simla Hills : Matiana. Bashahr; Sangla.

A* var. **apicalis** Breit. Kol. Rundsch. 1914. 160; Bänn. Kol. Rundsch. x. 1923. 140; id. Ent. Mitt. 1925. 195; Andr. FBI. 114 and 123.	As type form.
B* var. **nigrocollis** Andr. AM. (9). xv. 1925. 312; Bänn. Ent. Mitt. 1925. 195; Andr. FBI. 115 and 123.	As type form.
A Neb. **yunnana** Bänn. Kol. Rundsch. 1928. 4.	Yunnan.
var. **pallidipennis** Bänn. Kol. Rundsch. 1928. 4.	Yunnan.

NEOBLEMUS* Jeannel.

Neoblemus Jeann. AM. (9). xii. 1923. 409; id., Mon. des Trechinae (i), Ab. xxxii. 1926. 432.

B* **andrewesi** Jeann. AM. (9). xii. 1923. 410 and 413. f. 9; id. Mon. (i) 436. ff. 222-3.	Dehra Dun. Haldwani.
A* **bedoci** Jeann. AM. (9). xii. 1923. 410 and 412. f. 8; id. Mon. (i). 436. ff. 217-21.	Tonkin.
B* **championi** Jeann. AM. (9). xii. 1923. 410 and 411; id. Mon. (i). 434.	Sikkim : Darjiling Gopaldhara, Namsoo. Eastern Duars. U. P. : Dehra Dun, Mussoorie, Almora, Upper Gumti Valley.

NOTIOPHILUS* Duméril.

Notiophilus Duméril Zool. Analyt. 1806. 194; Dej. Sp. ii. 1826. 276. Lac. 43; Putz., Note sur les Notiophilus, Mém. Liége 1866. 153; Spaeth Verh. Zool.-bot. Ges. Wien. 1899. 510; Dupuis Gen. Ins Fasc. 134, Notiophilinae, 1912. 3; Andr. FBI. 135.

O* **orientalis** Chaud. BM. 1850. i. 428; Putz. Note 163; Spaeth. Verh. Zool.-bot. Ges. 1900. 50; Dupuis Gen. Ins. Fasc. 134. t. f. 14; Andr. EMM. 1926 66; id. FBI. 136. f. 26.	Kashmir : Lidar Valley Tanin. U. P. : Almora, Upper Gumti Valley, East Ramganga Valley, Konain, Pilibhit. Simla Hills : Matiana. Murree. Kurram Valley : Parachinar.

B* **radians** Andr. EMM. 1926. 66 and 70; Andr. FBI. 136 and 138. — Kumaon : Gori Valley.

ODACANTHA* Paykull.

Odacantha Payk. Faun. Suec. i. 1798. 169; latr. et Dej. HN. et Ic. 127; Dej. Sp. i. 1825. 174; Schm. Goeb. FCB. 21; Lac. 74.

Ophionea Klug. Ent. Bras. Spec. 1821. 298 (part); Bedel BF. 1910; 72; Andr. AM. (9). iii. 1919. 475.

Casnonia Latr. et Dej. HN. et Ic. 130 (Part); Dej. Sp. i. 1825. 170; Schm. Goeb. FCB. 18; Chaud. BM. 1848. i. 44; Lac. 72.

Arame Andr. AM. (9). iii. 1919. 477; id. AM. (9). xiv. 1924. 592.

B* **aegrota** *(Casnonia)* Bates TES. 1883. 278; id. AG. xxxii. 1892. 382; Andr. AM. (9). iii. 1919. 475. — Japan.

G **albicolon** *(Casnonia) Bates* AG. xxxii. 1892. 381. — Karen Hills.

O **apicallis** Chaud. BM. 1872. i. 408. — Siam., Singapore.

V* **bimaculata** *(Casnonia)* Redt. HK. 498. t. 23. f. 2; Chaud. BM. 1850. i. 25; id. BM. 1862. ii. 283; Bates CRB. 1891. 335; id. AG. xxxii. 1892. 381; Andr. MB. 116. — Throughout Northern India, extending southwards to Belgaum (Andrewes) and Malabar (Chaudoir). Ceylon : Hambegamuwa. Bhamo., Cambodia. Java., Borneo. Philippine Is.

G* **celebensis** *(Casnonia)* Gestro AG. 1875. 854; Chaud. BM. 1877. i. 267. — Celebes.

N **chaudoiri** *(Ophionea)* Boh. ER. 2; Chaud. BM. 1862. ii. 283; Redt. RN. 4; Bedel BF. 1910. 71; Andr. TES. 1923 (1924). 459. — Indo-China. China.

O *beauchenei (Ophionea)* Fairm. AF. 1888. 333; Maindr. BF. 1910. 35.

X **var. blaisei** Bedel BF. 1910. 71. — Tonkin.

X	**var. rubricollis** Bedel BF. 1910. 72.	China.
	distigma *(Casnonia)* × Chaud. BM. 1850. i. 26; id. BM. 1872. i. 407; Bates AG. xxxii. 1892. 381.	Bhamo., Katha. Malay Peninsula.
P*	*bimaculata (Casnonia)* | | Schm. Goeb. (not Redt.) FCB. 18; Chaud. BM. 1848. i. 47; Andr. TES. 1923. 3.	Sumatra. Borneo.
Z*	*subapicalis (Casnonia)* Oberth. NLM. v. 1883. 216; Andr. AM. (10). ii. 1928. 588.	
G	**dubia** Gestro AG 1879. 558.	New Guinea.
O	**fulvipennis** Chaud. BM. 1872. i. 407; Bates TES. 1883. 278.	Celebes ? Hongkong.
O*	**fuscipennis** *(Casnonia)* Chaud. BM. 1850. i. 26; id. BM. 1862. ii. 290; id. BM. 1877. i. 266; Gestro AG. 1875. 852 and 855; Fairm. AF. 1888. 334; Bates AG. xxxii. 1892. 282; Andr. AM. (9). xix. 1927. 106; id. TES. 1928. 9.	Malacca. Tonkin. Siam., Sumatra. Celebes., China. Japan.
B*	*flavicauda (Casnonia)* Bates TES. 1873. 303; id. AF. 1889. 279; id. AG. xxxii. 1892. 382; Chaud BM. 1877. i. 267. Andr. AM. (9). xix. 1927. 106.	Formosa.
G'*	*litura (Casnonia)* † Bouch. (not Schm. Goeb.) AF. 1903. 173.	
G*	**graciliceps** *(Casnonia)* Bates AG. xxxii. 1892. 381; (?) Schm. Goeb. FCB. t. 1. f. 9; Andr. TES. 1923. 5 and 56.	Caclutta. "Between Sorbhog and Golakganj". Burma : Bhamo, Teinzo, Toungoo, Tharrawaddy, Rangoon., Cambodia. Java.
O*	**latifascia** *(Casnonia)* Chaud. BM. 1872. i. 404; id. BM. 1877. i. 267.	"East Indies". China., Java.
P*	**litura** Schm. Goeb. FCB. 22; Chaud. BM. 1872. i. 405 and 407; id. BM. 1877. i. 266; Bates TES. 1883. 278; Andr. TES. 1923. 4.	Tharrawaddy. Java. Japan.
B*	**macra** *(Arame)* Andr. AM. (9). iii. 1919. 478.	Colombo. Peradeniya.
	baemorrhoidalis (Casnonia) † Bates (not Motch) AM. (5). xvii. 1886. 199.	

O*	**metallica** *(Casnonia)* Fairm. AF. 1888. 334.	Tonkin., China.
A*	**obesa** *(Arame)* Andr. AM. (9). xii. 1923. 220.	Malay Peninsula. Cochin-China., Java Verlaten I. (Krakatau Group) Sumatra.
G	**opacipennis** *(Casnonia)* Gestro AG. 1888. 107; Bates AG. xxxii. 1892. 381.	Dacca. Bhamo.
L	**pilifera** *(Casnonia)* Nietn. AM. (3). ii. 1858. 179; Chaud. BM. 1872. i. 407; Bates AG. xxxii. 1892. 381; Lesne Miss. Pav. 78; Bedel BF. 1910. 71 (note); Andr. AM. (9). xix. 1927. 106.	"South India". Ceylon: Vavuniya, Colombo, Anuradhapura. Bhamo., Siam. Laos., Cambodia.
J*	**punctata** *(Casnonia)* Nietn. AM. (3). ii. 1858. 178; Chaud. BM. 1872. i. 407; Andr. AM. (9). xix. 1927. 106.	Sardah. Sunderbans. Pupri., Calcutta.
M*	*haemorrhoidalis (Casnonia)* Motch. BM. 1864. ii. 219; Chaud. BM. 1872. i. 404; id. BM. 1877. i. 266; Bates AG. xxxii. 1892. 382; Andr. AM. (9). xix. 1927. 106; id. TES. 1928. 8.	Simla., Puri., Chanda. Tranquebar. Tinnevelly. Trichinopoly. Colombo., Mergui. Siam., Annam. Celebes. Philippine Is.
H*	**stigma** *(Arame)* Andr. Spol. Zeyl. 1923. 242.	Ceylon.
A*	**tenuis** Andr. AM. (9). xviii. 1926. 280.	Sumatra., Borneo.
A*	var. **linea** Andr. AM. (9). xviii. 1926. 281.	Banguey I.
N*	var. **rubripes** Andr. AM. (9). xviii. 1926. 281.	Sumatra.
P*	**tetraspilota** *(Casnonia)* Schm. Goeb. FCB. 19; Chaud. BM. 1848. i. 49; id. MB. 1872. i. 407; Andr. TES. 1923. 3.	Garo Hills. Dawna Hills. Tenasserim. Andaman Is.
O*	**virgulifera** *(Cosnonia)* Chaud. BM. 1872. i. 403; id. BM. 1877. i. 267.	Rangoon. Siam.
G*	**xanthe** *(Casnonia)* Bates AG xxxii. 1892. 382.	Pusa., Sunderbans. Pyinmana. Tharrawaddy. Rangoon., Insein.

OESYPERUS* Andrews.

Oesyperus Andr. AM. (9). xii. 1923. 444.

A* **planus** Andr. AM. (9). xii. 1923. 446. — Shembaganur.

I* **pygmaeus** Andr. AM. (9). xii. 1923. 446. — Palni Hills.

A* **unctulus** Andr. AM. (9). xii. 1923. 445. — Kodaikanal.

OMOPHRON* Latreille.

Omophron Latr. Hist. Nat. Crust. et Ins. iii. 1802. 89; Dej. Sp. ii. 1826. 257; Lac. 42; Chaud., Note monographique sur le genre Omophron, RMZ. 1868. 54; Gestro AG. 1888. 174; id. AG. xxx. 1892. 961; Hayward Psyche 1902. 304; Rouss. Gen. Ins. Fase. 83, Omophroninae, 1908. 2; Bänn. DEZ. 1915. 225; id. DEZ. 1918. 97; Andr. AM. (9). iv. 1919. 294; Sem. RRE. xviii. 1922. 39; Andr. MB. 4; id. FBI. 139.

Scolytus || F. (not Geoff). Ent. Syst. i. (1). 1792. 180 (part).

Epactius [Schneid. N. Mag. Ent. 1791. 23 (note)] Bedel Fn. med. 1895. 14.

Epacter Sem. RRE. xviii. 1922. 38.

Andromophron Sem. RRE. xviii. 1922. 40.

Obligemophron. Sem. RRE. xviii. 1922. 40.

Q **aequalis** Moraw. Beitr. 6; Bates TES. 1873. 229; id. TES. 1883. 216; Bänn. DEZ. 1918. 108. — Japan.

A* **affinis** Bänn. DEZ. 1918. 105; Andr. FBI. 142 and 148. t. 4. f. 1. — Berhampur. Chota Nagpur. Nowatoli.

O* **axillaris** Chaud. Mon. 59; Bänn. DEZ. 1918. 103; Andr. MB. 4; id. EMM. 1926. 66; id. FBI. 143 and 160 t. 5. f. 3. — Bengal. Kumaon : Almora, Swal River basin, Sarju Valley. Haldwani. Kotkhai (Simla).

B* **bicolor** Andr. AM. (9). iv. 1919. 292; id. FBI. 143 and 159. t. 5. f. 2. — Kangra., Calcutta. Daca.

B* **brettinghamae** Pasc. Journ. Ent. i. 11860. 38; Chaud. Mon. 55; Bänn. DEZ. 1918. 102; Andr. — Dacca. Burma : Teinzo.

	TES. 1919. 195; Sem. RRE. xviii. 1922. 40. *(Oligomophron)*; Andr. FBI. 143 and 158.	
G*	*levigatus* Gestro AG. 1888. 172; Bates AG. xxxii. 1892. 269; Bänn. DEZ. 1918. 102.	
A*	**chelys** Andr. AM. (9). vii. 1921. 404; Sem. RRE. xviii. 1922. 40 *(Andromophron)* ; Andr. FBI. 143 and 156. f. 27.	Sikkim : Gopaldhara.
	gemmeus × Andr. TES. 1921. 190; id. FBI. 143 and 157. t. 5. f. 1.	Naga Hills. Manipur., Almora.
B*	*gemma* || Andr. (not Casey) AM. (9). vii. 1921. 406.	Dehra Dun. Haldwani., Tonkin.
O*	**guttatus** Chaud. Mon. 62; Bänn. DEZ. 1918. 99; Sem. RRE. xviii. 1922. 38 (*Epacter)*; Andr. FBI. 141 and 145. t. 3. f. 4.	Berhampur. Chota Nagpur : Nowatoli., Ramnad.
O*	**interruptus** Chaud. Mon. 62; Bänn. DEZ. 1918. 99; Andr. FBI. 141 and 144. t. 3. f. 3.	Dacca.
A*	**lunatus** Bänn. DEZ. 1918. 104; Andr. MB. 4; id. FBI. 141 and 146. t. 3. f. 5.	Chota Nagpur : Nowatoli., Motinala. Mandla.
O*	**maculosus** Chaud. BM. 1850. i. 424; Bänn. DEZ. 1918. 99; Andr. TES. 1921. 189; id. FBI. 142 and 151. t. 4. f. 3.	Dacca., Pusa. Jamna River. Fyzabad.
G*	**oberthuri** Gestro AG. xxx. 1892. 962; Bänn. DEZ 1918. 102; Andr. EMM. 1926. 66; id. FBI. 143 and 155. t. 6. f. 6.	Sikkim : Gopaldhara, Namsu, Dikchu (Tista Valley), Pedong, Sarju Valley., Almora. Haldwani. U. P. : Ramnagar (Fatehpur).
C*	**pictus** *(Scolytus)* Wied. Zool. Mag. ii. 1. 1823. 69; Bänn. DEZ. 1918. 99 and 106; Andr. TES. 1921. 174 and 189; id. FBI. 142 and 150. t. 2. f. 2.	Bengal. Pusa.
O*	**porosus** Chaud. Mon. 61; Bänn. DEZ. 1918. 99 and 103; Andr. FBI. 141 and 144. t. 3. f. 2.	Dacca.
O*	**rotoundatus** Chaud. BM. 1852. i. 101; id. Mon. 55; Bänn. DEZ. 1915. 232; id. DEZ 1918. 103; Andr. AM. (9). iv. 1919. 292; Sem. RRE. xx. 1926. 40; Andr. FBI. 143 and 161 t. 5. f. 4.	Mussoorie., Annam. Quetta., Western China., Turkestan. Persia., Mesopotamia. Armenia.

O*	**salgonensis** Chaud. Mon. 57. Bänn. DEZ. 1918. 103.	Cochin-China.
B*	**smaragdus** Andr. AM. (9). vii. 1921. 405; id. FBI. 142 and 152. t. 4. f. 4.	Almora : Bhatkot, Ranikhet.
G*	**striaticeps** Gestro AG. 1888. 173; Bates AG. xxxii. 1892. 269; Bänn. DEZ. 1918. 99 and 106; Andr. FBI. 142 and 149. t. 4. f. 2.	Burma : Myn Kyan. Laos.
L	**suturalis** Gistl. Naturg. Thierreichs 1848. 112; Andr. FBI. 162.	Bangal.
B*	**testudo** Andr. AM. (9). vi. 1919. 293; id. FBI. 141 and 147. t. 3. f. 6.	Pegu. Indo-China.
B*	**virens** Andr. FBI. 142 and 153. t. 4. f. 5.	Haldwani.
C*	**vittatus** *(Scolytus)* Wied. Zool. Mag. ii. 1. 1823. 69; Chaud. Mon. 56; Bänn. DEZ. 1918. 99 and 106; Andr. TES. 1921. 174 and 188; id. FBI. 142 and 154. t. 2. f. 1.	Bengal.

OMPHRA* Reiche.

Omphra Reiche AF. 1842. 330; Lac. 94; Chaud. RMZ. 1872. 141; Sloane PLS. NSW. 1914. 570 and 580; And MB. 53.

Helluo † Latr. et Dej. (not Bon.) HN. et Ic. 94; Dej. Sp. i. 1825. 283 (part).

J*	**atrata** *(Helluo)* Klug Jahrb. Ins. 1834. 72; Andr. RIM. xxii. 1921 616; Chopra RIM. xxvi. 1924. 115. ff. 1-6 (biology); Andr. MB. 54; id. AM. (9). xix. 1927. 101.	Bengal. ? Barkuda I. (Lake Chilka). ? C. P. : Jubbulpore, Majgaon. Nilambur., Dindigul.
O*	**complanata** Reiche AF. 1842. 342; Andr. RIM. xxii. 1921. 345.	Janakpur. Simla.
O*	*brevis* Chaud. BM. 1850. i. 36; id. RMZ. 1872. 141.	Barkuda I. (Lake Chilka). Nagpur., Bombay. Nasik., Ratnagiri. Belgaum., Madura. Pondicherry.
C*	**hirta** *(Galaerita)* F. Syst. Eleuth i. 1801. 214; Latr. et Dej. HN. et Ic. 95. t. 7. f. 1; Dej. Sp. i. 1825. 284; Klug Jahrb. Ins. 1834. 71; Hope Col. Man. ii. 1838.	Throughout South India and Ceylon, extending

	98; Motch. Etud. Ent. 1855. 54; Redt. RN. 5; Bates CRB. 1891. 336; Heyne-Tasch. EK. 14. t. 3. f. 7; Andr. TES. 1921. 163; id. TES. 1923 (1924). 460.	northwards to Pondicherry and the Nilgiri Hills. Chota Nagpur : Tetara (Bates).
J*	*rufitarsis (Carabus)* Illig. Mag. Ins. 1802. 164 and 490; Andr. AM. (9). xixi. 1927. 99.	
J*	**pilosa** *(Helluo)* Klug Jahrb. Ins. 1834. 71; Redt. RN. 5; Chaud. RMZ. 1872. 141; Putz. CRB. 1875. 45; Bates CRB. 1891. 336; Heyne-Tasch. EK. 14; Andr. TES. 1923 (1924). 460; id. MB. 53; id. AM. (9). xix. 1927. 101.	Throughout India and Ceylon.
K	*attelaboides (Galerita)* || F. Syst. Eleuth. i. 1801. 214. (not Spec. Ins. i. 1781. 305); Schaum. SEZ. 1847. 49; Erichs. SEZ. 1847. 141; Schaum. SEZ. 1848. 334; Motch. Etud. Ent. 1855. 54; Andr. TES. 1921. 163.	
O*	**rotundicollis** Chaud. RMZ. 1872. 140.	"East Indies".
J	**rufipes** *(Helluo)* Klug. Jahrb. Ins. 1834. 72; Chaud. RMZ. 1872. 141; Bates AM. (5). xvii. 1886. 200; Andr. AM. (9). xix. 1927. 101.	Throughout Ceylon.

ONYCHOLABIS* Bates.

Onycholabis Bates TES. 1873. 329.

B*	**acutangulus** Andr. AM. (9). xii. 1923 682.	Assam : Yambung. West Almora. Haldwani. Dehra Dun. Siwalik Hills.
G*	**melitopus** Bates AG. xxxii. 192. 371.	Himalayan tract from Simla Hills to Bhutan.
X*	*oberithüri (Cardiomera)* Maindr. BF. 1899. 155; Bedel Fn. med. 1902. 215 (note 2); Maindr. BF. 1905. 94.	Karen Hills. ? Taoo (Tenasserim). Laos.
O*	**sinensis** Bates TES. 1873. 329.	China.

OODES* Dejean.

Oodes Dej. Sp. ii. 1826. 374; Lac. 229; Chaud. BM. 1857. ii. 25; id. Mon. des. Oodides (i) AF. 1882. 341; Sloane PLS. NSW. 1910. 442; Andr. TES. 1919. 92.

B* **bivittatus** Andr. Spol. Zeyl. 1924. 134.	Ceylon.
J* **coelestinus** Chaud. Mon. 363; Andr. AM. (9). xixi. 1927. 109.	Borneo.
G* **cribristernis** Bates AG. xxxii. 1892. 323.	Karen Hills.
O* **parallelus** Laf. AF. 1851. 271. (note 5); Chaud. Mon. 347; Bates AG. xxxii. 1892. 323; Andr. AM. (9). xiv. 1924. 589.	Khasi Hills. Saidpur., Dacca. North Kanara. Ceylon : Mamaduwa, Anuradhapura, Horawupotana, Rangoon, Palon, Tharrawaddy. Malacca.
G* **peguensis** Bates AG. xxxii. 1892. 323.	Tharrawaddy. Prome., Palon.
S* **piceus** Nietn. JASB. xxv. 1856. 526; id. AM. (2). xix. 1857. 377; Andr. AM. (9). xix. 1927. 106.	Assam., Pusa. Tranquebar.
O* *vilis* Chaud. BM. 1857. ii. 32; id. Mon. 369; Bates AM. (5). xvii. 1886. 75.	Pondicherry. Ceylon : Colombo, Galle. Burma., Siam. Indo-China., Java. Sumatra. Philippine Is. Celebes., China.
G* **rhodopus** Bates AG. xxxii. 1892. 324.	Sikkim : "Sookia Pokri to Pokribong". Trichinopoly. Burma : Karen Hills, Bhamo, Palon, Malewoon.
O* **siamensis** Chaud. Mon. 358. Bates AF. 1889. 267; Vuillet. Ins. ii. 1912. 19.	Siam. Cochin-China. Cambodia., Sumatra. Borneo.
G **subcoriaceous** Chaud. Mon. 362.	Malay Peninsula.
O* **subolivaceus** Laf. AF. 1851. 271 (note 4); Bates CRB. 1891. 329.	Nepal. Calcutta.

O* *chalceus* Chaud. BM. 1857. ii. 30; id. Mon. 357.	Saran., Chapra.
M* *nepalensis* Motch. Etud. Ent. 1858. 171; id. BM. 1864. ii. 353; Andr. TES. 1928. 15.	Pusa. Chota Nagpur : Tetara. Dehra Dun. Bareilly., Fyzabad. Nagpur., Belgaum. North Kanara. Ceylon : Vavuniya, Colombo.
M* **sulcatus** Eschsch. Zool. Atl. v. 1833. 28; *(Systolocranius)* Chaud. Mon. 335; Andr. TES. 1928. 20.	Philippine Is. ? Borneo.
B* **taprobanae** Andr. Spol. Zeyl. 1923. 231.	Ceylon : Vavuniya, Anuradhapura.
	Laos., Japan.
B* **vicarius** Bates TES. 1873. 254; Chaud. Mon. 361.	Khasi Hills.
C* **virens** Wied. Zool. Mag. ii. 1. 1823. 50; Andr. TES. 1921. 169.	Sibsagar., Bengal. Chapra.
O* *vurians* Chaud. Mon. 352.	Ceylon : Kotte Tharrawaddy. Java., Philippines Is.
O* **westermanni** Laf. AF. 1851. 271 (note 2); Chaud. BM. 1857. ii. 31; id. Mon. 368; Bates AG. xxxii. 1892. 324.	Nepal terai : Kumdhik, Calcutta. Diamond Harbour.
O *hispanicus* Dej. Sp. ii. 1826. 379 (part).	Dacca., Saran., Pusa.
O* *villis* † Bates (not Claud) TES. 1873. 254.	Madras : Tranquebar, Madras town, Genji. Ceylon : Holupahani, Murunkan, Kotte, Dehiwela., Prome. Tharrawaddy. Toungoo. Indo-China. Java., Celebes. Japan.
B* **xanthochilus** Andr. Spol. Zeyl. 1923. 232.	Balaghat. Ceylon : Kala-Oya River, Vavuniya.

OPHIONEA* Klug.

Ophionea Klug. Ent. Bras. Spec. 1821. 298 (part); Eschsch. Zool. Atl. ii. 1829. 5; Schm. Goeb. FCB.

20; Chaud. BM. 1848. i. 31 and 43; Lac. 73; Maindr. BF. 1910. 34; Bedel BF. 1910. 72; Sloane PLS. NSW. 1917. 415; Andr. AM. (9). iii. 1919. 475.

Casnoidea Cast. Etud. Ent. 1834. 40.

E **bakeri** *(Casnoidea)* Dupuis AB. 1913. 270. — Philippine Is.

X *bhamoensis* † Maindr. (not Bates) BF. 1910. 35.

(O*) **fcrsteri** Bouch. AF. 1903. 172; Maindr. BF. 1910. 35. — Borneo., Sumatra.

X **gestroi** Mainder. BF. 1910. 34. — New Guinea.

L **indica** *(Attelabus)* Thunb. Nov. Ins. Spec. iii. 1784. 68. f. 81; Andr. AM. (9). iii. 1919. 476 (note); id. RIM. xxii. 1921. 345; id. MB. 116. — Throughout South Eastern Asia and the Malay Archipelago from Japan in the North to the Guinea in the South.

C* *cyanocephala (Cicindela)* F. Suppl. Ent. Syst. 1798. 60; (*Odacantha)* id. Syst. Eleuth. i. 1801. 229; Herbst Nat. Syst. Ins. Käf. x. 1806. 223. *(Casnonia)* Latr et Dej. HN. et Ic. 130. t. 7. f. 6; Dej. Sp. i. 1825. 173; *(Colliuris)* Brullé HN. iv. 1834. 139. t. 4 f. 3; Schm. Goeb. FCB. 20; Motch. Etud. Ent. 1855. 52; Bates TES. 1873. 303; Gestro AG. 1875. 855; Bates AM. (5). xvii. 1886. 199; id.AF. 1889. 279; id. CRB. 1891. 335; id. AG. xxxii. 1892. 380; Heyne-Tasch. EK. 13. t. 3. f. 13; Maindr. BF. 1910. 35; Andr. AM. (9). iii. 1919. 475; id. TES. 1921. 161; id. TES. 1923. 4.

P* **interstitialis** Schm. Goeb. FCB. 20; Gestro. AG. 1875. 855; Bates AM. (5). xvii. 1886. 99; id. AF. 1889. 279; id. AG. xxxii. 1892. 380; Bouch. AF. 1903. 172; Maindr. BF. 1910. 35; Andr. TES. 1923. 4.

V* *cyanocephala* † Redt. (not F.) RN. 4; Andr. TES. 1923 (1924). 460.

Pusa. Madras : Genji, Pondicherry. Ceylon : Colombo, Peradeniya. Mandalay., Rangoon. Tharrawaddy. Martaban., Siam. Indo-China., China. Malay Peninsula and Archipelago.

P* **nigrofasciata** Schnm. Goeb. FCB. 21; Lac. Atl. t. 3. f. 2 (as *O. Cyanocephala* F.); Bates AG. xxxii. 1892. 380; Bouch. AF. 1903. 172; Maindr. BF. 1910. 35; Andr. TES. 1921. 161; id. TES. 1923. 4. — Khasi Hills. Patkai Hills. Sardah.

P* **nigrofasciata**— Ceylon : Henaratgoda, Peradeniya. Burma. : Tharrawaddy, Thagata, Palon. Kuala Lumpur. Singapore. Indo-China., Siam. Java., Sumatra. Borneo., Nias I. Buru., Hainan.

G* var. **bhamoensis** Bates AG. xxxii. 1892. 380. Sikkim : Gopaldhara. Bhamo.

OPHONISCUS* Bates.

Ophoniscus Bates AG. xxxii. 1892. 337; Andr. AM. (9). xii. 1923. 446.

B * **compositus** *(Curtonotus)* Walk. AM. (3). ii. 1858. 204; Bates AM. (5). xvii. 1886. 211; Andr. TES. 1919. 189; id. AM. (9). Xii. 1923. 446. Kodaikanal. Trichinopoly. Madura. Ceylon: Nuwara Eliya Pattipola, Galle, Dikoya.

B* *ferreus (Siopelus)* Bates AM. (5). xvii. 1886. 76.

G* **cribrifrons** Bates AG. xxxii. 1892. 338. Burma: Bhamo, Pyaung-Gyaung. Cochin-China.

G* **hypolithoides** Bates AG. xxxii. 1892. 338. Kawkareik.

G* **iridulus** Bates AG. xxxii. 1892. 337. Kamrup., Siliguri. Dehra Dun. Purnea. Parasnath Mt. Karen Hills. Rangoon., Palon. Laos., Tonkin.

OPHONUS Stephens.

O. *Ophonus* Steph. Ill. Brit. Ent. i. 67; Dej. Sp. iv. 1829. 193; Lac. 298; Tchitch. HER. xxxv. 1901. 234.

Ps. *Pseudophonus* Motch. Ins. Sib. 1844. 196; Tchitch. HER. xxxv. 1901. 236.

P. *Pardileus* Des Gozis SEG. 1882. 289; Tchitch. HER. xxxv. 1901. 237; Andr. MB. 35.

EP. **calceatus** *(Carabus)* Dufts. Fn. Austr. ii. 1812. 81; *(Harpalus)* Dej. Sp. iv. 1829. 320; De. et Boisd. Icon. Col. Eur. iv. 1835. 180. t. 191. f. 3; Tchitch. HER. xxxv. 1901. 237 (note 47); id. HER. xxxvii. 1906. 254; Andr. MB. 35. — Baluchistan: Nushki. Central and Southern Europe. Central Asia. Siberia., China.

D′ *nonsignatus (Anisodactylus)* Kryn. BM. v. 1832. 72.

calcitrapus (Harpalus) Motch. Ins. Sib. 1844. 216. t. 9. f. 13.

J* Ps. **griseus** *(Carabus)* Panz. Fn. Germ. 1797. 38. I; Dej. Sp. iv. 1829. 251; Bates TES. 1873. 260; Putz. CRB. 1875. 49; Har. Abh. nat. Ver. Bremen iv. 1875. 285; Heyd. DEZ. 1879. 331; Tchitch. Ab. xxix. 1897. 45; Andr. AM. (9). xix. 1927. 99. — Throughout the palaearctic region. Tonkin. Yunnan. Japan.

P. **indicus** *(Harpalus)* Bates CRB. 1891. 332; Andr. MB. 35; id. EMM. 1926. 68. — Throughout the Himalayan tract from Kashmir to Bhutan.

A* P. **meridianus** *(Harpalus)* Andr. AM. (9). xii. 1923. 443. — Nasik, Nilgiri Hills. Kodaikanal. Madura.

B'* O. **pangoides** Reitt. DEZ. 1913. 649. — Bashahr : Poo.

B* P. sinicus *(Harpalus)* Hope TES. 1845. 14; Tchitch. HER. xxxvii. 1906. 251 and 253; Andr. TES. 1919. 176; id. MB. 35. — Murree (Bates) Tonkin., Formosa. Yunnan., China.

rugicollis (Platus) Motch. Etud. Ent. 1860. 5; Har. Abh. nat. Ver. Bremen. iv. 1875. 285; Bates TES. 1876. 3; Heyd. DEZ. 1879. 332; Bates TES. 1883. 236; Tchitch. HER. xxxbv. 1901. 237 (note 47); id. HER. xxxvii. 1906. 251; Andr. TES. 1928. 23. — Japan., Korea.

Q *Japonicus (Harpalus)* Moraw. Mel. Biol. Acad. Sci. Petr. iv. 1862. 245; id. BAP. x. 1863. 327; id. Beitr. 69; Bates TES. 1873. 261; Bates SYM. 7; Tchitch. AB. xxix. 1897. 46.

Q P. **tridens** *(Harpalus)* Moraw. Mel. Biol. Acad. Sci. Petr. iv. 1862. 245; id. BAP. v. 1863. 326; id. Beitr. 69; Bates TES. 1873. 261; Putz. CRB. 1875. 45 and 49; Heyd. DEZ. 1879. 331; Bates TES. 1883. 236; id. Entom. 1890. 212; id. Entom. 1891. Suppl. 9; Tchitch. Ab. xxix. 1897. 46; id. HER. xxxv. 1901. 237 (note 47); id. HER. xxxvii. 1906. 247; Andr. MB. 35. — Kulu (Bates). Indo-China. Yunnan. China. Japan. Korea.

ORTHOGONIUS* Macleay.

Orthogonius Macl. Ann. Jav. 1825. 26; Dej. Sp. i. 1825. 279; Schm. Goeb. FCB. 55; Lac. 269; Chaud., Essai monographique sur les Orthogoniens, AB. xiv. 1871. 98; Andr. MB. 58.

Apsectra Schm. Goeb. FCB. 61.

Haplopisthius Chaud. BM. 1850. ii. 434.

Maraga Walk. AM. (3). ii. 1858. 203; Waterh. EMM. x. 1873. 17.

C* **acrogonus** *(Carabus)* Wied. Zool. Mag. i. 3. 1819. 167; Dej. Sp. v. 1831. 398; Lac. Atl. t. 10. f. 7 (front leg); Chaud. Mon. 104: Heyne-Tasch. EK. 23. t. 3. f. 47; Bouch. AF. 1903. 173; Andr. TES. 1921. 166. — Kedah. Java. Sumatra. Borneo.

B* *brunnilabris* Macl. Ann. Jav. 1825. 27; Andr. TES. 1919. 165.

O* **acutangulus** Chaud. BM. 1878. ii. 5. — Ceylon : Kelani Valley, Colombo, Kandy, Kotte Dehiwela, Labugama, Deniyaya.

B* *parallelus* † Bates (not Chaud.) AM. (5). xvii. 1886. 201; Andr. TES. 1921. 149.

O* **aequistriatus** Jord. NZ. 110. — Cilolo.

C* **alternans** *(Plochionus)* Wied. Zool. Mag. ii. 1. 1823. 52; Dej. Sp. i. 1825. 280; id. Sp. v. 1831. 397; Brullé HN. iv. 1834. 225. t. 8. f. 1; Desmarest, Voy. La Bonite. 1841. 291. 2. f. 1; Schm. Geob. FCB. 60; Chaud. Mon. 102; Bates AG. xxvii. 1889. 110; id. AG. xxxii. 1892. 399; Lesne Miss. Pav. 79; Andr. TES. 1919. 165; id. TES. 1921. 170 ; id. TES. 1923. 34. — Tezpur., Sibsagar. Kamrup., Margherita. Khira., Shan Hills. Burma : Teinzo, Katha. Siam., Tonkin. Laos., Java. Sumatra., Borneo. Philippine Is. New Guinea.

duplicatus † Bates (not Wied.) AG. xxvii. 1889. 110 (♀); id. AG. xxxii. 1892. 399; Andr. TES. 1921. 166.

Y* **andrewesi** van Emden DEZ. 1928. (1929). 381. — Philippine Is.

O* **angustus** Chaud. Mon. 114. — ? Malay region or S. Africa.

A* **assmuthi** Wasm. Ent. Mitt. 1920. 80. — Khandala.

O* **baconi** Chaud. Mon. 109; Bates AG. xxxii. 1892. 401.	Bengal., Almora. Bhamo.
O* **crassicrus** Chaud. Mon. 105; Bouch. AF. 1903. 173.	Java., Borneo.
O* **davidi** Chaud. BM. 1878. ii. 3; Bates AG. xxvii. 1889. 110.	China.
P* **deletus** Schm. Goeb. FCB. 56; Chaud. Mon. 113; Heyne-Tasch. EK. 23; Andr. TES. 1923. 31; id. MB. 117.	Surada. Chandkhira. Manipur. Karen Hills.
O* *melanarius* Chaud. Mon. 113.	Palon.
G* *quadricollis* Bates AG. xxxii. 1892. 400.	Tharrawaddy. Penang., Langkawi. I. Laos., Sumatra.
G* **dispar** Bates AG. xxxii. 1892. 401.	Karen Hills. Thagata., Tenasserim.
G* **doriae** Putz. in Chaud. Mon. 104 (note).	Borneo.
S* *collaris* Dohrn SEZ. 1891. 253; Andr. TES. 1921. 149; id. TES. 1922. 246.	
C* **duplicatus** *(Carabus)* Wied. Zool. Mag. i. 3. 1819. 166; Dej. Sp. i. 1825. 279; Andr. TES 1921. 165; id. MB. 59; id. FBI. f. 7. B.	Throughout Northern India and southwards to Central Provinces. Manipur.
P* *puncticollis* Schm. Goeb. FCB. 57; Chaud. Mon. 112; Bates CRB. 1891. 338; id. AG xxxii. 1892. 399; Andr. TES. 1923. 32.	Burma : Karen Hills, Bhamo, Teinzo, Rangoon, Martaban, Tenasserim.
P* *profundestriatus* Schm. Geob. FCB. 58; Bates AF. 1889. 280; id. AG. xxvii. 1889. 110; Andr. TES. 1921. 179; id. TES. 1923. 32.	Malay Peninsula. Laos. Cochin-China (Bates). Java (Wiedemann).
O* **femcralis** Chaud. BM. 1848. i. 99; id. Mon. 111; Andr. MB. 59.	Nilgiri Hills. Pondicherry. Vizagapatam. Surada., Saidapet. Travancore. Trivandum. Pallode., Tenmalai.

O*	**fugax** Chaud. Mon. 108.	Ceylon: Niroddumunai, Diyatalawa, Nowalapitiya, Woodside Urugalla, Kandy.
Z*	**hageni** Oberth. NLM. v. 1883. 222; Bates AG. xxxii. 1892. 404.	Sumatra.
O*	**hirtus** Chaud. Mon. 103.	Penang.
H*	**hopei** Gray in Griffiths, Anim. Kingd. xiv. 1832. 273. t. 13. f. 4. ; Chaud. Mon. 13; Heyne-Tasch. EK. 23; Andr. TES. 1919. 204.	Throughout the Malay Peninsula. Siam., ? India (Gray).
O	*malabariensis* Gory AF. 1833. 196.	? Malabar (Gory).
O*	**hypocrita** Chaud. Mon. 102.	Philippine Is. ? Java (Chaudoir).
O*	**inops** Chaud. Mon. 111.	Moluccas.
O*	**insularis** Chaud. Mon. 106.	Penang., Borneo.
O*	**intermedius** Chaud. Mon. 102; Andr. AM. (9). xx. 1927. 272.	Java.
B*	*alternans* † Macl. (not Wied.) Ann. ♂ Jav. 1825. 27.	
	macleayi × Andr. TES. 1919. 166; id. TES. 1921. 170.	
O	**longicornis** Chaud. Mon. 109.	Siam.
E*	**lucidus** Bates CRB. 1891. 337; Andr. MB. 59.	Chota Nagpur : Konbir, Tetara, Ranchi., Raniganj. Bombay : Igatpuri, Belgaum, North Kanara., Surada.
E	var. **nitidulus** Bates CRB. 1891. 337.	Chota Nagpur : Konbir, Tetara.
O*	**luzonicus** Chaud. Mon. 123.	Philippines Is.
O*	**mellyi** *(Haplopisthius)* Chaud. BM. 1850. i. 432; id. Mon. 101.	Assam : Margherita, Sibsagar, Barapani, Jorhat, Patkai Hills, Garo Hills, Lakhimpur. Chandkhira. Bengal., Bhamo.
	duplicatus † Bates (not wied.) AG. xxvii. 1889. 110 ( ); id. AG. xxxii. 1892. 399; Andr. TES. 1921. 166.	

O*	**mniszechi** Chaud. Mon. 101; Bates Entom 1891. Suippl. 9.	Kulu (Bates). Malay Peninsula. Langkawi. I. Borneo.
O*	**mouhoti** Chaud. Mon. 107.	Tharrawaddy. Paungde. Indo-China.
P*	**opacus** Schm. Goeb. FCB. 60; Andr. TES. 1923. 34.	Garo Hills. Sikkim : Gopaldhara, Pashok, Kurseong. Lebong. Shan States. Andaman Is. Maymyo. Karen Hills. Tenasserim.
P*	*angusticollis* Schm. Goeb. FCB. 61; Chaud. Mon. 122; Andr. TES. 1923. 35.	
G*	*apiculatus* Bates AG. xxxii. 1892. 403.	
O*	**ovatus** Chaud. BM. 1878. ii. 4.	Celebes.
O*	**paralleius** Chaud. Mon. 109.	Ceylon : Madulsima, Diyatalawa, Halupahani.
O*	**parvus** Chaud. Mon. 112.	Nilgiri Hills.
O*	**piceus** Chaud. Mon. 122.	Kedah., Malacca. Negri Sembilan.
B*	**picilabris** Macl. Ann. Jav. 1825. 27; Andr. TES. 1919. 165.	Malay Peninsula. Java.
O*	*femoratus* Dej. Sp. i. 1825. 281; Chaud. Mon. 122; Bouch AF. 1903. 173.	Sumatra.
O*	**picipennis** Chaud. Mon. 100.	Cambodia.
B*	**planigerus** *(Maraga)* Walk. AM. (3). ii. 1858. 204; Chaud. Chaud. Mon. 121; Waterh. EMM. x. 1873. 17; Bates AM. (5). xvii. 1886. 211; Andr. TES. 1919. 189.	Ceylon : Kekirawa, Mullativa, Koggala, Habarane, Anuradhapura.
O	**pelitus** Chaud. Mon. 105.	Malacca., Borneo.
O*	**punctulatus** Chaud. Mon. 110.	"East Indies."
G*	**rufiventris** Bates AG. xxxii. 1892. 402; Andr. TES. 1923. 34.	Karen Hills. Kengtung.
A*	**saundersi** Andr. AM. (9). xviii. 1926. 282.	Kedah., Signapore.

O*	**schaumi** Chaud. Mon. 112; Wasm. Zoo. Jahrb. xvii. Syst. 1902. t. 4. f. 12. a-d. and t. 5. f. 2. (larva).	Ceylon : Colombo, Peradeniya Kandy, Dehiwela, Galgamuwa.
G*	**schraumi,** † Bouch. (not Chaud.) AF. 1903. 173.	Sumatra.
	schmidt-goebeli × Chaud. Mon. 99.	Tenasserim.
P*	*duplicatus (Apsectra)* † Schm. Goeb. (not Wied.) FCB. 61; Andr. TES. 1921. 165; id. TES. 1923. 35.	
P*	**sulcatus** Schm. Geob. FCB. 59; Chaud. Mon. 110; Bates AG. xxxii. 1892. 403; Andr. TES. 1923. 33.	Tenasserim. Cambodia.
O*	*crenaticrus* Chaud. Mon. 113.	
O*	**suturalis** Chaud. Mon. 104.	Malay Peninsula. Sumatra. Mentawei Is. Borneo.
Z	**termiticola** Wasm. Zool. Jahrb. xvii. Syst. 1902. 141 and 145.	Perak.
V*	**xanthomerus** Redt. RN. 12; Chaud. Mon. 124; Andr. TES. 1923 (1924). 465.	Hongkong.

ORTHOTRICHUS* Peyron.

Orthotrichus Peyr. AF. 1856. 717; Andr. MB. 47.

Trichotarus Motch. BM. 1864. ii. 327.

G*	**alternatus** Bates AG. xxxii. 1892. 368.	Naga Hills. Burma : Senmigion, Prome, Katha, Rangoon.
O	**cymindoides** *(Anchomenus)* Dej. Sp. v. 1831. 371; Bates TES. 1873. 281.	Japan., Egypt. Syria.
E*	**indicus** Bates CRB. 1891 334; id. AG. xxxii. 1892. 368; Andr. MB. 47.	Throughout North India to the Central Provinces. Nilgiri Hills. Manipur. Karen Hills. Laos. Tonkin. Yunnan.

OXYCENTRUS* Chaudoir.

Oxycentrus Chaud. BM. 1854. i. 345; Bates TES. 1876. 3; Tchitch. HER. xxxv. 1901. 245.

Agreuter Schm. Goeb. in litt. FCB. (cover).

G* **acutulus** Bates AG. xxxii. 1892. 344. — Thagata (Tenasserin).

O* **angustus** Bates TES. 1876. 3 (note); id. AF. 1889. 268; id. AG. 1889 xxvii. 103. id. AG. xxxii. 1892. 344. — Burma : Bhamo, Teinzo, Rangoon, Moulmein, Pyinmana, Victoria Point.

melas (Agreuter) Schm. Goeb. FCB. t. 3. f. 2; Andr. TES. 1923. 58. — Siam. Indo-China.

G* **angustus** † Bouch. (not Bates) AF. 1903. 172. — Sumatra.

B* **argutoroides** *(Harpalus)* Bates TES. 1873. 261; id. TES. 1876. 3. t. 1. f. 3; id. TES. 1883. 208; Tchitch. HER. xxxv. 1901. 245 (note 73). — Japan.

O* **borneensis** Bates TES. 1876. 4 (note). — Borneo.

X* **foveicollis** Bates AF. 1889. 268; Lesne Miss. Pav. 75. — Laos., Cambodia. Palon., Rangoon.

G* **Omaseroides** Bates AG. xxxii. 1892. 344. — Haflong., Calcutta.

O* **parallelus** Chaud. BM. 1854. i. 347. — Pusa., Chapra. Saran., Dehra Dun. Haldwani.

OXYGNATHUS* Dejean.

Oxygnathus Dej. Sp. ii. 1826. 473; Lac. 198; Putz., Mon. des Clivina et genres voisins, Mém. Liége ii. 1846. 522; id., Postscr. ad Cliv. Mon., Mém. Liége xviii. 1863. 5; id., Rév. Gén. des Clivinides, AB. x. 1867. 23; Bates AG. xxxii. 1892. 274; Andr. FBI. 340.

I* **aboranus** Andr. FBI. 341 and 342. t. 9. f. 5. — Assam : Rotung and Kalek.

C* **elongatus** *(Scarites)* Wied. Zool. Mag. ii. 1. 1823. 38; Dej. Sp. ii. 1826. 474; Dej. et Boisd. Ic. et HN. i. 210. t. 22. f. 5; Brullé HN. v. 1835. 67; Putz. Postscr. 5. t. 1. f. 1; Bates AG. xxxii. 1892. 274; Andr. TES. 1921. 168; id. FBI. 341. f. 49. — Bengal (Wiedemann). Toungoo., Prome. Tharrawaddy. Paungde. Rangoon. Palon.

OXYLOBUS* Chaudoir.

Oxylobus Chaud. BM. 1855. i. 5; id., Mon. des. Scaritides (i); AB. 1879. 129; Andr. MB. 8; id. FBI. 292.

I*	**alternans** Andr. FBI. 295 and 302. t. 7. f. 6.	Travancore : Wallardi.
O*	**alveolatus** Chaud. Mon. 134; Andr. FBI. 296 and 308. t. 8. f. 3.	Balasore. Vizagapatam.
B*	**armatus** Andr. FBI. 297 and 320. ff. 45. and 45a.	Madura., Ramnad.
S*	**asperulus** Chaud. BM. 1857. ii. 58; id. Mon. 133; Andr. TES. 1922 245; id. Spol. Zeyl. 1924. 129; id. FBI. 296 and 311. t. 8. f. 6.	Nilgiri Hills. Pillur., Kodaikanal. Horseley Konda.
B*	*quddricollis* † Bates (not Chaud.) AM. (5). xvii. 1886. 71.	Yercaud., Madura. Shembaganur. Dhoni Forest (S. Malabar). Mysore : Nandidrug. Ceylon : Colombo. Diyatalawa, Nedukeni, Boralesgamuwa, Heneratgoda, Peradeniya.
A*	var. **amyntas** Andr. MB. 70. t. 1. f. 2; id. FBI. 296 and 313.	C. P. : Majgaon, Motinala, Mukhi.
B*	**bipunctatus** Andr. FBI. 296. and 309.	Dapoli.
A*	**dekkanus** Andr. FBI. 294 and 299. t. 7. f. 3.	Bangalore.
H*	**dentatus** Andr. FBI. 295 and 301.	Madras.
A*	**dispar** Andr. FBI. 295 and 303.	Belgaum. North Kanara. Castle Rock. Travancore.
Q	**dissors** Tchitch. HER. xxviii. 1894. 224; Andr. FBI. 322.	Tonkin (Tchitcherin)
A*	**follis** Andr. FBI. 294 and 299. t. 7. f. 7.	Kodaikanal. Horseley Konda. Sirumalai. Travancore : Pirmaid and Wallardi.

O* **feveiger** Chaud. Mon. 133; Andr. FBI. 296. and 310. t. 8. f. 4.	Madras. Nilgiri Hills. Mysore : Chikkaballapura.
A* **inaequalis** Andr. FBI. 294 and 300. t. 7. f. 5.	North Kanara.
O* **lateralis** *(Scarites)* Dej. Sp. i. 1825. 400; Chaud. BM. 1855. i. 8; id. Mon. 131; Andr. FBI. 297 and 316. t. 9. f. 2.	Poona. Nilgiri Hills: Kallar, Ootacamund. Coimbatore. Konakuppam. Manaparai. Alagar Hills. Kodaikanal Pondicherry. Nagalapuram (Chingleput). Tinnevelly. Madras. Trichinopoly. Bangalore. Madura. Ceylon : Kala Oya R., Niroddumunai.
B* var. **designans** *(Scarites)* Walk. AM. (3). ii. 1858 203; Chaud. Mon. 133; Bates AM. (5). xvii. 1886. 210; Andr. TES. 1919. 186; id. FBI. 297 and 318.	Ceylon.
A* var. **ingens** Andr. FBI. 297 and 318.	Poona., Trichinopoly. Tinnevelly. Madura.
A* **lirifer** Andr. FBI. 295 and 307.	Ghozeh (Belgaum district). Ramandrovy (Madras).
A* **mahratta** Andr. FBI. 294 and 297. t. 7. f. 1. 0	Sampgaon and Sanikop (Belgaum district).
O* **meridionalis** Bates CRB. 1891. 325 (note); Andr. AM. (9). xvii. 1926. 253; id. FBI. 294 and 298. t. 7. f. 2.	"South India."
B* **montanus** Andr. FBI. 296 and 315. t. 9. f. 1.	Anaimalai Hills.
B* **nanus** Andr. FBI. 296 and 315.	Nilgiri Hills.

B* **ovalipennis** Andr. Spol. Zeyl. 1924. 130; id. FBI. 297 and 319.	Ceylon : Koggala.
C* **porcatus** *(Scarites)* F. Suppl. Ent. Syst. 1798. 43; Hope Col. Man. ii. 1838. 95 (No. 10); Motch. Etud. Ent. 1855. 40; Heyne-Tasch. EK. 20. t. 3. f. 32; Andr. TES. 1921. 157; id. MB. 8; id. FBI. 295. and 305. t. 8. f .1.	Sahibganj. Tinapahar., Giridih. Konbir., Tetara. Chakardharpore. Barkul.
O* *costatus* Chaud. Mon. 134; Bates CRB. 1891 325; Andr. RIM. xxii. 1921. 339.	Punjab : Baddia. C. P. : Chitrakot, Jubbulpore. Motinala, Majgaon. Belgaum. Nilgiri Hills: Kallar, Pillur. Malabar. Horseley Konda. Coimbatore. Maulinga (Kanara). Vizagapatam. Barkuda I. (Lake Chilka). Shevaroy Hills. Madura. Shembaganur. Ceylon : Anuradhapura.
Q var. **minor** Tchitch. HER. xxviii. 1894. 227.	North Kanra.
B* *costatus* † Bates (not Chaud.) AM. (5). xvii. 1886. 71; Andr. AM. (9). iii. 1919. 469; id. FBI. 295 and 306.	Ceylon : Colombo, Belihul Oya, Weligama, Bentota, Vavuniya, Anuradhapura.
A* var. **obliteratus** Andr. FBI. 295 and 307. t. 8. f. 2.	Shembaganur.
O* **punctatosulcatus** Chaud. BM. 1855. i. 6; id. Mon. 131; Bates CRB. 1891. 325; Andr. FBI. 295 and 304.	Nepal (Chaudoir). Tetara., Ranchi. Chakardharpore. Motinala., Dapoli.
A* var. **fugiens** Andr. FBI. 295 and 305.	Berar. Pondicherry. Horseley Konda. Bangalore.

A* **pygmaeus** Andr. FBI. 295 and 307. — South Mysore.

O* **quadricollis** Chaud. BM. 1855. i. 7; id. Mon. 130; Heyne-Tasch. EK. 20; Andr. FBI. 296 and 314. t. 8. f. 7. — Malabar. Nilgiri Hills: Ouchterlony Valley, Gudalur, Coonoor.

A* **rugatus** Bänn. Ent. Blätt. 1928. 64; Andr. FBI. 297 and 319. — Madras.

A* **rugiceps** Andr. FBI. 296 and 310. t. 8. f. 5. — Belgaum. North Kanara.

L **sculptilis** *(Scarites)* Westw. Arcan. Entom. i. 1842. 88. t. 23. f. 1; Chaud. BM. 1855. i. 9; id. Mon. 133; Bates AM. (5). xvii. 1886. 210; Andr. TES. 1919. 180; id. FBI. 322. — Described as coming from Van Diemens Land but presumably Indian.

OXYODONTUS* Chaudoir.

Oxyodontus Chaud. Mém. sur les Coptodérides, AB. xii. 1869. 239.

O* **tripunctatus** Chaud. Mém. 239. — Celebes.

PACHYTRACHELUS* Chaudoir.

Pachytrachelus Chaud. BM. 1852. i. 85; Lac. 259; Andr. MB. 39.

Batoscelis [Schm. Geob. in litt. FCB. (cover);] Lac. 261.

? Systenognathus Putz., Postscr. ad Cliv. Mon., Mém. Liége xviii. 1863. 18.

O* **cribriceps** Chaud. BM. 1852. i. 86; Bates CRB. 1891. 333. — Kurseong. Murshidabad., Dacca. Berhampur. Dinapore., Pusa. Gwalior. Chota Nagpur : Konbir, Ranchi. Nagpur., Ratnagiri. Belgaum.

O* **oblangus** *(Agonoderus)* Dej. Sp. v. 1831. 813; Bates AG. xxxii. 1892. 343; *(Discoderus)* Vuillet Ins. ii. 1912. 19; Heller. DEZ. 1916. 276; Sloane. PLS. NSW. 1920. 320; Andr. MB. 39. — Throughout South East Asia, but not extending to Japan. Australia.

C* *angulatus (Carabus)* F. Syst. Eleuth. i. 1801. 203. (not Spec. Ins. i. 1781. 302); Schaum SEZ. 1847. 48; Motch. Etud. Ent. 1855. 46; Andr. TES. 1919. 125; id. TES. 1921. 162.

politus (Batoscelis) Schm. Goeb. FCB. t. 2. f. 8; Andr. TES. 1923. 57.

M *ceylonicus (Batascelis)* Motch. BM. 1861. i. 103; Andr. AM. (9). xvii. 1926. 254; id. TES. 1928. 23.

L ? **porosus** *(Systenognathus)* Putz. Postscr. 19. t. 1. ff. 2-4. — India.

PARENA* Motchulsky.

Parena Motch. Etud. Ent. 1859. 31; Chaud. BM. 1877. i. 207; Andr. Ins. Samoa iv. 1927. 13; id. TES. 1928. 16.

Phloeodromius W. Macl. TES. NSW. ii. 1871. 85; Sloane PLS. NSW. 1898. 493; Andr. AM. (9). iii. 1919. 483; id. TES. 1921. 179.

Crossoglossa Chaud. Mon. des Callidides, AB. xv. 1872. 177; Alluaud. BF. 1917. 85.

M* **bicolor** Motch. Etud. Ent. 1859. 32; Andr. TES. 1928. 16. — Java.

B* **cavipennis** *(Crossoglossa)* Bates TES. 1873. 316. — China., Japan.

O* **fasciata** *(Crossoglossa)* Chaud. Mon. 179. — Moluccas., Java. Boreno., Sumbawa.

Y* **hastata** *(Phloeodromus ?)* Heller PJS. xix. 1921. 528. t. 1. f. 2. — Philippine Is.

B* **laesipennis** *(Crossoglossa)* Bates TES. 1873. 317. — Japan.

O* **mellea** *ICrossoglossa)* Chaud. Mon. 179. — Moluccas. Andonare I.

B* **monostigma** *(Crossoglossa)* Bates TES. 1873. 316. — Japan.

O* **nigrolineata** *(Plochionus)* Chand. BM. 1852. i. 44; id. Mon. 180; Bates CRB. 1891. 339. — Bengal., Pusa. Dehra Dun.

B* *latecincta (Crossoglossa)* Bates TES. 1873. 315; id. AF. 1889. 282; Andr. TES. 1921. 179. — Kalka. Chota Nagpur : Konbir Pachmarhi.

B* *latecincta (Crossoglossa)*— Bassein Fort (Bombay)., Belgaum. North Kanara. Madras., Nilambur. Tinnevelly. Ceylon: Mamaduwa. Tharrawaddy. Andaman Is. Annam., Tonkin. China., Japan. Krakatau.

M' **plagiata** *(Phloeodromius)* W. Macl. PLS. NSW. 1876. 167; Sloane PLS. NSW. 1920. 322. New Guinea. Australia.

A* **rubripicta** Andr. TES. 1928. 17. Kallar (Nilgiri Hills). Nilambur., Bhamo. Momeit. Langkawi. I. Cochin-China.

Y* **sellata** *(Phloeodromus ?)* Heller PJS. xix. 1921. 526. t. l. f. 1. Philippine is.

O* **testacea** *(Crossoglossa)* Chaud. Mon. 178. Dacca "Nepal" Dapoli., Belgaum. Sumatra.

PAROPISTHIUS* Casey.

Paropisthius Casey Mém. Col. ix. 1920. 148 (note); Bånn. Ent. Mitt. 1925. 187; Sem. RRE. xx. 1920. 38; Andr. FBI. 103.

Opisthius Dupuis Gen. Ins. Fasc. 126, Opisthiinae, 1912. 2 (part).

O **davidis** *(Elaphrus)* Fairm. AB. 1887. 82; Sem. RRE. xx. 1926. 40. Yunnan.

B* **indicus** *(Opisthius)* Chaud. AF. 1863. 449; Dupuis Gen. Ins. Fasc. 126. t. ff. 1. and 8-10; Andr. TES. 1919; 195; Sem. RRE. xx. 1926. 38; Andr. FBI. 101. f. 18. Tibet : Tsanpo Valley, Doshongla. Nepal : Kalapokri, Sikkim, Darjiling, Simana, Nagri Spur, Mungphu, Khamba-Jong, Guentok, Senchal, Ghoom, Phallut.

PATELLUS* Chaudoir.

Patellus Chaud. Mon. des Oodides (ii). AF. 1882. 551. — Burma.

O* **drimostoides** Chaud. Mon. 553.

PAUSSOTROPUS* Waterhouse.

Paussotropus Waterh. TES. 1877. 3.

B* **parallelus** Waterh. TES. 1877. 3; Andr. TES. 1919. 197. — Batchian.

PENTAGONICA Schmidt-Goebel.

Pentagonica Schm. Goeb. FCB. 47; Lac. 133; Schaum. BEZ. 1863. 74; Bates TES. 1873. 321; Chaud. BM. 1877. i. 212; Sloane PLS. NSW. 1898. 494 and 513; Dupuis Gen. Ins. Fasc. 145, Pentagonicinae. 1913. 2.

Rhombodera || Reiche (not Burm.) RZ. 1842. 313; Lac. 139.

Didetus Lac. Trans. Am. Phil. Soc. x. 1853. 377; Lac. 399.

Elliolia Nietn. JASB. xxv. 1850. 524; id. AM. (2). xix. 1857. 375.

Trichothorax Montr. AF. 1860. 235.

Xenothorax Woll. Col. Hesp. 1867. 15.

Wakefieldia Broun Man. New Zeal. Col. 1880. 62.

A* **batesi** × Andr. TES. 1923. 24. *nigripennis* † Batest (not Bates 1873) AG. xxxii. 1892. 426. — Karen Hills. Kuala Lumpur. Java., Sumatra. Sebesi (Krakatau group).

D* **biangulata** Dupuis AB. 1912. 312; id. Gen. Ins. Fasc. 145. t. f. 1. — Formosa.

P* **erichsoni** Schm. Goeb. FCB. 48; Andr. TES. 1923. 25. — Ceylon

M* *marginata* Motch. BM. 1861. i. 105. t. 9 f. 3; Andr. TES. 1928. 7. — Sikkim: Gopaldhara, Kurseong., Manipur. Burma (Schmidt-Goebel). Selangor., Pahang. Langkawi I. Singapore.

M* *marginata* Motch.—	Hongkong. Borneo. Great Sangir. Batchian.
D* **formosana** Dupuis AB. 1912. 316; id. Gen. Ins. Fasc. 145. t. ff. 2. and 8.	Formosa.
D* **horni** Dupuis AB. 1913. 175.	Ceylon : Weligama, Kandy.
S* **pallipes** *(Elliotia)* Nietn. JASB. xxv. 1856. 525; id. AM. (2). xix. 1857. 376; Schaum BEZ. 1863. 75; Andr. AM. (9). xix. 1927. 106.	Kandy., Colombo. Nuwara Eliya. Pahang. Kuala Lumpur. Singapore. Philippine Is.
M* *transparipes* Motch. Etud. Ent. 1859. 29; Bates AM. (5). xvii. 1886. 209; Andr. TES. 1928. 16.	Sumatra. Sumbawa.
P* **ruficollis** Schm. Goeb. FCB. 48; Bates AG. xxxii. 1892. 426; Dupuis Gen. Ins. Fasc. 145. t. ff. 5. and 9-11; Andr. TES. 1923. 23.	Patkai Hills. Nilgiri Hills. Aratapara. Peradeniya.
A* *dichroa* Sloane PLS. NSW. 1903. 636.	Rangoon., Kedah. Tonkin., Sumatra. Borneo., Java. Gilolo., Timor. Andonare I., Allor I. Philippine Is. Australia.
X* **ruficollis** † Bates (not Shm. Goeb). AF. 1889. 286; Andr. TES. 1923. 23.	Annam.
D* **semisuturalis** Dupuis AB. 1912. 314; id. Gen. Ins. Fasc. 145. t. f. 3.	Patkai Hills. Formosa.
J* **suturalis** *(Rhombodera)* Schaum BEZ. 1863. 75; Andr. AM. (9). xix. 1927. 108.	Hongkong.
Y* **varicornis** Heller DEZ. 1916. 269.	Philippine Is.

PERICALUS* Macleay.

Pericalus Macl. Ann. Jav. 1825. 15; Schm. Goeb. FCB. 85; Lac. 147; Schaum BEZ. 1860. 190; Chaud. BEZ. 1861. 123; Bates EMM. vi. 1869. 71.

Pericallus Dupuis AB. 1913. 82.

Coeloprosopus Chaud. BM. 1842. 839.

B*	**cicindeloides** Macl. Ann. Jav. 1825. 16. t. 1. f. 2; Brullé HN. iv. 1834. 230; Dupuis AB. 1913. 82; Andr. TES. 1919. 143.	Java., Sumatra. Perak. Palawan.
A*	**depressus** Andr. AM. (9). xvii. 1926. 285.	Malay Peninsula. Sumatra., Borneo.
E*	**distinctus** Dupuis AB. 1913. 83.	Burma. Andaman Is.
B*	**dux** Andr. AB. 1920. 25.	Laos.
O*	**figuratus** Chaud. BEZ. 1861. 124.	Celebes.
D*	**formosanus** Dupuis AB. 1913. 83.	Formosa.
B*	**funestus** Andr. AM. (9). xviii. 1926. 284.	Sumatra.
J*	**gratus** Schaum. BEZ. 1861. 124; Andr. AM. (9). xix. 1927. 108.	Celebes.
H*	**guttatus** Chevr. Mag. Zool. 1832. Cl. ix. t. 46; Brullé HN. iv. 1834. 231; Chaud. BEZ. 1861. 123; Dupuis. AB. 1913. 82; Andr. TES. 1919. 206.	Java. Sumatra.
A*	var. **violaceus** Andr. AM. (9). xvii. 1926. 284.	Malay Peninsula. Sumatra.
J*D*	**laetus** Schaum BEZ. 1860. 190; Andr. AM. (9). xix. 1927. 108.	Borneo. Celebes.
J*	*adonis* Schauf. HER. xxi. 1887. 104; Andr. AM. (9). xix. 1927. 109.	
Y*	**levifrons** Heller DEZ. 1916. 273.	Philippine Is.
O*	**longicollis** Chaud. AB. xii. 1869. 159.	Malacca., Perak. Borneo.
P*	**ornatus** Schm. Goeb. FCB. 86; Bates AG. xxxii. 1892. 411; Dupuis AB. 1913. 83; Andr. TES. 1923. 49.	Assam : Khasi Hills, Naga Hills, Garo Hills, Lakhimpur, Dejoo, Rotung, Kobo. Sikkim : Kurseong, Gopaldhara. Chandkhira N. Shan States. Karen Hills.

P*	**ornatus** Schm. Goeb.—	Ruby Mines. Tenasserim. Siam., Indo-China.
Y*	**philippinus** Heller DEZ. 1916. 274. t. 3. f. 1.	Philippine Is.
O*	**picturatus** Chaud. AB. xii. 1869. 160.	Celebes.
B*	**quadrimaculatus** *(Catascopus)* Macl. Ann. Jav. 1825. 15; *(Coeloprosopus)* Chaud. BM. 1842. 840; id. BEZ. 1861. 123; Andr. TES. 1919. 141.	Perak., Pahang. Malacca., Siam. Java.
L	*quadrisignatus (Catascopus)* Cast. AF. 1832. 392.	Sumatra., Borneo. Lombok.
A*	**spiniger** Andr. AM. (9). xviii. 1926. 286.	Batchian.
O*	**tetrastigma** Chaud. BEZ. 1861. 123.	Perak., Singapore. Sumatra., Borneo.
O*	**undatus** Chaud. BM. 1848. i. 111; id. BEZ. 1861. 123.	Philippine Is.
J*	**xanthopus** Schaum BEZ. 1860. 191; Andr. AM. (9). xix. 1927. 108.	Perak. Borneo.

PERIGONA* Castelnau.

Perigona Cast. Etud. Ent. 1835. 151; Lac. 134 (note); Chaud. BM. 1871. ii. 281; Putz. AG. 1873. 218; Fvl. RE. 1907. 97.

Nestra Motch. BM. 1851. ii. 506; id. Etud. Ent. 1859. 37.

Trechicus Lec. Trans. Am. Phill. Soc. x. 1853. 386; Lac. 393 (note 2).

Spathinus Nietn. AM. (3). ii. 1858. 428.

Tachynotus Motch. Bull. Mose. 1861. i. 100; Putz. SEZ. 1870. 9; Andr. TES. 1928. 6.

Pentoplogenius Moraw. Beitr. 24; Fvl. RE. 1889. 100.

Siltopia Cast. Trans. Roy. Soc. Vict. viii. 1868. 127.

Extromus Pér. Trans. S. Afr. Phil. Soc. vii. 1896. 586.

B*	**acupalpoides** Bates TES. 1883. 264; id. AG. xxxii. 1892. 378; Csiki Ann. Mus. Hung. 1924. 172.	U. P. : Ramnagar. (Fatehpur). Kurseong. Burma : Karen Hills, Tikekee,

B*	**acupalpoides** Bates—	Thagata, Meetan. Formosa., Japan.
J'	**angustata** Fvl. RE. 1907. 104.	Java., Sumatra.
B'	**astrolabica** Csiki Ann. Mus. Hung. 1924. 172.	New Guinea.
G*	**bigener** Bates AG. xxxii. 1892. 379.	Karen Hills. Perak.
B'	**biroi** Csiki Ann. Mus. Hung. 1924. 173.	New Guinea.
M*	**castanea** *(Tachynotus)* Motch. BM. 1861. i. 100. t. 9. f. 1; Jean. AM. (9). xii. 1923. 394; Andr. TES. 1928. 6.	Colombo.
E*	**convexicollis** Putz. AG. 1875. 729.	Johore.
B'	**erimae** Csiki Ann. Mus. Hung. 1924. 173.	New Guinea.
Q	**exigua** *(Pentoplogenius)* Moraw. Beitr. 25; Bates TES. 1873. 282; Fvl. RE. 1889. 100; id., RE. 1907. 98.	Japan.
B'	**lebiodes** Csiki Ann. Mus. Hung. 1924. 174.	New Guinea.
	litura *(Trechus)* Perr. Ann. Soc. Linn. Lyon. xi. 1864. 72; Fvl. RE. 1882. 252; (No. 60. *Perigona ?)* AF. 1889. 274; Fvl. RE. 1903. 224; Andr. TES. 1921. 178.	Cambodia. Cochin-China. Sumatra. New Guinea. New Caledonia.
O*	**livens** Putz. AG. 1873. 225; id. AG. 1875. 730; Fvl. RE 1907. 99.	? "Coromandel."
B'	**ludovici** Csiki Ann. Mus. Hung. 1924. 174.	New Guinea.
B'	**malayensis** Csiki Ann. Mus. Hung. 1924. 172.	Malacca.
G(E)	**minor** Putz. AG. 1875. 734.	Borneo.
O*	**nigriceps** *(Bembidium)* Dej. Sp. v. 1831. 44; ? Nietn. AM. (3). ii. 1858. 429 (*Spathinus*); Fvl. RE. 1889. 99; K. Dan Münch. Kol. i. 1903. 249; Fiori Riv. Col. Ital. i. 1903. 24; Fvl. RE. 1907 102; Alluaud. AF. 1916. 59; Sloane PLS NSW. 1920. 322; Csiki Ann. Mus. Hung. 1924. 171; Clerm. BF. 1925. 249; Alluaud AG. lii. 1928. 76.	Cosmopolitan, but not yet recorded from Central or South America.
C'	*umbripennis (Trechicus)* Lec. Trans. Am. Phil. Soc. x. 1853. 386.	
B*	*fimicola (Trechus)* Woll. Ins. Mad. 1854. 63; id. Col. Hesp. 1867. 27; Bates AM. (5). xvii. 1886. 156; Andr. TES. 1919. 181.	

B* *jansoniana (Trechus)* Woll. AM. (3). i. 1858. 19.

O *atriceps (Nestra)* Fairm. AF. 1869. 184.

B* *japonica (Trechicus)* Bates TES. 1873. 281; Putz. AG. 1875. 734; Bates TES. 1876. 4; id. TES. 1883. 208; id. CRB. 1891. 326.

G(E) *beccarii* Putz. AG. 1875. 732; Bates AM. (5). xvii. 1886. 156; id. AG. xxxii. 1892. 379.

O* *discalis* Chaud. RMZ. 1876. 353 (not 553).

B* *tachyoides* Bates TES. 1883. 265; Andr. AM. (9). xx. 1927. 272.

B* *suffusa* Bates AM. (5). xvii. 1886. 151.

B* *pusilla (Extromus)* Pér. Trans. S. Afr. Phil. Soc. vii. 1896. 587; id. Ann. S Afr. Mus. iii. 1904. 296.

A *australica* Sloane PLS. NSW. 1903. 635.

M* **nigricollis** *(Nestra)* Motch. BM. 1851. ii. 506; Putz. AG. 1873. 222; id. AG. 1875. 728; Andr. TES. 1928. 3. — South India. Ceylon., Tenasserim. Java., Sumatra. Borneo.

M* **nigrifrons** *(Nestra)* Motch. Etud. Ent. 1859. 38. t. 1. f. 1; Putz. AG. 1873. 220; id. AG. 1875. 727; Bates AM. (5). xvii. 1886. 149; Andr. TES. 1928. 3 and 18. — Ceylon : Nuwara Eliya Galle Point, Bogawantalawa.

B' **papauana** Csiki Ann. Mus. Hung. 1924. 173. — New Guinea.

G* **plagiata** Putz. AG. 1875. 734; Andr. TES. 1921. 178; Csiki Ann. Mus. Hung. 1924. 172; Andr. AM. (10). ii. 1928. 588. — Assam : Rotung, Kobo., Dehra Dun. Saharanpore. Nilgiri Hills., Kandy. Matale. Burma : Bhamo, Katha, Myitkyina, Meetan, Thagata, Insein., Perak. Singapore., Siam. Indo-China. Philippine Is. Java., Sumatra. Ternate., Aru Is. Kei Is., New Guinea. Japan.

B* *discipennis* Bates TES. 1883. 265; AM. (9). xx. 1927. 271.

B* *ruficollis* † Bates (not Motch.) AM. (5). xvii. 1886. 149; id. AG. xxvii. 1889. 104; id. AG. xxxii. 1892. 378.

X* *nana* Bates AF. 1889. 273; Andr. AM. (9). xvii. 1926. 258.

J' *annamita* Fvl. RE. 1907. 104; Andr. TES. 1928. 3.

J' **procera** Fvl. RE. 1907. 103. — Java.

M* **ruficollis** *(Nestra)* Motch. BM. 1851. ii. 506; Putz. AG. 1873. 222; id. AG. 1875. 728; Andr. TES. 1928. 2. and 3. — Singbhum. Tenasserim. Tavoy., Perak.

E* *luzonica* Putz. AG. 1875. 728; Andr. AM. (10). ii. 1928. 587. — Singapore., Siam. Sumatra., Borneo. Batchian. Great Sangir. Philippines Is.

B* **sinuata** Bates TES. 1882. 265. — Japan.

B* **sinuaticollis** Bates AM. (5). xvii. 1886. 149. — Ceylon : Dikoya.

G **subcordata** Putz. AG. 1875. 730. — Kei Is.

G **subcyanescens** Putz. AG 1875. 732. — New Guinea

G **suturalis** Putz. AG. 1875. 728. — Celebes. New Guinea.

PERILEPTUS* Schaum.

Perileptus Schaum. Nat. Ins. Deutschl. i. 1860. 663; Motch. BM. 1864. ii. 190; Putz. DEZ. 1870. 9; Jeann., Mon. des Trechinae (1), Ab. xxxii. 1926. 402.

Blemus † Dej. (not Steph). Sp. v. 1831. 33; Cast. Etud. Ent. 1834. 94; id. Hist. Nat. Col. i. 1840. 155.

Ochthephilus || Nietn. (not Muls.) JASB. xxvi. 1857. 136; id. AM. (2). xx. 1857. 275; Jeann. AF. 1921 (1922). 165.

Pyrrhotachys Sloane PLS. NSW. 1896. 374; Jeann. Mon. (1). 405.

B* **cameroni** Jeann, AM. (9). xii. 1923. 397 and 407. f. 7; id. Mon. (1). 427. f. 205. — Kaligad (Dehra Dun). Shugnu (Manipur).

S* **ceylanicus** *(Ochthephilus)* Nietn. JASB. xxvi. 1857. 136; id. AM. (2). xx. 1857. 276. Jeann. AM. (9). xii. 1923. 397 and 401. f. 3. id. Mon. (1). 416. f. 191; Andr. AM. (9). xix. 1927. 106. — Poona. Ceylon : Negombo.

A* **denticollis** Jeann. AM. (9). xii. 1923. 397 and 406; id. Mon. (1). 426. ff. 202-4. — Yunnan.

B* **imaicus** Jeann. AM. (9). xii. 1923. 397 and 405. f. 6; id. Mon. (1). 425. f. 201. — Sikkim : Gopaldhara Rongni River.

B* **imaicus** Jeann.— Manipur : Shugnu. Eastern Duars. U. P. : Almora, Upper Gumti Valley, Binal Gad, Saiya. Garhwal : Siuni Gad.

B* **indicus** Jeann. AM. (9). xii. 1923. 397 and 399. f. 2; Andr. EMM. 1926. 68; Jeann. Mon. (1). 417. f. 192. Sikkim : Gielle Khola (Tista Valley). Garo Hills. Manipur : Shugnu. Eastern Duars. Haldwani. Dehra Dun. Chakrata., Almora. Sarju Valley. Tanakpur., Ranikhet.

B* **japonicus** Bates TES. 1873. 296; Jeann. AM. (2). xii. 1923. 397; id. Mon. (1). 414 ff. 188-9. Japan.

A'* **melas** Jeann. Mon. (1). 416. Philippine Is.

A* **platypterus** Jeann. AM. (9). xii. 1923. 397 and 403. f. 5; id. Mon. (1). 423. f. 200. Java.

A* **pusillus** Jeann. AM. (9). xii. 1923. 397 and 402. f. 4; id. Mon. (1). 417. ff. 198-9. Tonkin.

B* **robustus** Jeann. AM. (9). xii. 1923. 397 and 398. f. 1; id. Mon. (1). 415. f. 190. Almora., Ranikhet. Garhwal : Siuni Gad.

PERIPRISTUS* Chaudoir.

Peripristus Chaud., Mém. sur les Thyréoptérides, AB. xii. 1869. 135.

C* **ater** (*Thyreopterus*) Cast. Etud. Ent. 1835. 149; Schm. Goeb. FCB. 79; Chaud. Mém. 136; Bates. AF. 1889, 283; id. AG. xxvii. 1889. 110; id. AG. xxxii. 1892. 408; Andr. TES. 1923. 46. Kobo. Sikkim : Gopaldhara, Kalimpong, Pashok. Nilgiri Hills. Anaimalai Hills. Cochin State : Parambikulam. Burma : Bhamo, Katha, Shwegu, Shwegyin, Malewoon, Karen

G* **ater** (*Thyreopterus*)— | Hills, Thagata, Tenasserim. Nicobar Is. Malay Peninsula. Siam., Indo-China. Sumatra., Java. Borneo.

D* var. **schenklingi** Dupnis AB. 1912. 288; id. AB. 1913. 82. | Formosa.

PERONOMERUS* Schaum.

Peronomerus Schaum AF. 1853. 440; Chaud., Mon. sur les Panagéides, AB. 1878. 162.

B* **auripilis** Bates TES. 1883. 235. | Japan.

J*D* **fumatus** Schaum. AF. 1853. 440; Chaud. BM. 1861. ii. 354; Redt. RN. 9; Chaud. Mon. 162; Bates TES. 1883. 234; Andr. TES. 1923 (1924). 462; id. AM. (9). xix. 1927. 107. | Dacca (Chaudoir) Tonkin., Hongkong. Japan. Philippine Is. Sumatra.

O* *aeratus* Chaud. BM. 1861. ii. 354.

Y* *nigrinus* † Heller (not Bates) DEZ. 1916. 276; Andr. AM. (9). xvii. 1926. 254.

F' **insularis** Schoenf. Ent. Nachr. xvi. 1890. 168. | Liu Kiu Is.

B* **nigrinus** Bates TES. 1873. 245; Chaud. Mon. 164; Bates TES. 1883. 235. | Japan.

PHEROPSOPHUS Solier.

Pheropsophus Sol. AF. 1833. 461; Lac. 99; Chaud., Mon. des Brachynides, AB. 1876. 16; Arrow TES. 1901. 193; Andr. BM. 55; Hubthl. DEZ. 1914. 437.

Stenaptinus Maindr. BF. 1906. 15; Hubthl. DEZ. 1911. 547.

O* **aptinoides** Chaud. Mon. 19; Heller. AB. 1903. 246. | East Indies.

O* **assimilis** Chaud. Mon. 38. | Siam., China.

L **australis** Cast. Trans. Roy. Soc. Vict. viii. 1868. 109; Arrow. TES. 1901. 205. | New Guinea. Australia.

M' *papuensis* W. Macl. PLS. NSW. 1876. 166; id. PLS. NSW. 1886. 138; Heller. Abh. Mes. Dresd. xiii. 3. 1910. 7.

Y	**baliothorax** Heller Abh. Mus. Dresd. xiii. 3. 1910. 6.	New Guinea.
Y	var. **aptinomorphus** Heller Abh. Mus. Dresd. xiii. 3. 1910. 7.	New Guinea.
W*	**bimaculatus** *(Carabus)* L. Mantissa. 1771. 532; F. Syst. Ent. 1775. 243; Oliv. Ent. iii. 1795. 35 65. t. 2. f. 16; Dej. Sp. i. 1825. 299; Hope Col. Man. ii. 1838. 100; Motch. Etud. Ent. 1855. 55; Redt. RN. 5; Chaud. Mon. 34; Rates AM. (5). xvii. 1886. 200; Andr. TES. 1919. 120; id. TES. 1921. 153; id. TES. 1923 (1924). 460.	Poona. Anaimalai Hills. Kodaikanal. Pondicherry. Tinnevelly. Madras., Coimbatore. Ramnad., Chingleput. Madura., Trichinopoly. Mysore : Chikkaballapura. Travancore. Ceylon : Kitugalle, Udagama, Weligama, Colombo, Elephant Pass, Avisawella.
B*	var. **posticalis** Arrow TES. 1901. 204; Andr. TES. 1919. 198.	Kodaikanal.
X*	**cardoni** Maindr. BF. 1898. 131.	Chota Nagpur : Nowafully Nowatoli C. P. : Baster (from stomach of black ibis).
O*	**catoirei** *(Brachinus)* Dej. Sp. i. 1825. 301; Putz. CRB. 1875. 45; Chaud. Mon. 24 (part); Bates Entom. 1891. Suppl. 10; id. AG. xxxii. 1892. 391; Arrow TES. 1901. 202; Andr. MB. 55.	Throughout India, westward to Kashmir, Sind, and the Khyber Pass. Ceylon : Hambamtota, Kalupahani, Murunkan. Burma : Kachin, Teinzo. Andaman Is.
B*	**chaudoiri** Arrow TES. 1901. 203; Andr. MB. 56.	Nepal., Almora.
O*	*catoriei* † Chaud. (not Dej.) Mon. 24 (part).	C. P. : Nagpur, Mandla, Rajadhar. Belgaum., Surada. Madras., Ramandrug.

B*	**chaudoiri**—	Trichur. Trichinopoly Ceylon : Vavuniya, Nedunkeni.
P*	**consularis** *(Brachinus)* Schm. Goeb. FCB. 75; Bates CRB. 1891. 336; id. AG. xxxii. 1892. 392; Andr. TES. 1923. 44; id. MB. 116.	Assam : Silonibari, Sibsagar, Siliguri, Rotung., Dacca. Sardah., Chapra.
O*	*stenoderus* Chaud. BM. 1850. i. 77; id. Mon. 41; Arrow. TES. 1901. 204; Andr. MB. 56.	Spiti : Gharry. Kangra., Dehra Dun. Ranikhet. Chota Nagpur : Tetara, Konbir. Surada. Karen Hills., Bhamo.
O*	**discicollis** *(Brachinus)* Dej. Sp. i. 185. 300; Chaud. Mon. 21; Arrow. TES. 1901. 203. *affinis (Brachinus)* Dej. Sp. i. 1825. 301.	Nagpur., "Deccan." Madras., Red Hills. Pondicherry. Trichinopoly. Ceylon : Murunkan, Anuradhapura.
O*	**emarginatus** Chaud. Mon. 20.	Philippine Is. Formosa.
O*	**fumigatus** *(Brachinus)* Dej. Sp. i. 1825. 307; Chaud. Mon. 40.	Philippine Is.
X	**gironieri** Eyd. et Soul. RZ. 1839. 264; Desm. Voy. la Bonite 1841. 293. t. 2. f. 2; Chaud. Mon. 32.	Philippine Is.
B*	**heathi** Arrow TES. 1901. 205. t. 9. f. 1; Andr. TES. 1919. 198.	Burma : Moulmein, Yagyo.
C*	**hilaris** *(Carabus)* F. Suppl. Ent. Syst. 1798. 56; Schaum. SEZ. 1847. 49; Motch. Etud. Ent. 1855. 55; Chaud. Mon. 25; Gestro AG. 1882. 297; Bates AG. xxxii. 1892. 391; Arrow. TES. 1901. 202; Andr. TES. 1921. 159; id. MB. 55.	Throughout India. Baluchistan : Nushki, "Baluchistan-Afghanistan Boundary Commission". Minhla., Bhamo.
O*	var. **sobrinus** *(Brachinus)* Dej. Sp. ii. 1826. 462; Chaud. Mon. 25; Arrow. TES. 1901. 202.	Throughout India. Quetta. Socotra.

X*	*desbordesi* Maindr. BF. 1899. 16; Arrow. TES. 1901. 202; Maindr. BF. 1905. 95; Bedel. Fn. med. 1914. 310 (note 4).	
G*	**infantulus** Bates AG. xxxii. 1892. 390.	Karen Hills. Plapoo (Tenasserim).
O*	**javanus** *(Brachinus)* Dej. Sp. i. 1825. 305; Redt. RN. 5; Chaud. Mon. 42; Heyne-Tasch. EK. 14; Bouch. AF. 1903. 173; Andr. TES. 1923 (1924). 460.	Margherita. Orissa : Balaghai. Malabar (Chaudoir). Throughout Burma and the Malay Peninsula.
P*	*interruptus* † Schm. Goeb. (not Dej.) FCB. 74; Andr. TES. 1923. 44.	Siam., Indo-China. China.
O*	*fimbriatus* Chaud. Mon. 42.	Japan.
O*	*agnatus* Chaud. Mon. 43; Bates AF. 1889. 281; id. AG. xxxii. 1892. 392; Lewis AM. (6). xvii. 1896. 331; Dupuis AB. 1913. 419; Arrow TZS. xx. 1915. 499.	Formosa., Java. Sumatra., Borneo. Philippine Is. New Guinea.
J*	*kuntzeni* Dupuis AB. 1914. 29; Andr. AM. (9). xix. 1927. 110.	
Q	**jessoensis** Moraw. Mel. Biol. Acad. Sci. Petr. iv. 1862. 238; id. BAP. v. 1863. 322; id. Beitr. 22; Bates TES. 1873. 305; Putz. CRB. 1875. 52; Har. Abh. Nat. Ver. Bremen iv. 1875. 285; Chaud. Mon. 35; Heyd. Dez. 1879. 329.	Japan. Manchuria. China.
X*	**krichna** Maindr. BF. 1906. 15.	Khasi Hills. Shillong.
O*	**lineifrons** Chaud. BM. 1850. i. 80; id. Mon. 24; Bates AB. 1892. 233; Arrow. TES 1901. 203.	Tezpur., ? Darjiling. Pusa., Chapra. Bengal., Allahabad. Mughalsarai. Lachhiwala., Simla. Chota Nagpur. Nagpur.
O*	**lissoderus** Chaud. BM. 1850. i. 79; id. Mon. 24; Andr. FBI. 21; f. 6b.	Tibet (Chaudoir) Throughout India and Ceylon.
B*	*catoirei* † Bates (not Dej.) AM. (5). xvii. 1886. 201.	

M	**marginicollis** Motch. Etud. Ent. 1853. 44; Chaud. Mon. 43; Fairm. RE. 1887. 312; Andr. TES. 1928. 23.	China.
Y	**minahassae** Heller AB. 1903. 247.	Celebes.
O*	**nanodes** Bates AG. xxxii. 1892. 391 (note)	Assam: Sadiya Panji, "between Pasighat and Kobo", Dosing, Yambung, "below Damda".
O*	**nebulosus** Chaud. Mon. 27.	Malewoon (Tenasserim).
D*	var. **formosanus** Dupuis AB. 1912. 322; id. AB. 1913. 81.	Perak. Siam.
P*	*marginalis (Brachinus)* † Schm. Goeb. FCB. 74; Chaud. Mon. 34; Bates AF. 1889. 281; id. AG. xxxii. 1892. 393; Lesne Miss. Pav. 79; Vuillet Ins. ii. 1912. 17; Andr. TES. 1921. 149; id. TES. 1923. 43.	Indo-China. Sumatra. Formosa. China., Japan.
X*	*bimaculatus* † lesne (not L.) Miss. Pav. 79.	
	stabilis × Andr. AM. (9). xiii. 1924. 470.	
B*	**nigricollis** Arrow TES. 1901. 203. t. 9. f. 2; Andr. TES. 1919. 198.	Mysore : Banglore, Chikkaballapura, Manganale, Shimoga.
B*	**occipitalis** *(Aptinus)* Macl. Ann. Jav. 1825. 28; Chaud. Mon. 42; Arrow TES. 1901. 204; Andr. TES. 1919. 168.	Manipur., Asasm. Lushai Hills. Bengal.
O*	*ambiguus (Brachinus)* Dej. Sp. i. 1825. 304; Chaud. Mon. 37; Bates AG. xxxii. 1892. 391; Heyne-Tasch. EK. 14.	Bihar., Orissa. Throughout Central and Southern India. Ceylon and Burma. Andaman Is. Malay Peninsula and Archipelago. Hongkong. Formosa.
O*	*fuscicollis (Brachinus)* Dej. Sp. i. 1825. 306; Putz. CRB. 1875. 45; Chaud. Mon 37; Bates AM. (5). xvii. 1886. 200; id. AG. xxxii. 1892. 391; Bouch. AF. 1903. 173.	
O*	*interruptus (Brachinus)* Dej. Sp. i. 1825. 306; Chaud. Mon. 37.	Bhutan., Dafla Hills. Lushai Hills. Shan Hills.
O*	*quadripustulatus* Chaud. BM. 1843. ii. 706; id. Mon. 37.	

O* **picicollis** Chaud. Mon. 44; Bates AG. xxxii. 1892. 392.	Mormugao. Bangalore. Trivandrum. Ceylon. Karen Hills. Bhamo., Palon. Rangoon. Andaman Is. Malay Peninsula Siam., Laos.
Y **prophylax** Heller AB. 1903. 247.	Sibsagar. Dafla Hills. Garo Hills. Chandkhira. Manipur.
A* **scythropus** Andr. AM. (9). xii. 1923. 221.	Shillong., Khasi Hills.
O* *melanholicus* † Chaud. (not Schm. Goeb.) Mon. 20; Andr. TES. 1923. 40.	Sikkim : Gopaldhara, Pashok, Kurseong, Mungphu, Pedong, Maria Basti. Manipur. Ruby Mines.
B* var. **dissolutus** Andr. AM. (9). xii. 1923. 222.	Laos.
O* **siamensis** Chaud. Mon. 29; Bates AF. 1889. 281.	Siam., Cochin-China.
B* **tripustulatus** *(Brachinus)* F. Ent. Syst. i. (1). 1792. 145; id. Syst. Eleuth. i. 1801. 218; Hope Col. Man. ii. 1838. 101; Moth. Etud. Ent. 1855. 55; Andr. TES. 1919. 124; id. TES. 1921. 155 and 159; id. RIM. xxii. 1921. 346; id. AM. (9). xiv. 1924. 592.	Barkuda (Lake Chilka) Ennur., Madras. Pondicherry. Travancore. Siam (Fabricius)
O* *marginalis (Brachinus)* Dej. Sp. i. 1825. 310; Heyne-Tasch. EK. 14. t. 2. f. 31; Andr. TES. 1921. 149.	
O* *amoenus* Chaud. BM. 1850. i. 78; id. Mon. 36.	
C* var. **annulus** *(Brachinus)* F. Syst. Eleuth. i. 1801. 217; Motch. Etud. Ent. 1855. 55; Chaud. Mon. 47; Andr. TES. 1921. 164.	Tranquebar.

B* var. **curtus** Arrow TES. 1901. 204. t. 9. f. 3; Andr. TES. 1919. 198. — North Kanara. Mahe., Malabar. Nilambur.

PHYSOCROTAPHUS* Parry.

Physocrotaphus Parry TES. 1849. 180; Lac. 181; Chaud. BM. 1862. ii. 303.

L **ceylonicus** Parry TES. 1849. 180. t. 18. f. 1; Chaud. BM. 1862. ii. 304; Bates AM. (5). xvii. 1886. 200. [Lac. Atl. t. 7. f. 1. is not *Physocrotaphus ceyloncius,* as stated, but *Helluodes taprobanae* Westw.] — Ceylon Dikoya.

PHYSODERA* Eschscholtz.

Physodera Eschsch. Zool. Atl. ii. 1829. 8; Schm. Goeb. FCB. 46; Lac. 130; van de Poll. NLM. xi. 1889. 251.

L **amplicollis** van de Poll NLM xi. 1889. 254; Heller PJS. xxiii. 1923. 305. — Java

Y **bifenestrata** Heller PJS. xxiii. 1923. 304 and 305. t. 1. ff. 5-6. — Borneo.

Z* **cupreomicans** *(Calleida)* Oberth. NLM. v. 1883. 218; Andr. AM. (10). ii. 1928. 590. — Sumatra.

L **cyanipennis** van de Poll NLM. xi. 1889. 253; Heller PJS. xxiii. 1923. 305. — Celebes.

dejeani Eschsch. Zool. Atl. ii. 1829. 8. t. 8. f 6; Schm. Goeb. FCB. 46; Lac. Atl. t. 4. f. 3; Mohnike Sitzungsber. niederrhein. Ges. Bonn. 1875. 154; Andr. TES. 1923. 22; Heller PJS. xxiii. 1923. 305. — Sikkim : Gopaldhara. Tenasserim. Andaman Is. Perak., Penang. Java., Sumatra. Borneo., Balabac I. Philippine Is.

Y* **eburata** Heller PJS. xxiii. 1923. 304 and 305. t. 1. f. 4. — Philippine Is.

L **eschscholtzi** Parry TES. 1849. 179. t. 18. f. 2; Mohnike Sitzungsber. niederrhein. Ges. Bonn. 1875. 154; Bates AM. (5). xvii. 1886. 209; id. AG. xxxii. 1892. 425; Heller PJS. xxiii. 1923. 305. — Khasi Hills. Chandkhira. Sikkim: Namsoo, Pashok., Peradeniya. Burma : Teinzo. Malay Peninsula. Laos., Tonkin., China. Sumatra., Java. Borneo., Philippine Is.

O* *davidis* Fairm. AB. 1887. 92; van de Poll NLM. xi. 1889; 252; Andr. TES. 1921. 150.

L **parvicollis** van de Poll NLM. xi. 1889. 252; Heller PJS. xxiii. 1923. 305. — Hongkong.

PLANETES* Macleay.

Planetes Macl. Ann. Jav. 1825. 28; Lac. 94; Bates TES. 1873. 304; Andr. MB. 52.

Heteroglossa Nietn. JASB. xxvi. 1857. 141; id. AM. (2). xx. 1857. 279.

B* **bimaculatus** Macl. Ann. Jav. 1825. 29. t. 1. f. 8; Schaum. SEZ. 1847 46; Chaud. RMZ. 1872. 139; Bates AG. xxxii. 1892. 388; Heyne-Tasch. EK. 14. t. 2. f. 29; Andr. TES. 1919. 169. — Assam (Bates). Karen Hills. Bhamo. Kuala Lumpur. Siam., Indo-China. Java., Sumatra.

J*S* **elegans** *(Heteroglossa)* Nietn. JASB. xxvi. 1857. 143; id. AM. (2). xx. 1857. 281; Andr. AM. (9). xix. 1927. 106. — Colombo.

J* **immaculatus** Schaum BEZ. 1863. 81; Chaud. RMZ. 1872. 140; Bates AF. 1889. 280; Vuillet Ins. ii. 1912. 17; Andr. AM. (9). xix. 1927. 108. — Dacca (Chaudoir). Malacca., Siam. Indo-China. Sumatra.

A* **indicus** Andr. AM. (9). x. 1922. 166. — Nagpur.

B* **muiri** Andr. AM. (9). xiii. 1924. 469 and 471. — Macao.

B* **puncticeps** Andr. AM. (9). iii. 1919. 480; id. TES. 1919. 169. — China. Japan.

bimaculatus ‡ Bates (not Macl.) TES. 1873. 304; Putz. CRB. 1875. 52; Heyd. DEZ. 1879. 329.

ruficeps × Schaum BEZ. 1863. 81; Chaud. RMZ. 1872. 139; Heyne-Tasch. EK. 14; Andr. MB. 53. — Dacca (Chaudoir). Dehra Dun. Surada., Madras,.

J*S* *bimaculatus (Heteroglossa)* || Nietn. (not Macl.) JASB. xxvi. 1857. 144; id. AM. (2). xx. 1857. 282; Andr. AM. (9). xix. 1927. 106. — Nilgiri Hills. Ceylon : Colombo, Hambegamuwa, Kotte. Tharrawaddy.

J*S* **ruficollis** *(Heteroglossa)* Nietn. JASB. xxvi. 1857. 144; id. AM. (2). xx. 1857. 282; Chaud. RMZ. 1872. 140; Vuillet Ins. ii. 1912. 17; Andr. AM. (9). — Dacca., Pusa. Colombo., Cochin-

xix. 1927. 106. — China., Cambodia.

Z* **secernendu** Oberth. NLM v. 1883. 217. — Sumatra., Borneo.

B* **simplex** Bates AM. (5). xvii. 1886. 199. — Peradeniya.

PLATIA* Chaudoir

Platia Chaud. Mém. sur les Thyréoptérides, AB. xii. 1869. 155.

O* **brachydera** Chaud. Mém. 157. — Batchian.

O* **laticeps** Chaud. Mém. 157. — Celebes. New Guinea.

O* **lineella** Chaud. Mém. 156. — Batchian., Kei Is.

papuana × Sloane PLS. NSW. 1917. 433; Andr. AM. (9). xix. 1927. 110. — New Guinea.

D* *lineella (Agonochila)* || Sloane DEZ. 1907. 182.

PLATYMETOPUS* Dejean.

Platymetopus Dej. Sp. iv. 1829. 68; Lac. 300; Bates AG. xxxii. 1892. 332; Tchitch. HER. xxxv. 1901. 232; Alluaud BF. 1917 (1918). 321 Andr. TES. 1919. 156; id. MB. 31.

C* **flavilabris** *(Carabus)* F. Suppl. Ent. Syst. 1798. 59; Hope Col. Man. ii. 1838. 90; Schaun. SEZ. 1847. 48; Motch. Etud. Ent. 1855. 43; Andr. TES. 1921. 160; id. RIM. xxii. 1921. 344; id. MB. 31. — India Ceylon.

S* *senilis (Ophonus)* Nietn. JASB. xxvi. 1857. 150; id. AM. (2). xx. 1857. 372; Bates AM. (5). xvii. 1886. 76; id. Entom. 1891. Suppl. 9; id. CRB. 1891. 329; Andr. AM. (9). xix. 1927. 107.

N* var. **thunbergi** *(Harpalus)* Quens in Schönh. Syn. Ins. i. 1. 1806. 188 (note); Dej. Sp. iv. 1829. 74; Bates TES. 1873. 271; Har DEZ. 1878. 66; Heyd. DEZ. 1879. 332; Bouch. AF. 1903. 171; Vuillet Ins. ii. 1912. 19; Heller DEZ. 1916. 276. Andr. AM. (9). xiii. 1924. 469. — Indo-China. Malay Peninsula and Archipelago. China. Japan.

B* *punctulatus (Harpalus)* Macl. Ann. Jav. 1825. 21; Andr. TES. 1919. 151.

J*	*obscuripes* Schauf. HER. xxi. 1887. 106; Andr. AM. (9). xix. 1927. 110.	Japan.
B*	var. **corrosus** Bates TES. 1873. 270; Har. Abh. nat. Ver. Bremen iv. 1875. 286; id. DEZ. 1878. 66; Heyd. DEZ. 1879. 332.	China. Formosa.
D*	*formosanus (Dioryche)* Dupuis AB. 1912. 324.	
X*	var. **punctulicollis** Bates AF. 1889. 269; id. AG. xxxii. 1892. 330.	Burma., Andaman Is. Malay Peninsula. Sumatra. Indo-China. Yunnan.
O*	**laticeps** Dej. Sp. iv. 1829. 76; Heller DEZ. 1916. 276.	Philippine Is.
A*	**pictus** Andr. Spol. Zeyl. 1923. 233.	Sitapur (U.P.). Belgaum. Coimbatore. Yemmiganur. Bangalore. Arabia.
O*	**quadrimaculatus** Dej. Sp. iv. 1829. 70; Dej. et Boisd. Ic. et HN. iv. 56. t. 175. f. 3; Bates AF. 1889. 271 Lesne Miss. Pav. 73.	Siam. Cochin-China. Annam., China.
S*	**rugosus** *(Ophonus)* Nietn. JASB. xxvi. 1857. 150; id. AM. (2). xx. 1857. 373; Bates CRB. xx. 1891.	Sikkim terai: Panighata. Almora. Fyzabad. Chota Nagpur : Konbir. Rambha (Ganjam). Barkuda I (Lake Chilka). Vizagapatam. Ennur. Nagalapuram (Chingleput). Madras., Mangalore. Nilgiri Hills. Mysore. Ceylon : Colombo, Niroddununal,

329; Andr. RIM. xxii. 1921. 343; Andr. RIM. xxii. 1921. 343; id. AM. xix. 1927. 106.

Varadua, Matale, Negombo, Nalanda, Kandy, Kotte.

X* **tritus** Bates AF. 1889. 270.

Victoria Point.
Tenasserim (Prague Mus.).
Cochin-China.
Hongkong.

PLATYTARUS* Fairmaire.

Platytarus Fairm. BF. 1850. 17; id. Fn. Ent. Fr. Col. i. 1851. 33; Shaum. Nat. Ins. Dentschl i. 1860. 294; Chaud., Genres aberrants du groupe des Cymindides, BM. 1875. ii. 11; Bedel Fn. med. 1906. 262.

O* **boysi** *(Cymindis)* Chaud. BM. 1850. i. 50; id. BM. 1875. ii. 11; Bates AG. xxxii. 1892. 416; Bedel Fn. med. 1906 262.

Simla., Pusa.
Nagpur., Belgaum.
Mysore: Nandidrug.
Ceylon :
Anuradhapura.
Kawkareik.
Indo-China.
Hongkong.

G **planulatus** *(Cymindoidea)* Bates AG. xxxii. 1892. 417; Bedel. Fn. med. 1906. 262.

Palon.

A* **porcatus** Andr. AM. (9). xii. 1923. 452.

Madras.
Horseley Konda.
Mysore : Nandidrug, Chikkaballapura.

PLOCHIONUS* Dejean.

Plochionus Latr. et Dej. HN. et Ic. 150; Dej. Sp. i. 1825. 250; Schm. Gocb. FCB. 42; Agassiz Nom. Zool. Index. Univ. 2nd. edn. 1848. 863; Lac. 135; Chaud. Mon. des Callidides, AR. xv. 1872. 168.

Plocionus (correct in 2nd edn. supra) Agassiz Nom. Zool. Index Univ. 1846. 299.

L **pallens** *(Carabus)* F. Syst. Ent. 1775. 244; Chaud. Mon. 168; Fvl RE. 1889. 100; Bedel Fn. Scine I. 1879. 114; Andr. TES. 1921. 153; Alluand AG. Lii. 1928. 66 and 86.

Cosmopolitan

O* *bonfilsi* Latr. et Dej. HN. et Ic. 153. t. 12. f. 5; Dej.

Sp. i. 1825. 251; Dej. et Boisd. Ic. et HN. i. 130. t. 16. f. 1; Brullé HN. iv. 1834. 224. t. 7. f. 6; Hope Col. Man. ii. 1838. t. 1. f. 6.

O *boisduvali* Gory AF. 1833. 189.

C' *valens* Lec. New sp. N. AM. Col. i. Smithson. Misc. Coll. 167. 1863. 5.

POGONOGLOSSUS* Chaudoir.

Pogonoglossus Chaud. BM. 1862. ii. 304; Bouch. AF. 1903. 178; Heller DEZ. 1916. 271.

Libresthis Schm. Goeb. in litt. FCB. (cover)

Carpaulum Sloane PLS. NSW. 1904. 536; id. DEZ. 1907. 184.

(seaports), but not yet (1928). recorded from India.
Java., China.

G* **carinipennis** Bates AG. xxxii. 1892. 388; Bouch. AF. 1903. 179. — Karen Hills.

G **chaudoiri** Gestro AG. 1875. 863; Bouch. AF. 1903. 179. — Cambodia.

D* **horni** Sloane DEZ. 1907. 184; Andr. AM. (9). xix. 1927. 110. — New Guinea.

O **intermedius** Bouch. AF. 1903. 180. — Sumatra.

O **montanus** Bouch. AF. 1903. 180. — Sumatra.

O **schaumi** Chaud. RMZ. 1869. 204; Bouch. AF. 1903. 178. — Mysol.

G* **sumatrensis** Gestro AG. 1875. 863; Bouch. AF. 1903. 179. — Sumatra.

O **sylvaticus** Bouch. AF. 1903. 180. — Sumatra.

Y* **tagalus** Heller DEZ. 1916. 271. t. 3. f. 2. — Philippine Is.

A* **truncatus** Andr. TES. 1921. 183; *(Libresthis)* Schm. Goeb. FCB. t. 2. f. 4; Sloane DEZ. 1907. 184; Andr. TES. 1923. 57. — Kumaon : River Sarda Gorge. Assam (Bates). Burma : Shwegu Meetan.

G* *validicornis* ‡ Bates (not Chaud.) AG. xxxii. 1892. 388; Andr. TES. 1921. 148. — Laos., Tonkin. Borneo (Bates).

M' **unicolor** *(Planetes)* W. Macl. PLS. NSW. 1886. 137; Sloane DEZ. 1907. 184.

O* **validicornis** Chaud. BM. 1862. ii. 304; Gestro AG. 1875. 862; Bouch. AF. 1903. 179; Andr. TES. 1921. 183.

POGONUS* Stephens.

Pogonus [Nicolai Diss. inaug. med. sist. Col. Sp. Agr. Halensis. 1822. 15] Steph. Ill. Brit. Ent. i. 1828. (May). 106; Dej. Sp. iii. 1828 (Nov.) 6; Lac. 368; Chaud., Essai monographique sure le groupe des Pogonides, AB. xiv. 1871. 23; Bedel AF. 1878. 251; Sem. HER. xxxiv. 1900. 578.

New Guinea.

B'* **biroi** Csiki Ann. Mus. Hung. 1907. 574; Andr. RIM. xxii. 1921. 341.

Java.
Sumatra.

Sunderbans.
Sorabkatti (Ganges delta). Balugaon.
Barkuda I (Lake Chilka).
Bombay.
North Kanara.
Madras. Malabar.
Ceylon:
Hambantota.
Fao (Persian Gulf).

E* **japonicus** Putz. CRB. 1875. 52; Bates TES. 1876. 5; id. TES. 1883. 266.

Japan.

O* **micans** Chaud. BM. 1842. 820; id. Mon. 27; Andr. RIM. xviii. 1919. 99; id. Ent. Mitt. 1927. 143.

Seistan : Labi-Baring.
Persia., Basra.
Trans-Caspia.

PRAKASHA* Andrewes.

Prakasha Andr. AM. (9). iii. 1919. 474.

G* **amariformis** *(? Platymetopus)* Bates AG. xxxii. 1892. 333; Andr. AM. (9). iii. 1919. 474.

Toungoo.
Kacokareik.

PRISTODACTYLA Dejean.

Pristodactyla Dej. Sp. iii. 1828. 82; Lac. 343; Bates TES. 1873. 273; id. TES. 1883. 254; id. AG. xxxii. 1892. 368.

O* **crenata** *(Calathus)* Putz. AB. 1873. 82; Bates AG. xxxii. 1892. 368.

O* **lacerans** Bates PZS. 1889. 214.

PRISTONYCHUS Dejean.

Pristonychus Dej. Sp. iii. 1828. 43; Lac. 341; Schauf., Monographische Bearbeitung der Sphodrini, Sitzungsb. GES. Isis 1864. 139; Andr. MB. 46.

Laemosthenes Schauf. Mon. 121.

Intisphodrus Schauf. Mon. 126.

Laemostenus Bedel AF. 1878. 250; id. Fn. Seine i. 1879. 104 Andr. TES. 1919. 91.

I* **alticola** Fairm. BF. 1889. 16. — Nainital., Ranikhet. Almora., Mussoorie. Karen Hills. Kashmir : Goorais Valley. Burma : Mt. Yeomatong

B* **brunneus** *(Sphoidrus)* Hope Zool. Misc. 1831. 21; Andr. TES. 1919. 175. — "Nepal."

O* **kashmirensis** Bates PZS. 1889. 214; Andr. AM. (9). iii. 1919. 475; id. TES. 1921. 148. — Kashmir : Goorais Valley, Lidar Valley, Killanmarg, Gulmarg, Sintan, Pir Panjal, Upper Munda. Simla., Kulu.

A* var. babaulti Andr. MB. 100. — Kashmir : Sonemarg, Nagaberan.

I* *cordicollis (Sphodrus)* ‡ Bates (not Chaud.) SYM. 11; Andr. AM. (9). xvii. 1926. 256. — Spiti : Tsho Ti, Tchary-Djoni, Pulga. Lahoul : Sisu, Keiling, Sumdeo, Gandla. Kulu : Kandi, Koty. Rotang Valley. Murree., Narkanda.

O **picescens** Fairm. AB. 1887. 95.

O* **spinifer** Schauf. Sitzungsb. Ges. Isis 1862. 66; id. Mon. 174; Andr. AM. (9). iii. 1919. 475; id. TES. 1921. 148.

PSEUDOZAENA* Castelnau.

Pseudozaena Cast. Etud. Ent. 1834. 55; Lac. 159; Andr. AM. (9). iv. 1919. 295; Bänn. DEZ. 1927. 188 and 192.

Hoplognathus || Chaud. (not Macl.) BM. 1848. i. 101; Lac. 159.

Picrus × Chaud. BM. 1854. i. 290; id. Révision du groupe des Ozénides, AB. xi. 1868. 45.

Acanthophthalmus Montr. Ann. Soc. Agr. Lyon (2). vii. 1855. 8.

J **orientalis** *(Ozaena)* Klug Jahrb. Ins. 1834. 81. t. 1. f. 8; Chaud. BM. 1848. i. 101; id., BM. 1854. i. 291; id. Rév. 45; Bouch. AF. 1903. 169; Andr. AM. (9). xix. 1927. 101; Bänn. DEZ. 1927. 185. f. 3. and 192.

L *megacephala* Cast. Etud. Ent. 1834. 54. f. 2. f. 4; Brullé HN. iv. 1834. 256.

O* *obscura (Picrus)* Chaud. Rév. 46; Andr. AM. (9). xiv. 1924. 587.

tricostata *(Acanthophthalmus)* Montr. Ann. Soc. Agr. Lyon (2). vii. 1855. 8; Fairm. AB. 1883 (ii). 3; Bänn. DEZ. 1927. 193.

O* *alternata* Bates PZS. 1877. 152. t. 24. f. 2.

O* var. **opaca** *(Picrus)* Chaud. Rév. 46; Dupuis. AB. 1913. 81; Andr. AM. (9). xiv. 1924. 587; Bänn. DEZ. 1927. 192.

A var. **tenebrosa** Sloane Rec. Austr. Musc. i. 1890. 102; Bänn. DEZ. 1927. 192.

PSILUS* Putzeys.

Psilus Putz. CRB. 1877. 46; Andr. FBI. 344 and 386.

L(E*) **doriae** *(Dyschirius)* Putz. AB. xvi. 1873. 14; Andr. AM. (9). xvii. 1926. 378; id. AM. (9). x. 1927. 266 and 267.

A* **fulvaster** Andr. AM. (9). xx. 1927. 266 and 267.

A* **malayicus** Andr. AM. (9). xx. 1927. 267.

G* **paradoxus** *(Ardistomis)* Putz. AB. xi. 1868. 21; Andr. TES. 1921. 176; id. AM. (9). xvii. 1926. 381; id. AM. (9). xx. 1927. 266; id. FBI. 387; f. 58.

S* *acutipalpis* Putz. CRB. 1877. 46.

X* *trapezicollis (Clivina)* Bates AF. 1889. 263; id. BF.

Yunnan.
"Himalayas"
Dharmasala.
Campbellpore.

Malacca., Java.
Sumatra.

Mentawei Is.
Borneo., Kei Is.

Bismark Archipelago.
Woodlark I.
Duke of York's I.
Solomon Is.

Formosa., Ceram.
Ternate.
Philippine Is.
Samar I.

New Guinea.

Sarawak.

Borneo.

Penang.

Burma.
Malay Peninsula.
Siam., Cambodia.
Cochin China.

1889. 240.

G'* *lignicola (clivina)* Bouch. AF. 1903. 169.

B* **porphyreus** Andr. Spol. Zeyl. 1923 227; id. AM. (9). xx. 1927. 267; id. FBI. 387 and 389.

PTEROSTICHUS Stephens.

P. *Pterostichus* Steph. Ill. Brit. Ent. i. 67 and 120; Dej. Sp. iii. 1828. 337; Andr. TES. 1991. 92.

Platysma Steph. Ill. Brit. Ent. i. 67 and 124; Dej. Sp. iii. 1828. 304; Tchitch. HER. xxv. 1902. 507; Andr. TES. 1919. 92.

S* *Steropus* Steph. Ill. Brit. Ent. i. 1828. (May). 116; Dej. Sp. iii. 1828. (Nov.). 292; Tchitch. HER. xxx. 1897. 331.

B* *Bothriopterus* Chaud. BM. 1838. i. 9; Tchitch. HER. xxxiv. 1900. 606.

E. *Euryperis* Motch. Käf. Russl. 1850. t. ix.

Feronia [Latr. in Cuvier's Règne Animal iii. 1817. 191]; Lac. 323; Tchitch. HER. xxxii. 1898. 533. Sumatra. Borneo.

Sb. *Stobeus* Fairm. AB. 1888. 8.

St. *Steropanus* Fairm. AB. 1888. 10; Tchitch HER. xxx. 1897. 278; id. HER xxxii. 1898. 67. Ceylon : Kotte.

O B **aeneocupreus** *(Platysma)* Fairm. AB. 1887. 95; Bates Entom. 1891. Suppl. 70; Tchitch. HER. xxxiv. 1900. 612. Yunnan. China.

Q *kanssuensis (Bothriopterus)* Tchitch. HER. xxiii. 1889; id. HER. xxviii. 1894. 379.

B*S **championi** Andr. EMM. 1926. 69 and 76; id. FBI. 12. f. 3b. Kumaon : Gori, Sunderdhunga, and Pinda-Valleys. Garhwal : "Painsur above Lohba."

O Sb. **collucens** *(Stobeus)* Fairm. AB. 1888. 8. Yunnan.

O E. **curtatus** *(Euryperis)* Fairm. AF. 1886. 312; Tchitch. HER. xxx. 1897. 300. Yunnan.

X **P. davidi** Tchitch. Ab. xxix. 1897. 71.

O **P. diversus** *(Omaseus)* Fairm. AF. 1886. 311; Tchitch. HER. xxx. 1897. 297. — Yunnan.

X* **P. dolens** Tchitch. HER. xxxiv. 1900. 469. — Yunnan.

O St. **forticornis** *(Steropanus)* Fairm. AB. 1888. 10; Tchitch. HER. xxx. 1897. 279. — Sikkim Yunnan.

B* **P. gagates** *(Platysma)* Hope Zool. Misc. 1831. 21; Andr TES. 1919. 174. — Nepal.

O P. **haesitatus** Fairm. AB. 1888. 11; Tchitch. HER. xxx. 1897. 305. — Yunnan.

X* **P. harmandi** Tchitch. HER. xxxiv. 1900. 467. — Sikkim : Darjeeling, Senchal Hills, Gopaldhara, Tonglu. Allahabad (Brit. Mus.).

B* **P. indicus** *(Omaseus)* Hope Zool. Misc. 1831. 21 (♂); Andr. TES. 1919. 173. — Nepal.

B* *aëratus (Omaseus)* Hope Zool. Misc. 1831. 21 (♀) ; Andr. TES. 1919. 173.

O. P. **licinoides** *(Steropus)* Fairm. AB. 1888. 10; Tchitch. HER. xxx. 1897. 294. — Yunnan.

Q P. **nowitzkii** Tchitch. HER. xxxii. 1899. 657. — Sikkim : Darjeeling

O S. **scuticollis** *(Steropus)* Fairm. AB. 1888. 11. — Gopaldhara, Tonglu.

O P. **simillimus** Fairm. AF. 1886. 312. — Yunnan.

A* **sintanus** Andr. Eos iii. 1927. 74. f. 8. — Yunnan.

O P. **stictopleurus** *(Omaseus)* Fairm. AB. 1888. 10; Techitch. HER. xxx. 1897. 299. — Kashmir. Yunnan.

O P. **yunnanus** *(Platysma)* Fairm. AB. 1887. 94; Bates Entom. 1891. Suppl. 70. — Yunnan. China.

RHOPALOPALPUS* Laferté.

Rhopalopalpus Laf. AF. 1851. 212. and 262; Lac. 224; Chaud. BM. 1857. ii. 11.

phopalistes Gemm. & Har. Mun. Cat. 230; Chaud., Mon. des Chleniens, AG. viii. 1876. 291.

Gorakhpur (U. P.). Haldwani.

V* **janthinus** *(Chalaenius)* Redt. HK. 500; Chaud. Mon. 292.

O* *poeciloides* Laf. AF. 1851. 262; Chaud. BM. 1857. ii. 11.

RISOPHILUS* Leach.

Risophilus Leach in Brewsters' Edinb. Eney. ix. (1). 1815. Ent. 81; Anr. TES. 1919. 91.

Demetrias Dej. Sp. i. 1825. 228; Latr. et Dej. HN. et. Ic. 169; Schm. Goeb. FCB. 34; Lac. 115.

Peliocypas Schm. Goeb. FCB. 33; Lac. 116; Andr. TES. 1923. 14.

Tetregonica Motch. Etud. Ent. 1859. 26; Bates AM. (5). xvii. 1886. 207; Andr. TES. 1923. 15.

X* **annamensis** *Demetrias)* Bates AF. 1889. 284; id. AG. xxxii. 1892. 421.

B* **catenatus** *(Tetragonica)* Bates AM. (5). xvii. 1886. 208.

X* **cordicollis** *(Demetrias)* Bates AF. 1889. 284.

B* **euproctoides** *(Tetragonica)* Bates AM. (5). xvii. 1886. 209.

M **fuscus** *(Tetragonica)* Motch. Etud. Ent. 1859. 28. t. 1. f. 2; Putz. CRB. 1875. 69; Bates AM. (5). xvii. 1886. 207; Andr. TES. 1928. 24.

P* **hamatus** *(Peliocypas)* Schm. Goeb. FCB. 35; Andr. TES. 1923. 13.

B* **himalayicus** Andr. AM. (9). xii. 1923. 688; id. EMM. 1926. 70.

Dehra Dun., Kheri. Ramnagar (U. P.). Siwaliks., Kulu. Gurdaspur. Rawalpindi. Kashmir.

Bengal., Bhamo. Pegu., Annam.

Ceylon : Bogawantalawa.

Annam.

Colombo., Kandy. Peradeniya.

Ceylon : Nuwara Eliya Dikoya, Bogawantalawa, Maskeliya.

Nilgiri Hills. Tenasserim.

Katrain.
Kumaon : Almora Sarju Valley, Berenag, Ranikhet, Bhatkot, Khoula, Swal River basin, Malwa Tal.

Ceylon : Horton Plains

B* **intermedius** *(Tetragonica)* Bates AM. (5). xvii. 1886. 208.
Nongpoh (Khasi Hills).

P* **luridus** *(Peliocypas)* Schm. Goeb. FCB. 35; Bates AG. xxxii. 1892. 421; Andr. TES. 1923. 13.
Sikkim : Gopaldhara, Kurseong.
Bhamo., Laos.

A* **macellus** Andr. AM. (9). xii. 1923. 454.
Tonkin., Formosa.

Palni Hills : Kodaikanal, Kukkal.
Trichinopoly.
Madura.

B* **marginicollis** *(Demetrias)* Bates. TES. 1883. 285.
Shembaganur.

B* **melleus** *(Tetragonica)* Bates AM. (5). xvii. 1886. 208.
Japan.

A* **psilus** Andr. AM. (9). xii. 1923. 455.
Colombo.

B* **repandens** *(Dromius)* Walk. AM. (3) iii. 1859. 51; Bates AM. (5) xvii. 1886. 210; Andr. TES. 1919. 191.
Kodaikanal.
Ceylon.

P* **signifer** *(Peliocypas)* Schm. Goeb. FCB. 35; Andr. TES. 1923. 12.
Dehra Dun.
Burma.
Cochin-China.

P* **suturalis** *(Perliocypas)* Schm. Goeb. FCB. 34; Andr. TES. 1923. 12.
Sumatra., Java.
Burma.

O* **uniformis** *(Perliocypas)* Fairm. AF. 1888. 334; Andr. TES. 1923. 15.
Tonkin.

SAROTHROCREPIS* Chaudoir.

Sarothrocrepis Chaud. BM 1850. i. 76; 129; Chaud. BM. 1876. ii. 80; Sloane PLS. NSW. 1898. 493; id. PLS. NSW. 1917. 422.

O* **m-nigrum** Jord. NZ. 106.
Tenimber.

SCALIDION* Schmidt-Goebel.

Scalidion Schm. Goeb. FCB. 63; Lac. 135; Andr. TES. 1923. 36.

P* **hilare** Shm. Goeb. FCB. 64; Andr. TES. 1923. 36.
Bhutan., Sonapur.
Khasi Hills.
Kobo., Harrawala.

O* **xanthophana** *(Lebia)* Bates PZS. 1888. 382; id. PZS. 1889. 218; Andr. TES. 1923. 37; id. AM. (9). xiv. 1924. 593.

O* var. **nigrans** *(Lebia)* Bates PZS. 1889. 219.

SCAPTERUS* Dejean.

Scapterus Dej. Sp. ii. 1826. 471; Lac. 197; Putz., Mon. des Clivina et genres voisins, Mém. Liége ii. 1846. 522; id., Postscr. ad. Cliv. Mon., Mém. Liége xviii. 1863. 16. t. 1. ff. 16-18; id., Rév. Gén. des Clivinides, AB. x. 1867. 7; Andr. FBI. 327 and 328.

C* **crenatus** *(Scarites)* F. Ent. Syst. i. (1). 1792. 95; Hope Col. Man. ii. 1838. 95 (No. 12); Motch. Etud. Ent. 1855. 40; Andr. TES. 1921. 155; id. FBI. 329 and 332; t. 2. f. 3. Dehra Dun., Burma. China.

O* **guerini** Dej. Sp. ii. 1826. 472; Dej. et. Boisd. et. HN. i. 204. t. 22. f. 3. Gray in Griffiths' Anim. King. xiv. 1832. 201. t. 8. f. 3; Guér Icon. Règne Anim. Ins. 1844. 21. t. 5. f. 3; Putz. Rév. 9; Bates AG. xxxii. 1892. 274; Andr. FBI. 329. and 330. China. "East Indies."

G* **riparius** Gestro AG. 1882. 299. fig.; Andr. FBI. 329 and 332. Dehra Dun. Chichawatni. Teinzo., Rangoon.

A* **stevensi** Andr. FBI. 329 and 33. t. 9. f. 4. Chittagong Hill Tracts. Minhla.

G* **sulcatus** Putz. Postscr. 17; Chaud. RMZ. 1863. 117; Putz. Rév. 9; Andr. FBI. 329 and 331; t. 9 f. 3.

G* *figuloides* Gestro AG. 1882. 301. fig.; Andr. AM. (10). ii. 1928. 586.

SCARITES* Fabricius.

Scarites F. Syst. Ent. 1775. 249; Bon., Obs. Ent. ii., Mém. Acad. Sci. Turin. xx. 1813. 463; Dej. Sp. i. 1825. 364; Schm. Goeb. FCB. 93; Putz., Mon. des Clivina et genres voisins, Mém. Liége ii. 1846. 521; Lac. 194; Chaud. BM. 1855; i. 51; Motch. Etud. Ent. 1857. 95; Chaud., Mon. des Scaritides (ii), AB. 1880. 63; Reitt. Best Tab. eur. Col. xxxix, Scaritini, 1899. 4; Tchitch. RRE. 1904. 257; Andr.

MB. 9; id. FBI. 210 and 225.

Parallelomorphus Motch. BM 1849. ii. 67; id. Etud. Ent. 1857. 96.

T. *Taeniolobus* Chaud. BM. 1855. i. 30; Motch. Etud. Ent. 1857. 95; Chaud. Mon. 20.

Pleurogenius Motch. Etud. Ent. 1857. 95.

Prosebus Gistl, Acht. Hundred und zwangi neue and unbeschriebene wirbellose Thiere, Vacuna ii. 1867. 562.

Scallophorites Motch. Etud. Ent. 1857. 95.

Harpalits Motch. Etud. Ent. 1857. 95.

D. *Distichus* Motch. Etud. Ent. 1857. 96; Chaud. Mon. 44; Reitt. Best. Tab. 8 ; Tchitch. RRE. 1904. 258 and 265; Andr. MB. 9.

Silonibari (North Lakhimpur).

Broscomorphus Motch. Etud. Ent. 1857. 96.

Pusa., Chapra. Sepaya., Dehra Dun.

Paramecomorphus Motch. Etud. Ent. 1857. 96.

Dinapore.

O* **acutidens** Chaud. BM. 1855. i. 89; id. Mon. 83; Vuillet Ins. ii. 1912. 17.

Cambodia (Vuillet). China., Japan.

Q **aterrimus** Moraw. Beitr. 21; Bates TES. 1873. 237; Chaud. Mon. 120; Tchitch. RRE. 1904. 265.

China., Japan.

O* **barbarus** Dej. Sp. i. 1825. 388; Chaud. Mon. 96; Bates AG. xxxii. 1892. 272; Bedel Fn. med. 1895. 38 (note 2); Andr. TES. 1923. 55; id. FBI. 230 and 252.

Dacca., Calcutta. Dinapore., Pusa. Balasore. Tharrawaddy. Rangoon., Moulmein.

G* **batesi** × Andr. FBI. 231, 236. f. 38 (3), and 258.

Burma : Teinzo.

denticulatus ‡ Bates (not Chaud.) AG. xxxii. 1892. 273.

Rajabhatkhawa, Buxa

B*T. **beesoni** Andr. FBI. 234 and 290.

Dinapore., Pusa. Chapra.

O* **bengalensis** Dej. S. ii. 1826. 468; Chaud. BM. 1855. i. 79; id. Mon. 89; Bates Entom. 1891. Suppl. 8; Tchitch. RRE. 1904. 263; Andr. FBI. 229 and 247.

Kashmir : Goorais Valley (Bates). Yang and tze-Kiang (Bates). Persia (Tchitcherin).

Manipur.

G*T. **birmanicus** *(Taeniolobus)* Bates AG. xxxii. 192. 270; Andr. FBI. 234 and 291.	Burma : Teinzo.
A*D **borneensis** *(Distichus)* Bänn. Ent. Blätt. 1928. 65.	Borneo.
O* **boucardi** Chaud. Mon. 98; Andr. AM. (9). iii. 1919. 469; id. FBI. 232, 236. f. 38. (8 and 9), and 264.	Assam : Dejoo. Manipur., Moulmein. Tavoy., Tenasserim.
O* **ceylonicus** Chaud. Mon. 85; Bates AM. (5). xvii. 1886. 72; Andr. FBI. 228 and 241.	Nagpur. Ceylon : Galle Point, Colombo, Hambantota, Kaduwela, Hanwella, Ranna, Kandy, Kalawewa. Tharrawaddy. Pyonchaung (Toungoo). Tenasserim. Penang., Laos. Annam.
O* var. **estriatus** Fairm. AB. 1887. 93; Andr. AM. (9). xii. 1923. 212; id. FBI. 242.	China. "Laos or Tonkin." Annam.
A* **corbetti** Andr. FBI. 230 and 252.	Tharrawaddy.
B* **cormoides** Andr. FBI. 231, 236. f. 38. (5), and 259.	Chapra.
A* **crassus** Andr. FBI. 233, 237 f. 39. (18), and 278.	Belgaum.
O* **cycloderus** Chaud. Mon. 112; Andr. FBI. 233, 237 f. 39 (17), and 275.	Sikkim : Pedong.
O* **cylindronotus** Fald. BM. 1836. 363 t. 6. f. 4; Motch. Käf. Russ. 1850. 19; Chaud. BM. 1855. i. 69; Motch. Etud. Ent. 1857. 96; Solskky FT. 54. t. 1. f. 5; Chaud. Mon. 71; Reitt. Best. Tab. 5; Tchitch RRE. 1904. 263; Andr. FBI. 228 and 238.	"Seistan Commission." Kirghisia. Transcaspia. Turcomania. Turkestan.
A* **dentatus** × Andr. FBI. 229 and 246. *praedator* ‡ Bates (not Chaud) AG. xxxii. 1892. 273 (Teinzo); Andr. MB. 1924. 10 (Kalka).	Tezpur., Sahibganj. Sukna., Kalka. Multan., Teinzo. Thaungyin Valley. Yenangyaung. Laos.

A* **derogatus** Andr.—	Cambodia. Cochin-China.
	Tezpur. Brahmaputra River
O* **denticulatus** Chaud. Mon. 98.	above Jorhat. Kaliganj., Pusa.
A* **derogatus** Andr. FBI. 231, 237. f. 39 (15), and 254.	Chapra., Delhi. Nainital., Bangalore.
O* D **dicaelus** *(Distichus)* Chaud. Mon. 52.	Pondicherry.
B* **dubiosus** Andr. FBI. 230, 236. f. 38. (11), and 261.	Bhamo., Penang. Singapore.
	Tezpur., Sibsagar. Calcutta.
O* **dyschromus** Chaud. BM. 1855. i. 78; id., Mon. 82; Andr. FBI. 228 and 243.	Dehra Dun. Delhi., Pegu.
eurytus Fisch. Ent. Russ. iii.1828. 119. t. 5. f. 3; Motch Käf. Russ. 1850. 19; Chaud. BM. 1855. i. 74; Motch. Etud. Ent. 1857. 96; Solsky FT. 53; Chaud. Mon. 70; Bedel Fn. med. 1895. 41; Reitt. Best. Tab. 4; Tchitch. RRE. 1904. 262; Andr. Ent. Mitt. 1927. 143; id. FBI. 228, 235, and 237. f. 39 (22).	"East Indies." "Seistan Commission." Baluchistan : Nushki. Afghanistan., Persia. Turkestan., Irak.
Q *bucerdo* Mén. Cat. rais. 1832. 102.	Arabia., Syria. Asia Minor.
J* *exasperatus* Klug in Symb. Phys. Decas iii. 1832. t. 23. f. 4; Andr. AM. (9). xixi. 1927. 100.	Caucasus.
	Transcaspia.
O* *laetus* Fald. BM. 1836. 361. t. 6. f. 5.	Transcaucasia. Kirghisia.
O* *oblongus* Chaud. BM. 1843. ii. 725.	
K' *chaudoiri* Ball. BM. 1870. ii. 326.	Egypt., Sudan.
B* **fletcheri** Andr. FBI. 231, 236. f. 38. (10), and 257.	
B* **granulatus** Andr. FBI. 231 and 256.	Brahmaputra River above Jorhat.
	Pusa.
O **guineensis** Dej. Sp. v. 1831. 484; Chaud. BM. 1855. i. 59; id. Mon. 107; Andr. FBI. 233, 236. f. 38. (4), and 274.	Gaolundo to Gauhati (Brahmaputra River) Sibsagar (Ind. Mus.).
J *asphaltinus* Klug. in Symb. Phys. Decas iii. 1832. t. 24. f. 1.	Delhi., Karachi. Poona (Oxford Mus.).

O *nitidus* Chaud. BM. 1843. ii. 722. — Arabia.

O* *boysi* Chaud. BM. 1855. i. 57; id. Mon. 107; Andr. AM (10). ii. 1928. 586. — Abyssinia. Sudan.

O *rocheti* Chaud. BM. 1855. i. 58. — Senegal.

O* **inconspicuus** Chaud. BM. 1855. i. 82; id. Mon. 97; Bates SYM. 3; Andr. MB. 10; id. FBI. 230 and 253. — Nepal., Sibsagar. Margherita. Srimangal., Sikkim. Calcutta., Tinpahar. Muzaffarpur. Pusa., Chapra. Kierpur (Purnea). Ranikhet., Rudpur. Tanakpur. Dehra Dun. Rawalpindi. Kohat. Jhelum Valley. Mhow. C. P. : Nagpur, Motinala, Mandla, Majhgaon, Rajadhar. Yercaud., Surada.

?X* **indus** Oliv. Ent. 36. 9. t. 1. f. 2; Bon., Obs. Ent. ii, Mém. Acad. Sci. Turin. xx. 1813. 469; Dej. Sp. i. 1825. 395; Chaud. BM. 1855. i. 84; id. Mon. 102; Redt. RN. 8; Bates AM. (5). xvii. 1886. 72; id. Entom. 1891. Suppl. 8; id. CRB. 1891. 325; Heyne-Tasch. EK. 20. t. 3. f. 33; Fletch Agr. Res. Inst. Pusa Bull. 9. 1919. 33. f. 25 (biology); Andr. RIM. xxii. 1921. 340; id. AG. xlix. 1922. 391; id. TES. 1923. (1924). 462; Id. MB. 10; id. FBI. 29. f. 8b, 232, and 269. f. 42. — Throughout India and Ceylon.

T* *mancus* Bon., Obs. Ent. ii., Mém. Acad. Sci. Turin xx. 1813. 473; Dej. Sp. i. 1825. 394; Chaud. BM. 1855. i. 83; id. Mon. 102; Andr. AG. xlix. 1922. 392.

I* *terricota* ‡ Andr. (not Bon.) RIM. xxii. 1921. 339.

A*D **laticeps** × Andr. FBI. 234 and 283.

parus ‡ Andr. (not Wied.) MB. 1924. 9. — C. P. : Motinala,

O* **liopterus** Chaud. Mon. 87; Andr. FBI. 231, 236. f.

38 (2), and 262. f. 40.	Majhgaon, Mandla.
O* **longiusculus** Chaud. Mon. 86; Heller PJS. xxiii. 1923. 298.	Madras., Tenmalai. "North India".
B*D. **macleayi** *(Distichus)* × Andr. TES. 1919. 162.	
indus ‡ Macl. (not Oliv). Ann. Jav. 1825. 25.	Philippine Is.
A*D. **mahratta** Andr. FBI. 233 and 281.	Java.
	Nagpur., Berar.
A* **mandarinus** Bänn. Ent. Blätt. 1928. 66.	Mandla., Poona. Belgaum.
A* **mandibularis** Andr. FBI. 231, 237. f. 39. (16), and 260.	North Kanara. Tonkin., China.
	Brahmaputra River
O* T. **modestus** *(Distichus)* Chaud. Mon. 57; Andr. FBI. 234 and 289.	above Jorhat. Tezpur., Calcutta. Tharrawaddy.
I* **nitens** Andr. FBI. 233, 237. f. 39. (14), and 276. f. 44.	East Indies.
	"Baluchistan-
O* **orthomus** Chaud. BM. 1855. i. 55; id. Mon. 88; Andr. FBI. 232, 237, f. 39 (23) and 266. f. 41.	Afghanistan Boundary Commission."
O* **parallelus** Dej. Sp. i. 1825. 382; Chaud. BM. 1855. i. 108; id. Mon. 86; Lesne Miss. Pav. 63.	Bhutan., Assam. Bengal., Pusa.
C*D. **parvus** Wied. Zool. Mag. ii. 1. 1823. 37; Chaud. Mon. 103; Andr. TES. 1921. 167 and 175; id. MB. 9; id. FBI. 234 and 283.	Siam., Cambodia. Cochin-China. Java.
O* *lucidulus (Distichus)* Chaud. Mon. 57; Bates AG. xxxii. 1892. 271; Lesne Miss. Pav. 63; *(Scaritodes)* Vuillet Ins. ii. 1912. 17.	Dacca., Sahibganj. Sunderbans. Chandipur (Orissa).
O* D. **picicornis** Dej. Sp. v. 1831. 493; Chaud. Mon. 55; *(Taeniolobus)* Péring. Trans. S. Afr. Phil. Soc. vii. 1896. 398; Andr. FBI. 234 and 282.	Tharrawaddy. Palon., Rangoon. Siam., Cambodia.
	Saidpur (Bengal). Dacca. Kierpur (Purnea). Kheri.

J* *troglodytes* Erichs. in Klug Doubl. Verz. senegall. Ins. Berlin 1842. 4; id. Arch. Naturg. ix. 1843. 214; Andr. AM. (9). xix. 1927. 102.

O var. **minor** Nietn. JASB. xxv. 1856. 389; id. AM. (2). xix. 1857. 244; Chaud. Mon. 56; Bates AM. (5). xvii. 1886. 72; Andr. AM. (9). xix. 1927. 107.

Ceylon : Matnaduwa, Weligama., Abyssinia., Sennaar. River Niger. Senegal., Angola. Zanzibar. Mozambique., Natal.

B* **pinguis** Andr. FBI. 229. 237. f. 39 (20), and 245.

Colombo.

L D. **planus** Bon., Obs. Ent. ii., Mém. Acad. Sci. Turin. xx. 1813. 470; Dej. Sp. i. 1825. 395; Dej. et Boisd. Ic. et HN. i. 198. t. 21. f. 3; Mèn. Cat. rais. 1832. 103; Chaud. BM. 1855. i. 46; id. Mon. 53; Reitt. Best. Tab. 8; Tchitch. RRE. 1904. 265; Andr. AM. (9). iii. 1919. 470; id. RIM. xviii. 1919. 99; id. TES. 1921. 146; id. AG. xlix 1922. 391; id. FBI. 233 and 279.

Murshidabad. Rajabhatkhawa (Buxa). Pusa., Chapra. Chandipur. Chakrata.

Seistan : Labi Baring and "Seistan Commission". Persia. Transcaucasia, Irak., Arabia., Syria. Mediterranean Basin., Asia Minor. Egypt., Sudan. Abyssinia. Mozambique.

O* *nitidus* Dej. Sp. v. 1831. 484; Chaud. Mon. 55.

V *punctatostriatus* Redt. in Russegg. Reise Eur. As. Afr. 1843. 979.

J* *bisquadripunctatus* Klug in Peters' Reise Mossamb. Zool. v. 1862. 158; Chaud. Mon. 55; Reitt. Best. Tab. 8; Andr. AM. (9). xix. 1927. 101.

Tezpur., Kaliganj. Sikkim., Dacca. Sahibganj. Chandipur (Orissa). Pusa., Chapra. Changa Manga. Rawalpindi. Spiti : Ghari.

O* **praedator** Chaud. Mon. 97; Bates AG. xxxii. 1892. 273 (except Teinzo); Andr. MB. 10 (except Kalka); id. FBI. 231 and 256.

O* D. **puncticollis** *(Taeniolobus)* Chaud. BM. 1855. i. 47; id. Mon. 55; Bates CRB. 1891. 325; Andr. TES. 1921. 146 and 168; id. MB. 9; id. FBI. 234 and 287.

C* **punctum** Wied. Zool. Mag. ii. 1. 1823. 38; Chaud. BM. 1855. i. 109; id. Mon. 127; Andr. TES. 1919. 162; id. TES. 1921. 168; id. FBI. 232, 236. f. 38 (7). and 272. f. 43.

O* *opacus* Chaud. BM. 1855. i. 88; id. Mon. 103; Bates Entom. 1891. Suppl. 8; Andr. TES. 1921. 168.

G*D **rectifrons** *(Distichus)* Bates AG. xxxii. 1892. 272; Andr. FBI. 234 and 286.

B* **retusus** Andr. FBI. 231 and 255.

O* **salinus** Dej. Sp. i. 1825. 385; Fisch. Ent. Russ. iii. 1828. 120; Dej. et Boisd. Ic. et HN. i. 197. t. 21. f. 2; Mén. Cat. rais. 1832. 102; Motch. Käf. Russ. 1850. 19; Chaud. BM. 1855. i. 72; Solsky FT. 55; Chaud. Mn. 78; Reitt. Best. Tab. 5; Tchitch. RRE. 1904. 262; Andr. FBI. 228 and 239.

O *subuleti* Fisch. Ent. Russ. iii. 1828. 121. t. 5. f. 2.

? *fischeri* Zubk. BM. 1833. 315; Solsky FT. 54.

O* *crenulatus* Chaud. BM. 1842. 811.

B* **scaevus** Andr. FBI. 231. 236. f. 38. (1), and 261.

P* **selene** Schm. Goeb. FCB. 94; Chaud. BM. 1855. i. 108; Andr. TES. 1923. 55; id. FBI. 230, 237. f. 39.

Kashmir : Baltal, Harwan (Pohru Valley).
North Kanara.
Kodaikanal.
Throughout Burma.
Tonkin.

Throughout India and Ceylon.

Nepal., Assam.
Patkai Hills.
Malda., Chapra.
Pusa.
"Hurdwar and Roorkee."
Almora., Tanakpur
Boxar (Nainital).
Dehra Dun.
Mothronwala.
Akalga H., Kulu.

Darjeeling, Kurseong.
Bhamo., Katha., Laos.
Tonkin. Yunnan.

Assam : Dilkusha.

Baluchistan: Nushki.
Persia., Turkestan.
Caucasus.
Transcaucasia.
Syria., Caspian basin.
Transcaspia.

Pusa.

Assam., Dacca.
Calcutta.
Khargpur.

Pusa., Lucknow.

(13). and 250.

O* *capito* Chaud. BM. 1855. i. 92; id. Mon. 95; Bates AG. xxxii. 1892. 273.

B* **semicircularis** Macl. Ann. Jav. 1825. 24; Chaud. BM. 1855. i. 109; id. Mon. 127; Andr. TES. 1919. 162; id. FBI. 229 and 249.

O* *subproductus* Chaud. Mon. 90; Lesne Miss. Pav. 63.

X* *mancus* ‡ Bates (not Bon.) AF. 1889. 261; Andr. TES. 1921. 175.

O* **semirugosus** Chaud. BM. 1855. 1. 90; id. Mon. 82; Bates AG. xxxii. 1892. 272; Lesne Miss. Pav. 63; Vuillet. Ins. ii. 1912. 17; Andr. FBI. 228, 237. f. 39. (12), and 240.

O* *rugipennis* Chaud. BM. 1855. i. 81.

B* **stenodes** Andr. FBI. 229, 237. f. 39. (21), and 248.

O* D. **striaticeps** *(Distichus)* Chaud. Mon. 52; Andr. FBI. 234 and 284.

O* **subcylindricus** Chaud. BM. 1843. ii. 730; id. BM. 1855. i. 85; de la Brûl. AF. 1875. 127; Chaud. Mon 101; Tchitch RRE. 1904. 264; Andr. Ent. Mitt. 1927. 143; id. FBI. 232 and 269.

O *longipennis* Chaud. BM. 1855. i. 84.

Patna., Madras.
Palni Hills.
Ceylon :
Anuradhapura.
Toungoo.
Tharrawaddy.
Rangoon., Palon.

? Rangoon.
Kedan., Singapore.
Siam.

Indo-China., Java.

Khasi Hills.
Bengal.
Nagpur (*ex* stomach of black ibis).
Nagbhir (Chanda).
Madras., Toungoo.
Kedah., Malacca.
Siam., Indo-China.
Philippine Is.

Multan.

Sahibganj.
Brahmaputra River ("Goalundo to Gauhati.")
Rangoon.

Baluchistan.
Afghanistan Boundary Commission.
Egypt., Syria., Irak.

Arabia.

Bengal., Pusa.
Jumnagnar and Dhikala (Garhwal).
Bhogpur (Bijnor).

semicylindricus Reitt. Best. Tab. 7.

O* **subnitens** Chaud. BM. 1855. i. 87; id. Mon. 103; Andr. FBI. 232 and 271.

L **sulcatus** Oliv. Ent. 36. 7. t. 1. f. 11; Dej. Sp. i. 1825. 375; Chaud. BM. 1855. i. 107; id. Mon. 80; Bates PZS. 1888. 369; id. Entom. 1891. Suppl. 8; id. AG. xxxii. 1892. 272; Tchitch. RRE. 1904. 263; Andr. MB. 10; id. FBI. 228, 237. f. 39 (19), and 242.

B* *geryon* Hope Zool. Misc. 1831. 21; Andr. TES. 1919. 170.

J* *chinensis* Erichs. in Nov. Act. Leop. Carol. Nat. xvi. Suppl. i. 1834. 220; Chaud. BM. 1855. i. 107; Andr. AM. (9). xix. 1927. 101.

B* **tauropus** Andr. FBI. 229, 236. f. 38. (6), and 244.

T* **terricola** Bon., Obs. Ent. ii., Mém. Acad. Sci. Turin xx. 1813. 471; Dej. Sp. i. 1825. 398; Dej. et Boisd. Ic. et HN. i 200. t. 21. f. 5; id. Fn. Gall. Rhén. ii. 1882. 125 (larva); Bedel Fn. med. 1895. 42; Tchitch. RRE. 1904. 264; Andr. AG. xlix. 1922. 391; id. FBI. 232 and 267.

T* *arenarius* Bon., Obs. Ent. ii., Mém. Acad. Sci. Turin. xx. 1813. 472; Dej. Sp i. 1825. 396; Dej. et Boisd. Ic. et HN. i. 199. t. 21. f. 4; Brullé HN. v. 1835. 57; Muls. et Mayet in Mém. Acad. Lyon xix. 1872. 325 (larva); Fvl. Annuaire Ent. ii. 1874. 109 (larva); Chaud. Mon. 100; Bates SYM. 4; Reitt. Best Tab. 7; Andr. AG. xlix. 1922. 392.

volgensis Fisch. Ent. Russ. ii. 1824. 31. 5. 26. f. 3.

laevigatus ‡ Fisch. (not F.) Ent. Russ. iii. 1828. 122.

Kaladnugi, (Nainital).
Siwaliks.
Bodiar (Haldwani).
Kulhal (Dehra Dun).
Kathian (Chakrata).
Lalitpur.
Tharrawaddy.

Sikkim : Pedong.
Tinne (Mishmis).
Nepal : Soondrijal.
Assam., Khasi Hills.
Sitapur (U. P.).
Kumaon : Pyura, Sladevi, Almora.
Garhwal : Majkhali.
Kangra.
Kulu.
Spiti : Manikarn.
Murree.
Punjab : Baddia.
Bhamo., Tavoy.
Java., Sumatra.
Indo-China., China.
Macao., Korea.
Manchuria.

Bengal : Sara Ghat (Ganges).

Thoughout the Southern half of the Palaearctic region, including the Mediterranean basin.
? Pondicherry (Lesne).
? Combodia (Lesne).
Abyssinia (Bedel).
"Seistan

O* *persicus* Chaud. BM. 1842. 813; Solsky FT. 55.

X *compressus* Coq. AF. 1858. 758; de la Brûl. AF. 1875. 127 (note 13).

B* *pacificus* Bates TES. 1873. 238; Putz. CRB. 1875. 49; Heyd. DEZ. 1879. 329; Chaud. Mon. 101; Lesne Miss. Pav. 63; Dupuis AB. 1913. 81.

B*D. **uncinatus** *(Distichus)* Andr. Spol. Zeyl. 1923. 225; id. FBI. 234 and 288.

Commission." "Baluchistan-Afghanistan Boundary Commission." Baluchistan : Nushki.

Ceylon : Hambegamuwa.

SELINA* Motchulsky.

Selina Motch. Etud. Ent. 1857. 110; Schaum BEZ. 1860. 172. t. 3. f. 11; Bates EMM. viii. 1871. 31; Alluaud BF. 1918. 222; Andr. MB. 51.

Pselaphanax Walk. AM. (3). iii. 1859. 52.

Steleodera Schaum BEZ. 1863. 74; Chaud. BM. 1872. i. 396.

Sphinctodera Fairm. RE. 1901. 128.

M westermanni Motch. Etud. Ent. 1857. 110. t. f. 6; Schaum BEZ. 1863. 74; Reitt. WEZ. ii 1883. 96; Bates AM. (5). xvii. 1886. 1999; Fowl. Fauna Br. India. Col. Gen. Intr. etc., 1912. 58. f. 26; Andr. AM. (9). xiii. 1924. 466 and 469; id. MB. 51; id. TES. 1928. 23.

B* *setosa (Pselaphanax)* Walk. AM. (3). iii. 1859. 52; Waterh. Aid xv. 1882. t. 120. fig.; Oberth. NLM. v. 1883. 223; Andr. TES. 1919. 194.

Z* *ritsemae* Oberth. NLM. 1883. 223; Andr. AM. (9). xiv. 1924. 592.

I' *natalensis (Ega)* Péring. Trans. S. Afr. Phil. Soc. vii. 1896. t. 10. f. 10; *(Pselaphanax)* id. Trans. S. Afr. Phil. Soc. x. 1898. 317; Alluaud. BF. 1918. 223. fig.

O *perrieri (Sphinctodera)* Fairm. RE. 1901. 128.

Dacca., Calcutta. Madnapur. Sardah., Dehra Dun. Haldwani., Siwaliks. Nagpur., Seoni.

Mandla. North Kanara. Madras., Coimbatore. Tranquebar. Malabar. Chikkaballapura. (Mysore). Ceylon : Kandy, Niroddumunai, Kotte., Sumatra. Indo-China. Hongkong., Macao. Nairobi., Salisbury (Rhodesia)., Natal.

SERRIMARGO* Chaudoir.

Serrimargo Chaud. Mém. sur les Thyréoptérides, AB. xii. 1869. 134. — Madagascar.

grouvellei Bouch. BF. 1901. 90.

J* **guttiger** *(Thyreopterus)* Schaum BEZ. 1860. 189. t. 3. f. 5; Chaud. Mém. 135; Andr. AM. (9). xix. 1927. 108. — Sumatra.

O* **verrucifer** *(Thyreopterus)* Chaud. RMZ. 1869. 171; id. Mém. 135. — Perak., Kedah. Singapore., Borneo. Sumatra.

SFITAKANTHA* Andrewes. — Malay Peninsula.

Sfitakantha Andr. AM. (9). iii. 1919. 482.

P* **inpressa** *(Thyreopterus)* Schm. Goeb. FCB. 80; Chaud. AB. xii. 1869. 142; Bates AG. xxxii. 1892. 406; Andr. AM. (9). iii. 1919. 483; id. TES. 1923. 46. — Sadiya., Karen Hills. Penang., Perak. Laos., Java., Borneo. Sumbawa.

SIAGONA* Latreille.

Siagona Latr. Nouv. Dict. Hist. Nat. xxiv. 1804. Tab. Méth. 141; id. Gen. Crust. et Ins. i. 1806. 208; Bon., Obs. Ent. ii, Mém. Acad. Sci. Truin. xx. 1813. 457; Dej. Sp. i. 1825. 357; Lac. 162; Chaud. BM. 1850. i. 439; Jacq. du Val Gen. Col. Eur. i. 1857. 46; Chaud., Mon. des Siagonides, BM. 1876. i. 76; Péring. Trans. S. Afr. Phil. Soc. vii. 1896. 414; Bedel et Francois BF. 1897. 38; Bedel Fn. med. 1897. 106 (note 2); Gahan TES. 1900. 441; Horn DEZ. 1907. 428; Andr. MB. 13; Bänn. Ent. Blätt. 1928. 55; Andr. FBI. 174.

G* **angulifrons** Bates AG. xxxii. 1892. 284; Bänn. Ent. Blätt. 1928. 56. t. f. 13; Andr. FBI. 19. f. 5c. and d, 178, 182. f. 32 (3), and 184. — Mandalay. Tharrawaddy. Prome., Pyinmana. Palon., Rangoon. Tenasserim.

G* **angustipennis** Bates AG. xxxii. 1892. 286; Bänn. Ent. Blätt. 1928. 56; Andr. FBI. 180, and 200. t. 6. f. 3. — Palon and Tikekee (Pegu).

A* **apcalis** × Andr. TES. 1921. 181; Bänn. Ent. Blätt. 1928. 56; Andr. FBI. 180 and 195.

G* *cinctella* ‡ Bates (not Chaud.) AG. xxxii. 1892. 285; — Palon., Yenangyaung. Pyonchaung (Toungoo).

Andr. TES. 1921. 146.

O* **atrata** Dej. Sp. i. 1825. 360; Chaud. Mon. 85; Andr. TES. 1921. 181; Bänn Ent. Blätt. 1928. 56. t. f. 3; Andr. FBI. 179, 182. f. 32 (10), and 191.

Dacca., Sahibganj.
Narail., Pusa., Puri.
Balighai., Sylhet.
Cachar.
Burma (Chaudoir).

O* **baconi** Chaud. Mon. 89; Bates AG. xxxii. 1892. 284 and 285; Bänn. Ent. Blätt. 1928. 56; Andr. FBI. 179, 182 f. 32 (9), and 189.

Malda., Berhampur.
Narail., Sahibganj.
Darjiling., Pusa.
Katihar.

O* **cinctella** Chaud. Mon. 95; Andr. TES. 1921. 182; id. FBI. 180 and 197.

Kierpur (Purnea).
Chapra., Mhow.

X* **crassidens** Bates AF. 1889. 263; Bänn. Ent. Blätt. 1928. 56.

Rangoon.

A* **cyathodera** × Andr. AM. (9). vii. 1921. 409; Bänn. Ent. Blätt. 1928. 56; Andr. FBI. 178, 182. f. 32 (2), and 183 t. 5. f. 5.

Cochin-China.
Cambodia.

Bhamo.

C* *punctulata* ‡ Bates (not Chaud.) AG. xxxii. 1892. 285.

K* **depressa** *(Carabus)* F. Suppl. Ent. Syst. 1798. 56; *(Galerita)* id. Syst. Eleuth. i. 1901. 215; Bon., Obs. Ent. ii, Mém. Acad. Sci. Turin. xx. 1813. 458; Hope Col. Man. ii.1838. 99 (No. 5); Motch. Etud. Ent. 1855. 42: Andr. TES. 1921. 157; Bänn. Ent. Blätt. 1928. 56; t. f. 12; Andr. FBI. 179, 182. f. 32 (7), and 187, f. 33.

Nagpur.
Pondicherry.
Trichinopoly.
Tranquebar.
Madura.
Ceylon : Vavuniya.

J* *tomentosa (Scarites)* Herbst Nat. Syst. Ins. Käf. x. 1806. 260. t. 175. f. 12; Andr. AM. (9). xix. 1927. 98.

O* *germana* Chaud. Mon. 96.

O* var. **punctatissima** Chaud. Mon. 106; Andr. TES. 1921. 158; id. FBI. 181 and 189.

Trichinopoly.
Pondicherry.

O* **dilutipes** Chaud. BM. 1850; i. 440; id. Mon. 106; Bänn. Ent. Blätt. 1928. 56; Andr. FBI. 181, 182. f. 32 (8), and 202.

Simla., Mhow.
Belgaum.

A* var. **aligera** Andr. FBI. 179 and 203.

Belgaum.

A* **europea** Dej. Sp. ii. 1826. 468; Dej. et Boisd. Ic. et. HN. i. 191. t. 20. f. 2; Rambur Fn. Ent. Andalousic 1837. 41; Küst. Käf. Eur. xv. 1848. 7; Chaud. Mon. 91 (part); Andr. AM. (9). iii. 1919. 471; id.

India (Chaudoir).

RIM. xviii. 1919. 100; id. TES. 1921. 158; id. Ent. Mitt. 1927. 143; Bänn. Ent. Blätt. 1928. 56 and 59. t. f. 8; Andr. FBI. 180, 182. f. 32 (12), and 194.

depressa ‡ Bedel (not F.) AF. 1887. 195. (part); id. Fn. med. 1897. 108 (part).

Seistan : Labi Baring. Persia., Fao., Basra. Arabia. Caspian and Mediterranean basins. Abyssinia. Senegal (Chaudoir).

A***fabricii** × Andr. TES. 1921. 158; id. MB. 14; Bänn. Ent. Blätt. 1928. 56; Andr. FBI. 178, 182. f. 32 (4), and 185.

O* *europea* ‡ Chaud. (not Dej.) Mon. 91 (part).

X *depressa* ‡ Bedel (not F.) AF. 1887. 195 (part); Bates CRB. 1891. 326; Bedel Fn. med. 1897. 108 (part); Andr. AM. (9). iii. 1919. 470.

Throughout India and Ceylon.

C* **fiesus** *(Galerita)* F. Syst. Eleuth. i. 1801. 216. Dej. Sp. i. 1825. 363; Hope Col. Man. ii. 1838. 99 (No 7); Motch. Etud. Ent. 1855. 42; Chaud. Mon. 94; Bates AF. 1889. 264; id. AG. xxxii. 1892. 285; Vuillet Ins. ii. 1912. 17; Andr. TES. 1921. 164; Bänn. Ent. Blätt. 1928. 56. t. f. 11; Andr. FBI. 180, 182. f. 32. (13), and 196. f. 34.

O *dorsalis* Dej. Sp. v. 1831. 477; Chaud. Mon. 94; Bänn. Ent. Blätt. 1928. 59.

Throughout India and Burma. Malay Peninsula. Siam. Indo-China. Senegal. Tanganyika.

O* **induta** Chaud. Mon. 98; Bänn. Ent. Blätt. 1928. 56; Andr. FBI. 179. 181, and 190.

Sibsagar. Dunsiri Valley, (Dafla Hills)., Dacca.

A* **kulla** Andr. AM. (9). vii. 1921. 411; Bänn. Ent. Blätt. 1928. 56 and 59; Andr. FBI. 180, and 202. t. 6. f. 1.

Nagpur.

L **lunulata** Bon., Obs. Ent. ii., Mém. Acad. Sci. Turin. xx. 1813. 459; Bänn. Ent. Blätt. 1928. 63; Andr. FBI. 204.

? Ind. Or.

O* **obscuripes** Chaud. Mon. 86; Bänn. Ent. Blätt. 1928. 56. t. f. 4; Andr. FBI. 179, 182. f. 32 (11), and 193.

G* *subtilis* Bates AG. xxxii. 1892. 284; Andr. TES. 1921. 146.

Tharrawaddy. Paungde., Rangoon. Theinzeik., Palon. Senmigion. Tenasserim.

C* **plana** *(Galerita)* F. Syst. Eleuth i. 1801. 216; Hope Col. Man. ii. 1838. 99 (no. 6); Andr. TES. 1921. 163; Bänn. Ent. Blätt. 1928. 56; Andr. FBI. 180 and 198.

O* *plagiata* Chaud. Mon. 93.

Dacca. Satpara (Orissa). Barkuda I. (Lake Chilka)., Madras.

B* **plicata** Andr. FBI. 178 and 185.

A* **polita** × Andr. TES. 1921. 180; Bänn. Ent. Blätt. 1928. 56; Andr. FBI. 178, 182. f. 32 (5), and 192. t. 5. f. 6.

G* *atrata* ‡ Bates (not Dej.) AG. xxxii. 1892. 284; Andr. TES. 1921. 146.

O* **pubescens** Chaud. BM. 1850. i. 439; id. Mon. 95; Bänn. Ent. Blätt. 1928. 56; Andr. FBI. 179, 182. f. 32 (6), and 186.

A* var. **degener** Andr. FBI. 181 and 187.

pumila × Andr. TES. 1921. 164; Bänn. Ent. Blätt. 1928. 56. t. f. 7; Andr. FBI. 180, 182 f. 32 (14), and 199. t. 6. f. 2.

T* *plana* || Bon. (not F.), Obs. Ent. ii., Mém. Acad. Sci. Turin. xx. 1813. 458; Hope Col. Man. ii. 1838 99 (No. 6); Andr. AM. (9). iii. 1919. 470; id. AG. xlix. 1922. 390.

O* *depressa* ‡ Dej. (not F.) Sp. i. 1825. 361; id. Sp. V. 1831. 476; Chaud. Mon. 90.

O* **punctulata** Chaud. Mon. 99; Bänn. Ent. Blätt. 1928. 56; Andr. FBI. 178, 181 and 182. f. 32 (1).

A* **pygmaea** Andr. AM. (9). vii. 1921. 410; Bänn. Ent. Blätt. 1928. 56; Andr. FBI. 180 and 201.

A* **rubescens** Andr. FBI. 180 and 199.

I* **rustica** Andr. FBI. 181, 182. f. 32. (15), and 204.

O* **sublaevis** Chaud. Mon. 86; Bates AF. 1889. 264; Lesne Miss. Pav. 68; Bänn. Ent. Blätt. 1928. 56.

L **sulcicollis** Bon., Obs. Ent. ii., Mém. Acad. Sci. Turn. xx. 1813. 459; Bänn. Ent. Blätt. 1928. 63; Andr. FBI. 204.

SIMOUS* Chaudoir.

Trichinopoly.
Shembaganur.
Ceylon :
Hambegamuwa,
Murunkan, Wirawila.

"India".

Tharrawaddy.
Paungde.
Theinzeik.
Rangoon.

Insein.
Tikekee (Pegu).
Moulmein.

Nepal., Bankipur.
"N. W. India."

Nagpur., Belgaum.
North Kanara.
Dharwar.

Dacca., Calcutta.
Saidpur (Bengal).
Murshidabad.

Pusa., Chapra.
Kierpur (Purnea).
Dehra Dun.
Nagpur., Belgaum.

Madras.
Madura.

Dacca., Berhampore
Sahibganj., Pusa.

Belgaum.

Nagpur.

Simous Chaud., Mon. des Oodides (i), AF. 1882. 373.

O* **aeneus** *(Oodes)* Laf. AF. 1851. 270 (note 4); Chaud. Mon. 375.

B* **borneensis** Bates PZS.1889. 384; Andr. TES. 1919. 197.

O* **laevissimus** *(Oodes)* Chaud. Mon. 361; Andr. AM. (9). xiv. 1924. 588.

G* **lampros** Bates AG. xxxii. 1892. 322.

O* **lucidus** *(Oodes)* Chaud. RMZ. 1869. 119; id. Mon. 376; Bates AG. xxxii. 1892. 322; Lesne Miss. Pav. 70.

X* *aeneus* ‡ Bates (not Laf.) AF. 1889. 266; Andr. TES. 1921. 176.

O* **mouhoti** *(Oodes)* Chaud. RMZ. 1869. 119; id. Mon. 373; Lesne Miss. Pav. 70.

C* **nigriceps** *(Oodes)* Wied. Germ. Mag. iv. 1821. 114; Laf. AF. 1851. 270; Chaud. Mon. 375; Andr. TES. 1921. 167.

O* *pulcher (Oodes)* Dej. Sp. ii. 1826. 375.

SINURUS* Chaudoir.

Sinurus Chaud., Mém. sur les Thyréoptérides, AB. xii. 1869. 129.

G* **graciliceps** Bates AG. xxxii. 1892. 48.

G* **nitidus** Bates AG. xxxii. 1892. 407.

Pegu.

Malay Peninsula.
Siam., Cambodia.
Cochin-China.

? Ind. Or.

Java.

Sumatra., Borneo.

New Guinea.

Sibsagar.
Khasi Hills.
Ruby Mines., Bhamo.
Tharrawaddy.
Toungoo., Palon.
Rangoon.,
Cambodia.

Sikkim: Mungphu.
Rangoon.
Kawkareik., Papun.

Siam.
Indo-China.

Laos.
Cambodia.

Shillong.
Dunsiri Valley
(Dafla Hills).
Naga Hills.
Calcutta.
Berhampur.
Murshidabad.
Malda., Birbhum.
Chapra., Nainital.
Dehra Dun.

Khasi Hills.
Karen Hills.

O* **opacus** Chaud. Mém. 130; Bates AG. xxxii. 1892. 407.

SOMOTRICHUS* Seidlitz.

Somotrichus Seidl. Faun. Blat. Ed. ii. 1891. 7; Gangl. Käf. Mitteleur. i. 1892. 400; Bedel Fn. med. 1905. 243.

O* **unifasciatus** *(Lebia)* Dej. Sp. v. 1831. 389.

L *elevatus (Carabus)* || F. Ent. Syst. i. (1). 1792. 162 (not Mant. Ins. i. 1787 198); id Syst. Eleuth. i. 1801. 204; *(Lebia)* Brullé Silb. Rév. ii. 1834. 108; *(Lebia)* Schm. Goeb. FCB. 43; *(Lebia)* Chaud. BM. 1854. i. 133; *(Somoplatus)* Bedel Fn. Seine i. 1879. 114; id. Fn. med. 1905. 243. (note 3) and 244; Alluaud AF. lxxxv. 1916. 85; Andr. TES. 1921. 157; id. TES. 1923. 21.

B* *bicinctus (Coptodera)* Hope TES. iv. 1845. 15; Andr. TES. 1919. 178.

O* *massiliensis (Coptodera)* Fairm. AF. 1849. 419.

SPAROSTES* Putzeys.

Sparostes Putz., Rév. Gén. des Clivinides, AB. x. 1857. 27; Péring Trans. S. Afr. Phil. Soc. vii. 1896. 606; Andr. FBI. 343 and 344.

E* **brevicolli** Putz. Rév. 27.

O* **striatulus** Putz. Rév. 29; Andr. FB. 345. t. 9. f. 6.

Karen Hills.
Ruby Mines.

Palon., Andaman Is.
Perak., Java.
Sumatra., Borneo.,
Philippine Is.

Cosmopolitan but generally confined to seaports.

U. P. : Chakrata.
C. P. : Raipur.
Calcutta., Ceylon.
Batchian., Celebes.
Hongkong., Canton.
Mauritius., Rhodesia.
Algiers., Tarsus.

Marseilles.
Bordeaux., Paris.

Le Havre., Rouen.
Caen., Guadeloupe.

China.

Sikkim : Nurbong,
Darjeeling., Duars.
Pusa.
Kierpur (Purnea).
Chapra., Dehra Dun.
Lansdowne.
Pachmarhi.
Belgaum.
Nilgiri Hills.
Wynaad.

SPHODRUS* Schellenberg.

Sphodrus Schell. Helv. Ent. ii. 1806. 84/5' Dej. Sp. iii. 1828. 87; Lac. 340; Motch. BM. 1864. ii. 314; Schauf., Monographische Bearbeitung der Sphodrini, Sitzungsb. Ges. Isis 1864. 116; Andr. MB. 45.

O* **indus** Chaud. BM. 1852. i. 67; Schauf. Mon. 117; Bates SYM. 10; Andr. MB. 46. — Shan Hills. Shwegu., Katha. Rangoon. Moulmein. Tenasserim. Siam., Indo-China. Hongkong.

STENOLOPHUS* Latreille.

Stenolophus Latr. in Cuvier's Régne Anim (Ed. ii). vi. 1829. (April). 391; Dej. Sp. iv. 1829. (Dec.) 405; Lac. 303; Sloane, PLS. NSW. 1898 456; Tchitch. HER. xxxiv. 1900. 364; id. HER. xxxv. 1901. 246; Andr. MB. 40.

Egadroma Motch. Etud. Ent. 1855. 43; id. BM. 1864 ii. 201; Tchitch. HER. xxxiv. 1900. 364; id. HER. xxxv. 1901. 246.

A* **bajaurae** Andr. MB. 95. id. EMM. 1926. 69. — Dehra Dun. Kailana., Kasauli. Chamba., Murree. Baddia (Punjab).

G* **charis** Bates AG. xxxii. 1892. 349. — Pusa., Fyzabad. Kashmir., Kangra. Bajaura. Spiti : Mankaran. Kumaon : Sarju Valley. Bengal., Haldwani. Katha., Rangoon. Timor.

L **dingo** *(Harpalus)* Cast. Trans. Roy. Soc. Vict. viii. 1868. 197; Sloane PLS. NSW. 1920. 321; Andr. Ins. Samoa iv. 1927. 4.

D* *robustus Sloane* DEZ. 1907. 469; Andr. AM. (9). xix. 1927. 110.

V* *quinquepustulatus* ‡ Csiki (not Wied.) Denskschr. Akad. Wiss. Wien. Math. nat. Kl., xci. 1915. 164. — Langkawi I. Java., Mentawei Is. Sumatra. New Guinea. Australia. Tahiti.

discophorus Fisch. Ent. Russ. ii. 1824. 141. t. 26. f. 9; Dej. Sp. iv. 1829. 409; Dej. et Boisd. Ic. et HN. iv. 241. t. 198. f. 2; Andr. MB. 40.

G* **gonidius** Bates AG. xxvi. 1889. 104; id. AF. 1889. 272; id. AG. xxxii. 1892. 348; Bouch. AF. 1903. 172.

O **lucidus** Dej. Sp. iv. 1829. 419.

B* **opaculus** Bates AM. (5). xvii. 1886. 80.

B* **polygenus** Bates AM. (5). xvii. 1886. 79.

G* **rectifrons** Bates AG. xxxii. 1892. 348; Lesne Miss. Pav. 76.

C* **smaragdulus** *(Carabus)* F. Suppl. Ent. Syst. 1798. 60; id. Syst. Eleuth. i. 1801. 209; Dej. Sp. iv. 1829. 418; Hope Col. Man. ii. 1838. 93; Schaum. SEZ. 1847. 49; Motch. Etud. Ent. 1855. 43; id. BM. 1864. ii. 205; Bates AM. (5) xvii. 1886. 80; id. CRB. 1891. 333; id. AG. xxxii. 1892. 349; Bouch. AF. 1903. 173; Andr. TES. 1921. 160; Sloane PLS. NSW. 1920. 321; Andr. MB. 40.

B* *trechoides (Harpalus)* Hope TES. iv. 1845. 15; Andr. TES. 1919. 178.

B* *stolidus (Harpalus)* Walk. AM. (3). ii. 1858. 204; Bates AM. (5). xvii. 1886 80 and 221; Andr. TES. 1919. 189.

M* *apicalis (Egadroma)* Motch. BM. 1864. ii. 205; Andr. TES. 1928. 8.

M* *nitens (Egadroma)* Motch. BM. 1864. ii. 205; Andr. TES. 1928. 8.

Samoa.

Srinagar.
Mesopotamia.
Arabia., Siberia.
Turkestan.
Transcaspia.
Southern Europe.

Burma : Bhamo,
Tharrawaddy,
Toungoo-Katha,
Teinzo, Thagatta.
Cambodia.
Sumatra.

"East Indies".

Nuwara Eliya.

Nuwara Eliya.

Calcutta.
Sunderbans.
Chota Nagpur :
Barwa. Malabar.
Madras., Trichur.
Ceylon : Weligama,
Kandy, Kotte,
Colombo, Negombo.
Burma : Teinzo,
Karea Hills, Rangoon,
Panol, Kawkareik.
Singapore., Siam.
Indo-China.
Sumatra., Boreno.

B* *transmutans* Bates AM. (5). xvii. 1886. 80; id. Ag. xxxii. 1892. 349.

G* *cyanellus* Bates AG. xxvii. 1889. 103; id. AG. xxxii. 1892. 350; Bouch. AF. 1903. 172; Andr. TES. 1928. 8.

X* *chalceus* ‡ Lesne (not Bates) Miss. Pav. 76; Andr. AM. (9). xvii. 1926. 255.

C* var. **quinquepustulatus** *(Badister)* Wied Zool. Mag. ii. 1. 1823. 58; Dej. Sp. iv. 1829. 414; Bates TES. 1873. 270; Putz. CRB. 1875. 49; Bates AM. (5). xvii. 1886. 79; id. AF. 1889. 272; CRB. 1891. 333; id. AG. xxxii. 1892. 350; Bouch. AF. 1903. 173; Lesne Miss. Pav. 76; Sloane PLS. NSW. 1920. 31; Andr. TES. 1921. 171; id. AM. (9). xiii. 1924. 469.

China.

G* *rectifrons* ‡ Bouch. (not Bates) AF. 1803. 172.

M* **splendidus** *(Egadroma)* Motch. BM. 1864. ii. 205; Andr. TES. 1928. 8.

Throughout the whole of South East Asia, extending from Japan in the north to Queensland in the south.

G* *harpaloides* Bates AG. xxxii. 1892. 347.

STENOTELUS* Bouchard.

Stenotelus Bouch. AF. 1903. 174.

opacus Bouch. AF. 1903. 175.

Calcutta.
Sunderbans.
Mandalay.

Prome., Rangoon.
Tenasserim.

STEPHANA* Chaudoir.

Stephana Chaud. Mon. des Lébiides (ii); BM. 1871; i. 55.

O* **princesp** *(Lebia)* Chaud. BM. 1852. i. 42; id. Mon. 56.

Perak., Java.
Sumatra., Borneo.

STEVENSIUS* Jeannel.

Stevensius Jeann. AM. (9). xii. 1923. 432; id., Mon. des Trechinae (iii), Ab. xxxv. 1928. 90.

A'* **lampros** Jeann. AM. (9). xii. 1923. 433. f. 18; id. Mon. (iii). 92. ff. 1328-31.

Maflong (Khasi Hills). Kurseong.
Kumaon.
Thadyar (Jaunsar Himalayas).

STOMONAXELLUS* Tchitcherin.

Stomonaxellus Tchitch. HER. xxxv. 1901. 51.

Q **filicornis** Tchitch. HER. xxxv. 1901. 52. Deota (N. W. Himalayas).

STRIGIA* Brullé.

Strigia Brullé HN. vi. 1834. 382; Lac. 327.

Rathymus || Dej. (not Lep.) Sp. v. 1831. 783; Lac. 329.

Selenidia Motch. Etud. Ent. 1855. 45.

Rhathymus Chaud. BM. 1878. ii. 7; Tchitch. HER. xxxii. 1898. 536; id. HER. xxxiv. 1900. 448.

Parastrigia Maindr. BF. 1906. 251.

O* **atra** *(Rhathymus)* Chaud BM. 1878. ii. 8. Sikkim : Tonglu. Chota Nagpur : Barwa.

X* **maxillaris** Brullé HN. iv. 1834. 383. t. 15. f. 6; Chaud. BM. 1878. ii. 8; Bates CRB. 1891. 334; Andr. MB. 116.

C* **stigma** *(Carabus)* F. Syst. Eleuth. i. 1801. 192; Macl. Ann. Jav. 1825. 29; Hope Col. Man. ii. 1838. 83 (No. 121); Schaum SEZ. 1847. 46; Motch. Etud. Ent. 1855. 45; Chaud. RMZ. 1872. 140; id. BM. 1878. ii. 9; Andr. TES. 1921. 161. Coromandel. Koilpatti (Tinnevelly). Bengal. Chota Nagpur : Konbir. Delhi., Rawalpindi. Surada. Dacca. Mhow. Madura.

STYPHLOMERUS* Chaudoir.

Styphlomerus Chaud., Mon. des Brachynides. AB. 1876. 87.

Styphromerus Chaud. Mon. 88.

batesi *(Styphromerus)* × Chaud. Mon. 92; Lewis AM. (6). xvii. 1896. 331.

B* *bicolor (Crepidogaster)* ‡ Bates (not Boh.) TES. 1875. 307; id. TES. 1883. 208.

P* **fusciceps** *(Brachinus)* Schm. Goeb. FCB. 73; Chaud. Mon. 92; Bates AG. xxxii. 1892. 397; Andr. TES. 1923. 43; id. MB. 117.

Japan.

L *bicolor (Brachinus* || Boh. (not Burllé) ER. 3; Andr. AM. (9) xiii. 1924. 470.

dichrous (Brachinus) × Gemm. & Har. Mun. Cat. 105; Chaud. Mon. 92.

Surada., Mysore. Shwegu. Karen Hills.

Tharrawaddy. Prome., Rangoon.

Malay Penins la. Java., Formosa. Hongkong. Macao.

O* **ruficeps** *(Styphromerus)* Chaud. Mon. 91.

Dehra Dun. Nagpur. Coromandel. Madras., Nilambur. Nilgiri Hills. Palni Hills. Mysore : Chikkaballapura.

O* **timoriensis** *(Brachinus)* Jord. NZ. 105; Andr. TES. 1921. 149.

Sumatra., Timor. Andonare I.

SYRDENUS* Chaudoir.

Syrdenus Chaud., Essai monographique sur le groupe des Pogonides, AB. xiv. 1871. 34; Reitt., Best. Tab. eur. Col., lxv., Pogonini 1908. 5.

B* **grayi** *(Pogonus)* Woll. AM. (3). ix. 1862. 438; Chaud. Mon. 35; Bedel Fn. med. 1897. 90; Sem. RRE. 1903. 401 (Russian); Reitt. Best. Tab. 5; Alluaud AG. lii. 1928. 76.

T *fulvus (Pogonus)* Baudi BEZ. 1864. 206; Schaum BEZ. 1864. 206 (note); Chaud. Mon. 35; id. BM. 1878. ii. 80.

O* *extensus* Chaud. Mon. 35.

O *dilutus (Pogonus)* Fairm. iii. 1873. 333.

Bombay. Fao (Persian Gulf). Red Sea littoral. Cyprus. Sicily. Southern littoral of Mediterranean. Cape Verde. Canary Is.

SYSTOLOCRANIUS* Chaudoir.

Systolocranius Chaud. BM. 1857. ii. 23; id., Mon. des. Oodides (i), AF. 1882. 326; Alluaud BF. 1916. 227; Andr. MB. 29.

C* **linea** *(Oodes)* Wied. Germ. Mag. iv. 1821. 113; Chaud. Mon. 331; Andr. TES. 1921. 167; id. MB. 29.

O* *grandis (Oods)* Dej. Sp. ii. 1826. 376; Laf. AF. 1851. 269; Chaud. BM. 1887. ii. 24. — Calcutta., Birbhum. Malda., Murshidabad. Berhampur., Pusa. Chapra. Sitapur (U. P.). C. P. : Raipur, Chilpy, Jubbulpore, Baihar.

TACHYS* Stephens.

Tachys Steph. Ill. Brit. Ent. ii. 1828. 2 and 4; Dej. Sp. v. 1831. 33; Jacq. Duv. AF. 1852. 189; Motch. Etud. Ent. 1862. 27; Putz. AG. 1875. 737; Bates Biol. Centr. Amer. Ins. Col. i. (1). 1882. 139; Reitt. WEZ. 1884. 116; Net. Ent. Blätt. 1914. 173; Andr. MB. 18; id., Revision of the Oriental species of the genus Tachys, AG. li. 1925. 327.

Tachyta Kirby Faun. Bor. Amer. iv. 1837. 56; Schaum BEZ. 1863. 88.

Elaphropus Motch. BM. 1839. i. 73; Andr. AM. (9). iii. 1919. 472.

Pericompsus Lec. Ann. Lyc. N. Y. 1852. v. 191.

Tachylopha, Tachyura, Tachymenis (=Tachyta), Polyderis Motch. Etud. Ent. 1862. 27.

Barytachys Chaud. RMZ. 1868. 213; Bates Biol. Centr. Amer. Ins. Col. i. (1). 1882 140; Andr. Rev. 328.

Porotachys Net. Ent. Blätt. 1914. 174.

M* **acaroides** Motch. Etud. Ent. 1859. 39; Putz., AG. 1875. 740; Andr. Rev. 469; id. TES. 1928. 19.

G'* **acaroides** ‡ Bouch. (not Motch.) AF. 1903. 170. — Ceylon.

G* **acuticollis** Putz. AG. 1875. 740; Andr. Rev. 485 and 492. — Sumatra.

G* **aeneus** Putz. AG. 1875. 744; Andr. Rev. 401 and 417. — Aru Is. New Guinea.

var. **jordani** × Csiki Col. Cat. 97, Carab. 1928. 167. — Celebes., Tenimber.

O* *biplagiatus* || Jord. (not Dej.) NZ. 112; Andr. Rev. 401. and 418. — Sumbawa New Guinea.

G* **ambiguus** Andr. Rev. 382.

B* **amplians** Bates AM. (5). xvii. 1886. 155; Andr. AM. (9). iii. 1919. 472; id. Rev. 471 and 475. — Mentawei Is.

Ceylon : Kandy (in

S* **arcuatus** Putz. AG. 1875. 744; Andr. TES. 1922. 247; id. Rev. 404 and 438.

A* **babaulti** Andr. MB. 87. t. 2. f. 1; id. Rev. 407 and 452.

A* **barringtoni** Andr. Rev. 402 and 423.

A* **bathyglyptus** Andr. Rev. 343 and 367.

A* **belli** Andr. Rev. 400 and 411.

O* **bembidiiformis** Jor. NZ. 111; Andr. Rev. 402 and 422.

A* *helmsi* Sloane PLS. NSW. 1898. 476; id. PLS. NSW. 1921. 198. and 202.

M' **bifoveatus** W. Macl. TES. NSW. 1871. 117; Sloane PLS. NSW. 1921. 199 and 206; Andr. Rev. 471 and 476.

M' *ovatus* || W. Macl. (not Motch.) TES. NSW. 1871. 117; Sloane PLS. NSW. 1896. 369.

B* *perlutus* Bates TES. 1873. 299; Andr. AM. (9). iii. 1919. 472.

B' *unistriatus* || Reitt. (not Putz.) DEZ. 1887. 497.

unilineatus × Reitt. WEZ. 1888. 104; Net. Ent. Blätt. 1914. 174.

A* **blandus** Andr. MB. 85 t. 1. f. 8; id. Rev. 408 and 459.

E* *haemorrhoidalis* ‡ Bates (not Dej.) AB. 1892. 230.

A* **bombycinus** Andr. Rev. 382 and 385; id. EMM. 1926. 67.

description), Colombo (on label). Ceylon.

Lyallpur., Haldwani. Dehra Dun., Mhow. Nagpur., Motinala. Poona., Lonovla. Matheran., Madras., Nilambur.

Philippine Is.

Borneo.

Karachi.

Java., Andonare I.

Tenimber.

Tonkin., Formosa. Japan., Siberia.

Australia.

U. P. : Ranikhet. R. Sarda. Gorge, Haldwani, Dehra Dun., Siwalik Hills. Ranchi., Motinala. Nagpur. Hoshangabad. Mandla., Khandesh. Poona., Chipurupalle.

Sikkim : Gopaldhara. Eastern Duars. U. P. : Tenakpur, Swal River basin, Sarju Valley, R. Sarda

A* **borneensis** Andr. Rev. 401 and 418.	Gorge, Haldwani, Dehra Dun.
A* **brachys** Andr. Rev. 375 and 377.	Borneo. New Guinea.
B* *? atomarius* ‡ Bates (not Woll.) AM. (5). xvii. 1886. 152.	Bengal. Haldwani, Nilgiri Hills. Kodaikanal.
A* var. **decolor** Andr. Rev. 375 and 378.	? Ceylon., Selangor. Singapore. Nicobar Is.
J* **brunneus** Andr. Rev. 375 and 379.	Singapore. Formosa., China.
I* **buxans** Andr. Rev. 471 and 476.	Borneo.
A* **cameroni** Andr. Rev. 342 and 359.	Kodaikanal.
E* **cardoni** Andr. Rev. 343. and 364.	Singapore.
A* **castaneus** Andr. Rev. 395. and 396.	Chota Nagpur : Barwa
J* **ceylanicus** *(Bembidium)* Nietn. AM. (3). ii. 1858. 423; Andr. Rev. 404 and 436.	Sikkim : Gopaldhara. Almora.
M* *flaviculus* Motch. Etud. Ent. 1859. 39; Putz. AG. 1875. 746; Andr. TES. 1919. 198; id. TES. 1928. 18.	Sikkim : Gopaldhara, Gielle Khola (Tista Valley).
E* *anceps* Putz. AG. 1875. 742.	Eastern Duars. Calcutta., Almora. Dehra Dun.
B* *infans* Bates AM. (5). xvii. 1886. 154; id. AG xxxii. 1892. 294.	Mysore : Teppukadu. Ceylon : Kandy, Colombo, Trincomali. Burma : Senmigion, Karen Hills., Perak. Penang., Malacca. Annam., Sumatra. Java., Celebes.
B* var. **polyporus** Andr. Rev. 404 and 437.	Philippine Is. Hongkong.
B* **chalceus** Andr. Rev. 401 and 415.	Almora., Tanakpur. Haldwani.

A* **championi** Andr. Rev. 407 and 453.	Dehra Dun. Almora., Haldwani. Sikkim : Gopaldhara, Nurbong. Eastern Duars. Almora., Ranikhet.
B* **charactus** Andr. Rev. 407 and 451.	Haldwani. Kathgodam. Siwalik Hills. West Almora. Tanakpur. Haldwani.
A* **charis** Andr. Rev. 407 and 454.	Swal River basin. Prome.
E* *ornatus* ‡ Putz. (not Nietn., not Apetz.) AG. 1875. 741.	Bengal.
B* *scydmaenoides* ‡ Bates (not Nietn.) AM. (5). xvii. 1886. 154; id. CRB. 1891. 326.	Chota Nagpur : Konbir, Ranchi.
G* *nietneri* × || Bates (not Schaum). AG. xxxii. 1892. 295.	Kallikota., Nagpur. Matheran. Belgaum. Nilgiri Hills. Chipurupalle,
B* var. **rubescens** Andr. Rev. 407 and 455.	Mysore., Ceylon. ? Mandalay. Vizagapatam.
G* **cinctus** Putz., AG. 1875. 746; Andr. Rev. 342 and 357.	Ceylon : Colombo, Kotte, Anuradhapura.
X* **compactus** Andr. Rev. 406 and 449.	Amboina.
A* **comptus** Andr. AM. (9). x. 1922. 161; id. Rev. 461 (*not* t. 3. f. 10, see var. **pluripunctus).**	Tonkin. Mangaldai., Tezpur. Darjiling., Ranchi. Nagpur., Belgaum. Madras.
A* var. **borealis** Andr. Rev. 461 and 462; id. EMM. 1926. 67.	Ceylon : Anuradhapura. Sikkim : Gopaldhara. Sarju Valley (Almora).

B* var. **pluripunctus** Andr. Rev. 461 and 462. t.3. f. 10. — Dehra Dun.

G* **coracinus** Putz. AG. 1875. 739; Andr. Rev. 485 and 491. — Chota Nagpur : Barwa., Formosa. Almora.

G* *remotiporis* Bates AG. xxxii. 1892. 290. — Patkai Hills., Sadiya.

Burma : Meetan, Kawkareik, R. Mohnyin, Palon. Penang., Perak. Singapore., Siam. Tonkin., Java. Sumatra. Mentawei Is.

decoratus × Andr. Rev. 402 and 424. — Borneo., Celebes. Philippine Is.

J* *ornatus (Bembidium)* || Nietn. (not Apetz) AM. (3). ii. 1858. 426; Bates AM. (5). xvii. 1886. 154. — Ceylon : Colombo. Nalanda, Kandy.

A* **delicatus** Andr. Rev. 342 and 358.

A* **dentatus** Andr. Rev. 341 and 343. t. 3. f. 1. — Singapore.

Singapore., Siam. Cambodia. Cochin-China.

M* **dorsalis** Motch. BM. 1851, ii. 508; Andr. Rev. 497; id. TES. 1928. 4. — Java., Sumatra. Princes I.

B* **dulcis** Andr. Rev. 395. — ? Tenasserim.

A* **elegans** Andr. Rev. 463 and 465.

A* **erotyloides** Andr. Rev. 402 and 419. t. 4. f. 41. — Bareilly.

B* **eueides** Bates AM. (5). xvii. 1886. 153; Andr. Rev. 408 and 460. — Nilgiri Hills.

New Guinea.

G* **euryodes** Bates AG. xxxii. 1892. 289; Andr. Rev. 343 and 360. t. 4. f. 27; id. FBI. 16. f. 4d. — Ceylon : Dikoya.

Calcutta., Sardah. Haldwani. Ceylon : Anuradhapura. Karen Hills. Rangoon., Insein. Perak., Kuala Lumpur.

B*	**exaratus** Bates TES. 1873. 296; Andr. Rev. 388 and 392. t. 3. f. 4.	Singapore., Siam. Java., Sumatra. Boreno., Celebes.
B*	var. **curtus** Andr. 388 and 393.	Japan.
G*	**expansicollis** Bates AG. xxxii. 1892. 293; Andr. Rev. 382 and 383.	Japan.
M*	**fasciatus** *(Trechus)* Motch. BM. 1851. ii. 506; Andr. TES. 1922. 248; Jeann. AM. (9). xii. 1923. 394; Andr. TES. 1928. 3.	Karen Hills.
J*	*triangularis (Bebidium)* Nietn. AM. (3). ii. 1858. 422; Schaum BEZ. 1863. 91; Bates TES. 1873. 298; id. AM. (5). xvii. 1886. 151; id. AF. 1889. 274; Fvl. RE. 1903. 219; Sloane PLS. NSW. 1920. 321; id. PLS. NSW. 1921. 200 and 207; Andr. Rev. 340 and 350. t. 3. ff. 3 and 14. t. 4. ff. 19, 22, 30, 34, 35 and 38.	Throughout South-East Asia, including Japan., Australia New Caledonia. Arabia., Egypt. South Africa.
M'	*atriceps (Trechus)* W. Macl. TES. NSW. ii. 1871. 113; Sloane PLS. NSW. 1896. 359 and 372.	
O*	*trechiformis* Jord NZ. 111.	
I'	*migrator* Péring. Trans. S. Afr. Phil. Soc. vii. 1896. 593 and 595; id. Ann. S. Afr. Mus. xxiii. 1926. 648 and 650.	
B*	var. **incertus** Andr. Rev. 342 and 352 nom.	
	tropicus ‡ Bates (not Nietn.) AM. (5). xvii. 1886. 151.	
B*	**finitimus** *(Bembidium)* Walk. AM. (3). ii. 1858. 204; Bates AM. (5). xvii. 1886. 211; Andr. TES. 1919. 191; id. Rev. 407 and 455.	Ceylon : Dikoya.
A*	**florus** Andr. Rev. 402 and 426.	Ceylon : Anuradhapura.
M*	**fumigatus** Motch. BM. 1851. ii. 509; Andr. Rev. 466. t. 3. f. 17 and t. 4. f. 33; id. TES. 1928. 5.	Kuala Lumpur. Singapore., Java. Sumatra.
J*	*emarginatus (Bembidium)* Nietn. AM. (3). ii. 1858. 425; Putz. AG. 1875. 739; Bates AM. (5). xvii. 1886. 155; id. AF. 1889. 275; id. AG. xxxii. 1892. 295; Bouch. AF. 1903. 170; Dupuis AB. 1913. 426; Andr. RIM. xxii. 1921. 341.	Throughout South-East Asia.
J*	var. **scydmaenoides** *(Bembidium)* Nietn. AM. (3). ii. 1858. 427; Bates TES. 1883. 267 (part); Andr. Rev. 466.	

D*	var. **geminatus** Schaum. BEZ. 1860. 200; Putz. AG. 1875. 740; Andr. Rev. 466.	Ceylon. Japan.
	putzeysi Dupuis AB. 1913. 426.	Malay region.
J*	var. **socius** Schaum BEZ. 1863. 92; de la Brûl AF. 1875. 438 (note 82); Reitt. WEZ. 1884. 118; Net. Ent. Blätt. 1914. 175; Sloane PLS. NSW. 1921. 203; Andr. Rev. 466.	Egypt.
	var. **singalensis** × Csiki Col. Cat. 97. Carab. 1928. 180.	
E*	*pallidus* || Dupuis (not Chaud, not Reitt.) AB. 1913. 175 and 426; Andr. Rev. 466.	Ceylon.
B*	**fuscicauda** Bates TES. 1873. 298; Andr. Rev. 404 and 438.	
J*	**fusculus** Schaum BEZ. 1860. 200; Putz. AG. 1875. 741; Andr. Rev. 402 and 425.	China. Japan.
G*	**fusiformis** Andr. Rev. 402 and 420.	Tonkin. China.
G*	**gestroi** Andr. Rev. 382 and 384. t. 3. f. 5.	New Guinea.
G*	**glis** Andr. Rev. 472 and 481.	Mentawei Is.
A*	**gongylus** Andr. Rev. 406 and 450.	Celebes., Buru.
O*	**gradatus** Bates TES. 1873. 331; Andr. Rev. 404 and 434.	China.
G*	**haliploides** Bates AG. xxxii. 1892. 289; Andr. AM. (9). iii. 1919. 472; id. Rev. 471 and 478. t. 3. ff. 11 and 18; id. TES. 1928. 4.	China. India., Burma. Cambodia., Java. Celebes. New Guinea. Philippine Is.
A*	var. **nigrinus** Andr. Rev. 472 and 479.	Sikkim : Gopaldhara. Assam Valley. Kumaon : Almora. Upper Gumti Valley. Haldwani.
B*	var. **asthenes** Andr. Rev. 472 and 480.	Almora. Dehra Dun.

A* var. **diversus** Andr. Rev. 472 and 480.

A* **imperfectus** Andr. Rev. 401 and 413.

M* **impressipennis** Motch. Etud. Ent. 1859. 39; Putz. Ag. 1875. 745; Andr. Rev. 341 and 346. t. 3. ff. 2. and 13; id. TES. 1928. 18.

O* *dohertyi* Jord. NZ. 111.

A* *sinuaticollis* Sloane PLS. NSW. 1903. 579; id. PLS. NSW. 1921. 200 and 207.

M* **impressus** Motch. BM. 1851. ii. 508; Andr. Rev. 342 and 353; id. TES. 1928. 4.

E* **interpunctatus** Putz. AG. 1875. 741; Andr. Rev. 404 and 432. t. 3. f. 7.

A* **javanicus** Andr. Rev. 408 and 458.

J* **klugi** *(Bembidium)* Nietn. AM. (3). ii. 1858. 423; Bates AM. (5). xvii. 1886. 155; id. AG. xxxii. 1892. 291; Andr. Rev. 337, 395, 401 and 414. t. 3. f. 6.

L* *sulculatus* Putz. AF. 1875. 743; Bouch. AF. 1903. 170.

B* *euglyptus* Bates TES. 1883. 268; Andr. AM. (9). iii. 1919. 472.

E* *feanus* Bates CRB. 1891. 326; id. AG. xxxii. 1892. 292; Andr. MB. 19.

E* var. **sulcatopunctatus** Putz. AG. 1875. 742; Bates AG. xxxii. 1892. 292; Bouch. AF. 1903. 170 Andr. Rev. 414.

B* **laetificus** Bates TES. 1873. 298; Andr. Rev. 408 and 457.

B* **laotinus** Andr. Rev. 403 and 432.

M* **latissimus** Motch. BM. 1851. ii. 508; id. Etud. Ent. 1859. f. 4; Andr. Rev. 478; id. TES. 1928. 4.

M* var. **gracilis** *(Elaphropus)* Motch. Etud. Ent. 1862. 36; Andr. AM. (9). iii. 1919. 472; id. Rev. 477; id. TES. 1928. 4 and 20.

E* *subfasciatus* || Putz. (not Motch.) AG. 1875. 739.

Siwalik Hills.

Gahan (Simla Hills).

Eastern Duars.

Philippine Is.

Throughout South-East Asia (excluding Japan). Australia.

Calcutta.
Tenasserim.
Sumatra.

Celebes.

Java.

Throughout South-East Asia, including Japan.

Sumatra., Celebes.
Andonare I.
Philippine Is.

Japan.

Laos., Annam.

Burma.

Someswari River (Garo Hills).
Eastern Duars.
Sardah.

E* *unistriatus* Putz. AG. 1875. 740; Andr. Rev. 471 and 477.

G* *centractulus* Bates AG. xxxii. 1892. 290.

latus Peyron AF. 1858. 364. t. 9. f. 3; Bates AF. 1889, 275; id. Ag xxxii. 1892. 293; Andr. TES. 1928. 5.

M* *sulcatus* || Motch. (not Putz.) BM. 1851. ii. 509; Putz. AG. 1875. 740; Andr. TES. 1919. 199; id. TES. 1928. 5.

jureceki Reitt. Ent. Blätt. 1908. 50.

aratus × Andr. Rev. 400 and 408.

X* var. **vigens** Andr. Rev. 400 and 410.

B* var. **virgatus** Andr. 400 and 410.

C* **lissonotus** Andr. Rev. 342 and 356.

G* **loriae** Andr. Rev. 387 and 389.

A* **luteus** Andr. Rev. 387 and 390.

G* **luxus** Andr. Rev. 338 and 339.

A* **malabaricus** Andr. Rev. 405 and 443.

A* **malayicus** Andr. Rev. 485 and 487.

G* **mediocris** Andr. Rev. 405 and 442.

A* **micraulax** Andr. RIM. xxvi. 1924. 115; id. Rev. 395 and 397.

Diamond Harbour.
Calcutta.
Kierpur (Purnea).

Haldwani.
Belgaum.

Tenasserim.
Nicobar Is., Siam.
Cambodia., Java.
Celebes.

Calcutta., Sardah.
Pusa., Fyzabad.

U. P. : Almora, Tanakpur, R. Sarda Gorge, Kosi River (Ranikhet).
Haldwani.
Ceylon (F. Walker in British Mus.).
Bhamo.
Ruby Mines.
Kyanktan.
Moulmein., Tavoy.
Sumatra., Tonkin.
Kurdistan.
Asia Minor.

Tonkin., Cambodia.

Sardah., Pusa.

Java., Sumatra.

New Guinea.

Singapore.

Karen Hills.
Tonkin.

Cannanore.

Penang., Singapore.
Java.

New Guinea.

Assam : Siju Cave

A* **micraulax**—

B* **microscopicus** *(Tachyta)* Bates TES. 1873. 299; id. AM. (5). xvii. 1886. 153; Andr. Rev. 375 and 379.

A* **monostigma** Andr. Rev. 485 and 489.

A* **mus** Andr. Rev. 375 and 380. 5. 3. f. 15.

A* **nalandae** Andr. Rev. 405 and 441.

A* **nannodes** Andr. Rev. 406 and 447.

J* **nanophyes** Andr. Rev. 472 and 481.

L **nanus** *(Bembidium)* Gyllh. Ins. Suec. ii 1810. 30; Perris AF. 1862. 175. t. 5. ff. 510-16 (life history); Bates TES. 1883. 267; Fvl. RE. 1889. 97; Bedel Fn. med. 1896. 77; Andr. Rev. 485 and 486.

J *quadristriatus (Carabus)* || Illig. (not Schrank) Käf. Preuss. 1798. 188.

E' *minimus (Elaphrus)* Dufts. Faun. Austr. ii. 1812. 205.

L *inornatus (Bembidium)* Say Tr. Am. Phil. Soc. ii. 1825. 87.

B* *picipes (Tachyta)* Kirby Faun. Bor. Amer. iv. 1837. 56. t. 8. f. 6.

luticollis Motch. Käf. Russ. 1850. 8 (note 1).

rivularis Motch. Käf. Russ. 1850. 8 (note 2).

A* **nilgiricus** Andr. Rev. 406 and 446.

A* var. **unisculptus** Andr. Rev. 403 and 446.

G* **nitnes** Andr. Rev. 401 and 412.

B* **notaphoides** Bates AM. (5). xvii. 1886. 156; Andr. Rev. 395 and 398.

and Someswari River (Garo Hills), Janakmukh.
Sikkim : Gopaldhara. Chakrata.
Kumaon : Almora. Upper Gumti Valley. Kotkhai (Simla).

Japan.

Singapore.

Java.

Ceylon : Nalanda, Woodside-Urugalle.

Mussoorie.
Belgaum.
Nilgiri Hills.

Formosa.

Throughout the north temperate zone, including Japan.

Mysore. Nilgiri Hills.

Nilgiri Hills.
Ceylon : Kanthalay, Trincomali.

New Guinea.

G* var. **callispilotus** Bates AG. xxxii. 1892. 291; Andr. Rev. 395 and 399.

G* **obsolescens** Bates AG. xxxii. 1892. 289; Andr. Rev. 343 and 360; id. TES. 1928. 4.

G* **ocellatus** Bates AG. xxxii. 1892. 292; Andr. Rev. 400 and 410.

A* **ochrias** Andr. Rev. 374 and 376.

B* **opacus** Andr. Rev. 472 and 483.

B* var. **bicolor** Andr. Rev. 472 and 484.

A* **opalescens** Andr. Rev. 342 and 352.

L **ornatus** *(Bembidium)* Apetz Col. Brehm. 1854. 12; Schaum BEZ. 1863. 92; Bates AF. 1889. 275; id. AG. xxxii. 1892. 295; Kuntz. Mitt. Zool. Mus. Berl. ix. 1919. 154; Andr. RIM. xxii. 1921. 341; id. Rev. 401 and 416. t. 4. f. 24; id. FBI. 16. f. 4a.

J* *orientalis (Bembidium)* Nietn. AM. (3). ii. 1858. 425; Bates AM. (5). xvii. 1886. 156.

Tranquebar.
Ceylon : Kitugalle (in description), Dikoya (on label).

Sardah., Almora.
Tanakpur., Haldwani.
Burma., Teinzo.
Kyanktan., Toungoo.
Tonkin.

Karen Hills.
Tenasserim.
Dehra Dun.

Naga Hills.
Burma : Teinzo.

Eastern Duars.
U. P. : West Almora, Tanakpur, River Sarda Gorge, Haldwani, Mussoorie, Dehra Dun, Siwalik Hills.
Burma : Moulmein.

Sardah.

Sardah.

Ceylon : Negombo.

Dehra Dun., Ranchi.
Puri.
Barkuda I. (Lake Chilka).
Almora., Fyzabad.

Chilla (Hardwar).
Nagpur.
Hoshangabad.
Mandla., Madras.
Madura., Mysore Chikkaballapura.
Ceylon : Colombo, Hambegamuwa, Elephant Pass, Anuradhapura.
Philippine Is.

A* **orphninus** Andr. Rev. 341 and 345.

M* **ovatus** *(Lopha)* Motch. BM. 1851. ii. 509; Dupuis AB. 1913. 170; Andr. TES. 1919. 198; id. AM. (9). xiii. 1924. 468; id. Rev. 463. t. 3. f. 9; id. TES. 1928. 5.

D* *albicornis* Schaum. BEZ. 1860. 199; Motch. Etud. Ent. 1862. 27 and 55.

G* *mirabilis* Bates AG. xxxii. 1892. 294; Dupuis. AB. 1913. 175; Andr. TES. 1921. 146.

D* *mirandus* Dupuis. AB. 1913. 170.

J* var. **obliteratus** Andr. Rev. 463 and 465.

B* **pallescens** Bates TES. 1873. 297; Andr. Rev. 342 and 355.

A* **pallidicornis** Andr. Rev. 405 and 443.

G* **papuae** Andr. Rev. 402 and 421.

B* **peryphinus** Bates AM. (5). xvii. 1886. 153; Andr. Rev. 404 and 435.

G* **pictus** Andr. 405 and 439.

G* **plagiatus** Putz. AG. 1875. 745; Andr Rev. 369 and 370.

A* *doddi* Sloane PLS. NSW. 1903. 578; id. PLS. NSW. 1921. 199. and 206.

B* var. **sexmaculatus** Andr. Rev. 369 and 371.

var. **manilensis** × Csiki Col. Cat. 97. Carab. 1928. 192.

A* *longulus* || Andr. (not Bates) Rev. 368 and 371.

O* **poecilopterus** Bates TES. 1873. 331; id. AG. xxxii. 1892. 295; Andr. Rev. 405 and 440. t. 3. f. 16; id. EMM. 1926. 67.

Arabia., Egypt.
Sudan., Abyssinia.
Damaraland.
? Sicily.

Timor.

Assam Valley.
Patkai Hills.
Ceylon : Negombo.
Bhamo.
Kawkareik.
Moulmein.
Tavoy., Perak.

Penang.
Kuala Lumpur.
Singapore.

Sumatra.
Philippine Is.
Formosa., China.

Celebes.
? Hongkong.

Japan.

Khandesh.

New Guinea.

Ceylon : Kitugalle, Nalanda.

New Guinea.

Celebes., Sumbawa.

New Guinea.
Philippine Is.

Siam., Tonkin.
Formosa.

Philippine Is.

Garo Hills.
Sikkim :

X* *deliciolus* Bates AF. 1889. 274; id. AG. xxxii. 1. 92. 295. — Gopaldhara. Tarkhola (Tista Valley). Eastern Duars.

Sardah., Calcutta. Hooghly. Sunderbans. Almora., Ranikhet. Haldwani., Kheri. Dehra Dun. Sitapur (U. P.). Lyallpur. Burma : Bhamo, Teinzo, Prome, Rangoon, Tharrawaddy, Victoria Point. Kuala Lumpur. Singapore., Siam. Indo-China. Sumatra., Celebes. China.

M* **politus** Motch. BM. 1851. ii. 509; Putz. AG. 1875. 743; Bouch. AF. 1903. 170; Andr. TES. 1919. 199; id. TES. 1921. 146; id. MB. 19. id. Rev. 406 and 448. t. 3. f. 8 and t. 4. ff. 23, 29, 32, 36, 37, and 40; id. TES. 1928. 5.

J* *ebeninus (Bembidium)* Nietn. AM. (3). ii. 1. 58. 424; Motch. Etud. Ent. 1862. 55. — Throughout South-East Asia (except Japan).

E* *bioculatus* Putz. AG. 1 75. 743; Bates AM. (5). xvii. 1886. 153; id. AG. xxvii. 1889. 105; id. CRB. 1891. 326; id. AG. xxxii. 1892. 293.

G* var. **homostictus** Bates AG. xxxii. 1892. 293; Andr. Rev. 406 and 449.

A* var. **aspilotus** Andr. Rev. 406 and 449.

B* var. **constrictus** Andr. Rev. 406 and 449. — With type form.

A* **porosus** Andr. Rev. 470 and 472. — With type form.

G* **prolixus** Bates AG. xxxii. 1892. 288; Andr. Rev. 343 and 364. — Laos.

G* **psilus** Andr. Rev. 403 and 429. — Borneo.

A* **pulcher** Andr. Rev. 403 and 428. — Karen Hills.

B* var. **celebensis** Andr. Rev. 403 and 429.

A* **punctus** Andr. Rev. 471 and 473. var **mindorensis** × Csiki Col. Cat. 97. Carab. 1928. 193.

A* *variabilis* || Andr. (not Chaud.) Rev. 471 and 474.

J* **quadrillum** Schaum BEZ. 1 60. 201; Andr. Rev. 369. and 372; id. Ins. Samoa iv. 1927. 3.

G* *pictipennis* Putz. AG. 1875. 745; Bates AF. 1889. 274; Andr. TES. 1921. 178.

B* *spilotus* Bates AM. (5). xvii. 1886. 152.

A* var. **impictus** Andr. Rev. 369 and 373.

J* var. **languidus** Andr. Rev. 369 and 374.

A* **queensladicus** Sloane PLS. NSW. 1903. 577; id. PLS. NSW. 1921. 199 and 207.

G* *cruciger* || Putz. (not Bates) AG. 1875. 747; Andr. Rev. 368 and 369.

recurvicollis × Andr. Rev. 337 and 338.

B* *reflexicollis* || Bates (not Motch.) TES. 1883. 268.

G* **reticulatus** Andr. Rev. 387 and 388.

A* **rhombophorus** Andr. Rev. 387 and 390; id. EMM. 1926. 67.

A* var. **evanidus** Andr. Rev. 387 and 391.

New Guinea.
Philippine Is.
Sumatra., Celebes.
Philippine Is.
Philippine Is.
Chatrapur.
Rambha (Ganjam).
Barkuda I. (Lake Chilka).
Bombay.
Ceylon : Colombo, Weligama.
Siam., Annam.
Celebes., Batchian.
New Guinea.
Formosa., China.
Samoa.
Philippine Is.
Formosa.
Celebes.
New Guinea.
Australia.
Japan.
New Guinea.
Eastern Duars.
Sikkim : Gopaldhara, Namsoo, Gielle Khola and Tarkhola (Tista Valley).
U. P. : Almora Ranikhet, Tanakpur, River Sarda Gorge, Haldwani, Hardwar,

A* **saundersi** Andr. Rev. 472 and 482. — Dehra Dun. Lachiwala, Siwalik Hills, River Song.

B* **sericans** Bates TES. 1873. 297; Andr. Rev. 341 and 348. — With type form.

Penang., Singapore.

O* var. **vixstriatus** Bates TES. 1873. 331; Andr. Rev. 341 and 349. — Java., Sumatra. Sikkim.

M* **sericeus** Motch. BM. 1851. ii. 507; Andr. Rev. 343 and 362; id. TES. 1928. 3. — Japan.

G* *photinus* Bates AG. xxxii. 1892. 288. — China.

Bhamo., Karen Hills.

G* **serratus** And. Rev. 341 and 344. — Palon.

Kawkareik.

G* **singularis** Andr. Rev. 388. and 393. — Kuala Lumpur. Borneo.

A* **spurcus** Andr. Rev. 403 and 427.

Celebes.

A* **stevensi** Andr. Rev. 403 and 445. — Philippine Is.

Celebes.

Borneo.

Sikkim : Gopaldhara. Eastern Duars. Tehri State. U. P. : Mussoorie Almora, Nainital, Swal River, Basin, Haldwani, Khaula, Bhim Tal, Dehra Dun.

A* **striatifrons** Andr. Rev. 407 and 456. — Simla Hills : Barogh. Kangra.

X* **striatulus** Andr. Rev. 471 and 474. — Chota Nagpur : Barwa, Karen Hills.

G'* **subfasciatus** ‡ Bouch. (not Putz.) AF. 1903. 170.

Singapore.

A* **sundaicus** Andr. Rev. 343 and 362.

Tonkin.

M* **suturalis** Motch. BM. 1851. ii. 508; Putz. AG. 1875. 746; Andr. TES. 1919. 193; id. Rev. 403 and 430; id. TE. 1928. 4. — Sumatra.

A*	**tagax** Andr. Rev. 405 and 444; id. EMM. 1926. 67.	Java.
		Tenasserim. ? Ceylon (F. Walker in Brit. Mus.).
		Sikkim: Tarkhola (Tista Valley), Rangit River. U. P. : Almora, Nainital, Ranikhet, Haldwani, Upper Gumti Valley,
M	**tetraspilus** Solsky FT. 114; Reitt. DEZ. 1894. 33; Andr. Rev. 404 and 433.	Dehra Dun, Siwalik Hills, Mussoorie,
J*	**tostus** Andr. Rev. 408 and 459. *scydmaenoides* ‡ Bates (not Nietn.) TES. 1873. 299; id. TES. 1883. 267 (part).	Arni Gad, Terhi Garhwal, Chaubattia.
G*	**transumbratus** Bates AG. xxxii. 1892. 291; Andr. Rev. 342 and 350.	Lyallpur., Pusa. Turkestan.
A*	**triloris** Andr. Rev. 403 and 431.	China.
J*	**tropicus** *(Bembidium)* Nietn. AM. (3). ii. 1858. 421; Putz. AG. 1875. 746; Andr. Rev. 343 and 366. t. 4. f. 26; id. FBI. 16. f. 4c.	Burma : Semigion, Moulmein., Celebes.
		Timor.
M*	*cinctipennis* Motch. BM. 1861. i. 99; Andr. Rev. 498; id. TES. 1928. 6.	Ceylon : Colombo, Dikoya, Kandy,
B*	*subvillatus* Bates AM. (5). xvii. 1886. 151.	Trincomali,
J*	**truncatus** *(Bembidium)* Nietn. AM. (3). ii. 1858. 421; Motch. BM. 1861. i. 100; Fvl. RE. 1907. 101; Andr. Rev. 374 and 375. t. 4. ff. 20, 28 and 31.	Kanthalay, Bandarawella.
M*	*tenellus (Polyderis)* Motch. Etud. Ent. 1862. 35; Andr. Rev. 498; id. TES. 1928. 19.	Calcutta. Eastern Duars. Haldwani. Ceylon : Colombo. Shwegu.
M*	**umbrosus** Motch. BM. 1851. ii. 507; id. Etud. Ent. 1862. 32; Putz. AG. 1875. 745; Bates AM. (5). xvii. 1886. 151; id. AG. xxxii. 1892. 296; Andr. Rev. 485	Singapore. Cochin-China. Tonkin., Java.

and 488. t. 3. f. 12 and t. 4. ff. 21 and 25; id. TES. 1928. 3; id. FBI. 16. f. 4b.

M *parallelus* || Motch. BM. 1851. ii. 507 (not Käf. Russ. 1850. 8 (note 3)); Andr. TES. 1928. 4.

B* *extremus (Acupalpus)* Walk. AM. (3). ii. 1858. 204; Andr. TES. 1919. 190.

D* *nietneri (Tachyta)* Schaum BEZ. 1863. 88.

G* **unitarius** Bates AG. xxxii. 1892. 296; Andr. Rev. 402 and 425.

I* **vilis** Andr. Rev. 342 and 354. t. 4. f. 39.

B* **vixmaculatus** Andr. Rev. 407 and 452.

B* **wallacei** Andr. Rev. 485 and 490.

zonatus × Andr. Rev. 341 and 347.

G* *ephippiatus* || Putz. (not Say) AG. 1875. 747.

Sumatra., Borneo. Philippine Is.

Throughout South-East Asia, excluding Japan, but in India confined to the Himalaya tract.

Karen Hills., Palon. Tenasserim., Tavoy.

Kashmir

Laos.

New Guinea.

Celebes.

TANTILLUS* Chaudoir.

Tantillus Chaud. Mém. sur les Thyréoptérides, AB. xii. 1869. 126.

O* **brunneus** Chaud. Mém. 126; Bates AM. (5) xvii. 1886. 202.

B* **vittatus** Bates AM. (5). xvii. 1886. 202.

Ceylon : Dikoya, Pidurutalagalla.

Ceylon : Bogawantalawa.

TARIDIUS* Chaudoir.

Taridius Chaud., Genres aberrants du groupe des Cymindides, BM. 1875. ii. 7.

G **birmanicus** Bates AG. xxxii. 1892. 416.

O* **opaculus** Chaud. Mon. 8; Bates AG. xxxii. 1892. 416.

A* **stevensi** Andr. AM. (9). xii. 1923. 689.

Burma : Karen Hills, Teinzo.

North India (Chaudoir)., Bhamo. Shwegu.

Sikkim : Nurbong, Mahanadi, Valley,

TETRAGONODERUS* Dejean.

Tetragonoderus Dej. Sp. iv. 1829. 485; Schm. Goeb. FCB. 92; Lac 132; Chaud., Etude monographique des Masoréides, des Tetragonodérides, et du genre Nematotarsus, BM. 1876. ii. 33; Andr. MB. 60.

O* **arcuatus** Dej. Sp. iv. 1829. 495; Chaud. Mon. 38; Bates AG. xxxii. 1892. 415; Bedel Fn. med. 1905. 231; Andr. BM. 61.

Gopaldhara, Kalimpong, Kurseong, Haldwani. Tanakpur. Khondmal Hills.

E* *cadoni* Bates CRB. 1891. 338; id. AG. xxxii. 1892. 416; Andr. TES. 1921. 150.

Shillong. Chandpur (Assam). Sikkim : Kurseong, Gopaldhara. Baksa-Duar., Dacca. Calcutta., Sardah. Murshidabad., Pusa. Chapra. Bhogaon (Purnea). Sitamarhi. Muzaffarpur. Samastipur. Chota Nagpur : Konbir., Dehra Dun. Almora., Haldwani. Kangra., Bajaura. Lyallpur., Mandla. Chikalda., Bombay. Pondicherry. Karen Hills. Mandalay., Bhamo. Toungo., Rangoon. Palon., Laos., Egypt. Sudan., Mesopotamia.

B* **cursor** Bates AM. (5). xvii. 1886. 201.

C* **dilatatus** *(Bembidium)* Wied. Zool. Mag. ii. 1. 1823. 61; Dej. Sp. iv. 1829. 493; Chaud. Mon. 41; Bates CRB. 1891. 338; Andr. TES. 1921. 173.

Kandy.

Kurseong., Bengal. Puri., Almora. Belgaum. North Kanara. Madras., Palur (Arcot). Waltair. Bhawani Valley

O* **discopunctatus** Chaud. BM. 1850. i. 456; id. Mon. 48.

B* **fimbriatus** Bates AM. (5). xvii. 1886. 202.

P* *punctatus* ‡ Schm. Goeb. (not Wied). FCB. 92; Andr. TES. 1923. 54.

(Nilgiri Hills).

Pusa., Simla.

Belgaum.
North Kanara.

Nilgiri Hills : Hill Grove and Bhawani Valley.
Trichinopoly.
Mysore : Teppukadu, Manganali.
Ceylon : Kandy, Trincomali, Hambantota, Peradeniya, Tissa, Horawupotana.
Burma (Schmidt-Goebel).

M **notaphioides** Motch. BM. 1861. i. 99; Chaud. Mon. 54; Bates AM. (5). xvii. 1886. 201; Andr. TES. 1928. 24.

Berhampur., Puri.
Rambha (Ganjam).
Barkuda I. (Lake Chilka).
Bhandara.
North Kanara.
Madras.
Trichinopoly.
Palni Hills.
Tranquebar.
Ceylon : Colombo, Dikoya, Murunkam.

C* **punctatus** *(Bembidum)* Wied. Zool. Mag. ii. 1. 1823. 61; Dej. Sp. iv. 1829. 505; Chaud Mon. 48; Bates AG. xxxii. 1892. 415; Vuillet Ins. ii. 1912. 17; Andr. TES. 1921. 174.

Sardah., Dacca.
Calcutta., Pusa.
Haldwani.
Burma : Bhamo, Teinzo, Myenn Kyan.
Cambodia (Vuillet).

C* **quadrinotatus** *(Carabus)* F. Suppl. Ent. Syst. 1798. 55; id. Syst. Eleuth. i. 1801. 186; Dej. Sp. vi. 1829. 491; Motch. Etud. Ent. 1855. 50; Chaud. Mon. 41; Bates CRB. 1891. 338; Heyne-Tasch. EK. 15; Andr. TES. 1921. 157; id. RIM. xxii. 1921. 347; id. MB. 61.

Throughout India and Ceylon.

N **quadrisignatus** *(Carabus)* Quens. in Schönh. Syn. Ins. i. 1. 1806. 212 (note); Dej. Sp. iv. 1829. 491; Redt. RN. 15; Chaud. Mon. 41; Bates AF. 1889. 282; Andr. AM. (9). xiii. 1924. 470; id. TES. 1923 (1924). 467; id. MB. 61.

P* **rhombophorus** Schm. Goeb. FCB. 93; Chaud. Mon. 48; Bates AG. xxxii. 1892. 415; Andr. TES. 1923. 54.

C* **taeniatus** *(Bembidium)* Wied. Zool. Mag. ii. 1. 1823. 62; Andr. TES. 1921. 174.

O* *trifasciatus* Chaud. BM. 1850, i. 455; id. Mon. 48; Bates AG. xxxii. 1892. 416; Andr. AM. (9). xiv. 1924. 593.

THLIBOPS* Putzeys.

Thlibops Putz., Rév. Gén. des Clivinides, AB. x. 1867. 9; Andr. FBI. 328 and 334.

Y* **abbreviata** Heller PJS. xxiii. 1923. 297 and 298.

O **crenata** *(Scapterus)* Chaud. RMZ. 1863. 118; Putz. Rév. 10; Gestro. AG. 1882. 304; Heller PJS. xxiii. 1923. 297.

L(O*) **dohrni** *(Scapterus)* Chaud. RMZ. 1863. 117; Putz., Rév. 10; Gestro. AG. 1882. 304; Andr. TES. 1922 245; Heller PJS. xxiii. 1923. 298.

B* **filiformis** Andr. FBI. 335 and 337. f. 47.

B* **glaber** Andr. FBI. 335 and 338. f. 48.

Y* **glabriventris** Heller DEZ. 1916. 276; id. PJS. xxiii. 1923. 298.

Y* **integricollis** Heller PJS. xxiii. 1923. 296 and 298.

Y* **intermedia** Heller PJS. xix. 1921. 530; id. PJS. xxiii. 1923. 298.

Y* **minor** Heller DEZ. 1916. 275; id. PJS. xxiii. 1923. 298.

Y* **omega** Heller Abh. Mus. Dresd. vii. 8. 1899. 3; id. PJS. xxiii. 1923. 298.

X* **paviei** Lesne Bull. Mus. Paris ii. 1896. 239. f. 1; id. Miss. Pav. 63. t. 3. f. 2. and text-fig. 1; Heller PJS. xxiii. 1923. 298.

G* **puncticollis** Gestro AG. 1882. 302; Heller PJS. xxiii. 1923. 297; Andr. FBI. 335. f. 46.

Cambodia.
China.
India (Bates).

Sardah., Rangoon.
Tharrawaddy.
Palon., Martaban.
Tenasserim., Laos.

Bengal., Pusa.
Chapra.

Almora., Ranikhet.
Haldwani.
Tanakpur., Simla.
Burma : Teinzo,
Katha.

Philippine Is.

Cochin-China.

Java.

Dapoli.

Parli (Malabar).

Philippine is.

Philippine Is.

Philippine Is.

Philippine Is.

Philippine Is.

TILIUS* Chaudoir.

Tilius Chaud., Etude monographique des Masoréides, des Tetragonodérides, et du genre Nematotarsus, BM. 1876, ii, 71.

O* **holosericeus** *(Lionychus)* Chaud. BM. 1850. i. 68; id. Mon. 72; Bates CRB. 1891. 338. — Siam. Cochin-China.

TRECHODES Blackbrun.

Minhla.

Trechodes Blackb. Tr. Roy. Soc. S. Austr. xxv. 1901. 119; Jeann. AM. (9). xii. 1923. 414; id., Mon. des Trechinae (i), AB. xxxii. 1926. 484. — Berhampur., Simla. Kodaikanal. Trichinopoly.

A'* **bakeri** Jeann. Mon. (i) 491. f. 266.

G* **cauliops** *(Trechus)* Bates AG. xxxii. 1892. 298; Jeann. AM. (9). xii. 1923. 415 f. 10; id. Mon. (i). 490. ff. 262-3.

Philippine Is.

TRECHUS Schellenberg.

Karen Hills.

Trechus Schell. Helv. Ent. ii. 1806. 22/3; Dej. Sp. v. 1831. 2; Lac. 370; Pandellé, Rév mat. pur la faune des Col. de France, ii. 1867. 131; Putz., Trechorum oculatorum Monographia, SEZ. 1870. 7; Jeann. AF. xc. 1922. 170; id. AM. (9). xii. 1923. 415; Andr. MB. 19; Jeann., Mon. des Trechinae (ii) Ab. xxxii. 1927. 114; id., Mon. des. Trechinae (iii), Ab. xxxv. 1928. 790.

Blemus Steph. iii. Brit. Ent. i. 67. and 171.

A'* **bakeri** Jeann. AM. (9). xii. 1923. 416 and 426. f. 16; id. Mon. (ii) 157 and 165 ff. 545-8.

B* **cameroni** Jeann. AM. (9). xii. 1923. 416 and 422. f. 14; id. Mon. (ii). 157 and 162. ff. 540-3; id. AM. (10). i. 1928. 291. — Philippine Is.

A var. **bistriatus** Jeann. AM. (10). i. 1928. 291; id. Mon. (iii). 791. — Simla Hills :

B* **hingstoni** Jeann. AM. (10). i. 1928. 288. f. 2; id. Mon. (iii). 790. — Kotgarh, Narkanda.

A'* **imaicus** Jeann. AM. (9). xii. 1923. 416 and 421. f. 13; id. Mon. (ii). 156 and 161. ff. 537-9; id. AM. (10). i. 1928. 284. — Kangra.

O*	**indicus** Putz., Mon. 175; Andr. TES. 1922. 247; Jeann. AM. (9). xii. 1923 416. f. 11; id. Mon. (ii). 157. and 158; id. AM. (10). i. 1928. 284.	Sikkim : Gnatong. Sikkim : Tonglu, Karponang. Jalep-la. Kurseong. U.P.: Bodyar, Konain, Sijla Gad, Kanasar, Khedar Khud and Korawa Khud (all Chakrata District), Dhobi Ghat and Ringal Gad (Mussoorie district). Mohan Rau (Siwalik Hills.)
B*	var. **championi** Jeann. AM. (9). y. 1920. 109; Andr. TES. 1922. 248; Jeann. AM. (9). xii. 1923. 416 and 420. f. 12; Andr. MB. 20; Jeann. Mon. (ii). 157 and 158. ff. 528-32.	Aglar River (Tehri State). Simla Hills: Kotgarh. Yunnan. Sikkim : Gopaldhara, Darjiling. U.P.: Almora Upper Gumti Valley. Sunderdhunga Valley, Swal Riverbasin, Nainital, Chakrata, Ranikhet. Spiti: Pulga. Lahoul: Cimur. Dharmsala.
	var. **macroderus** Jeann. AM. (10). i. 1928. 290; id. Mon. (iii). 790.	Simla Hills: Kotgarh, Gahan. Mandi.
A	**macrops** Jeann. Mon. (ii). 157 and 160. ff. 533-6; id. AM. (10). i. 1928. 289.	Kashmir : Baltal, Sonemarg.
B*	**pumilio** Jeann. AM. (9). xii. 1923. 416 and 424. f. 15; id. Mon. (ii). 157 and 163. f. 544.	Kishtwar.

B* **thibetanus** Jeann. AM. (10). i. 1928. 284. f. 1; id. Mon. (iii). 791. — Kashmir : Sonemarg (?), Yunnan.

Kumaon : Pindar Valley.

Tibet : Tasam, Phuse la, Lamna la, Tashidzom, Kang la Pangla, Tuna, Gantsa, Chumbi Valley.
Sikkim : Tangu.

TRICHUS* Klug.

Trichis Klug in Symb. Phys. Decas iii. 1832. t. 21. f. a-c; Brullé HN. iv. 1834. 146; Chaud. BM. 1850. i. 65; Lac. 111; Chaud. Essai monographique sur le genre Cymindis proprement dit, BEZ, 1873. 114; Bedel Fn. med. 1906. 264.

J* **naculata** Klug in Symb. Phys. Decas iii. 1832. t. 21. f. 10; Brullé HN. iv. 1834. 147; Mars. Ab. xix. 1880. 164; Heyd. WEZ. 1905 95; Pic. WEZ. 1905. 273; Bedel Fn. med 1907. 265; Andr. AM. (9). xix. 1927. 100. — Sind. Fao (Persian Gulf). Asterabad., Suez. Port Said. Mediterranean region.

J* **pallida** Klug in Symb. Phys. Decas iii. 1832 t. 21. f. 9; Brullé HN. vi. 1834. 146; Bedel Fn. med. 1907. 265; Andr. AM. (9). xix. 1927. 100. — Karachi., Arabia. Sinai Peninsula. "Ethiopia."

TRICHISIA* Motchulsky.

Trichisia Motch. BM. 1864. ii. 331; Chaud., Mon. sur. les Panageides, AB. 1878. 164.

J* **cyanea** *(Isotarsus)* Schaum AF. 1853. 439; Chaud. BM. 1861. ii. 349; id. Mon. 165; Redt. RN. 9; Andr. TES. 1923 (1924). 462; id. AM. (9). xix. 1927. 107.

B'* *chinensis* Csiki Ann. Mus. Hung. 1907. 576; Andr. AM. (9). xvii. 1926. 251. — Hongkong.

O* **morio** *(Isotarsus)* Laf. AF. 1851. 221 (note 4); Chaud. BM. 1861. ii. 348; id. Mon. 165; Gardn. Ins. For. Rec. xiii. 1927. 66. t. 5. ff. 12-16 (Larva)

M *? cyanescens* Motch. BM. 1864. ii. 332; Andr. TES. 1928. 24. — Jorhat., Dacca., Pusa. Dehra Dun. Rawalpindi.

B'* **papuana** Csiki Ann. Mus. Hung. 1970. 576.

B' **quadrimaculata** Csiki Ann. Mus. Hung. 1907. 577. — Sumatra.

TRICHOCELLUS Ganglbauer.

Trichocellus Ganglb. Käf. Mitteleur. i. 1892. 366; Tchitch. Ab. xxix. 1897. 63; id., Mém. sur le genre Trichocellus, HER. xxxii. 1899. 444; id. HER. xxxiv. 1900. 52 and 366; id. HER. xxxv. 1901. 249; Andr. MB. 40.

I* **alticola** *(Dichirotrichus)* Bates PZS. 1878. 713; id. SYM. 5. t. 1. f. 6.; Tchitch. Mém. 458.

Q* **roborowskii** Tchitch. Mém. 457.

A* var. **lissonotus** Andr. MB. 41.

TRICHOTICHNUS Morawitz.

Trichotichnus Moraw. Beitr. 63; Tchitch. HER. xxxiv. 1900. 362; id. HER. xxxv. 1901. 244.

Asmerinx Tchitch. AF. 1898. 183; id. HER. xxxiv. 1900. 363; id. HER. xxxv. 1901. 134 and 245; id. HER. xxxvii. 1906. 278.

G* **birmanicus** Bats AG. xxxii. 1892. 342.

A* **javanus** Andr. AM. (9). xviii. 1926. 278.

B* **lamprus** *(Tachycellus)* Bates AM. (5). xvii. 1886. 80; Andr. AM. (9). iii. 1919. 475.

B* **liparus** Andr. EMM. 1926. 69 and 74.

B* **sumatrensis** Andr. AM. (9). xviii. 1926. 279.

TRIGONOGNATHA Motchulsky.

Trigonognatha Motch. Etud. Ent. 1857. 25; Moraw. Beitr. 56; Bates TES. 1883. 243; Tchitch. HER. xxix. 1895. 168; id. HER. xxxii. 1898. 67. and 77.

New Guinea.

New Guinea.

Pamir : Turkestan.

Tibet.

Kashmir : Baltal.
Leh.
Rukshu : Salt Lake
Nima-Mud.

Lebong., Mussoorie.
Karen Hills., Laos.

Java., Krakatau.
Buru.

Kodaikanal.
Trichinopoly.
Pondicherry.
Shembaganur.
Ceylon : Colombo,
Weligama,
Bandarawella.

Kumaon : West
Almora,
Sunderdhunga

Aurisma Fairm. AB. 1888. 9.

O **delavayi** *(Aurisma)* Fairm. AB. 1888. 9; Tchitch. HER. xxx. 1897. 282; id. HER. xxxii. 1898. 82.

G **feana** Bates AG. xxxii. 1892. 365; Tchitch. HER. xxxii. 1898. 82.

TRIGONOTOMA* Dejean.

Trigonotoma Dej. Sp. iii. 1828. 182; Chaud. BM. 1852. i. 71; Lac. 311; Chaud. Rév. des. Trigonotomides, AB. xi. 1868. 151; Tchitch. HER. xxxiv. 1900. 180; Kuntz. Ent. Rundsch. 1911. 182; id. Trig. 60.

?Q **adoxa** Tchitch. HER. xxxiv. 1900. 162; Kuntz. Trig. 61.

V **æthiops** Tchitch. HER. xxx. 1897. 264; id. HER. xxxii. 1898. 36; Kuntz. Trig. 63.

X* **aurifera** Tchitch. HER. xxxiv. 1900. 155; Lesne Miss. Pav. 77. t. 9 bis. f. 1; Kuntz. Ent. Rundsch. 1911. 182; id. Trig. 63.

G* **bhamoensis** Bates AG. xxvii. 1889. 105; id. AG. xxxii. 1892. 354; Kuntz. Ent. Rundsch. 1911. 182; id. Trig. 63 et seqq.

X* **cauta** Tchitch. HER. xxxiv. 1900. 159 and 185; Kuntz. Trig. 64 et seqq.

O* **chalceola** Bates TES 1873. 328; id. AF. 1889. 276; Tchitch. HER. xxxiv. 1900. 160; Kuntz. Trig. 61 et seqq.; Andr. AM. (9). xiii. 1924. 469.

A* **chalcodes** Andr. AM. (9). xii. 1923. 217.

G* **chrysites** Bates AG. xxxii. 1892. 354; Kuntz., Ent. Rundsch. 1911. 182; id. Trig. 63.

G **comotti** Gestro AG. 1882. 308; Kuntz., Trig. 63.

L **concinna** Cast. Etud. Ent. 1834. 77; Chaud. Rév. 159; Tchitch. HER. xxxiv. 1900. 183; Kuntz. Trig. 65 et. seqq.

O* **curtula** Chaud. Rév 160 Tchitch. HER. xxxiv. 1900. 160; Kuntz. Trig. 61 et seqq.

Valley, Gori River Gorge, Nainital.

Sumatra.

Yunnan.

Mt. Mooleyit (Tenasserim).

Cochin-China.

Burma.

Cochin-China.
Cambodia.
Laos.

Shillong., Manipur.
Burma : Bhamo, Teinzo.
Laos., Tonkin.
Formosa.

Ternate.

Cochin-China.
Yunnan.
China.

Laos.
Cochin-China.

Bhamo., Karen Hills.
Yunnan.

Minhla.

X* **cyanella** Tchitch. HER. xxxiv. 1900. 165; Kuntz. Trig. 61 et seqq.
Java.

O* **dohrni** Chaud. BM. 1852; i. 69; id. Rév. 159; Bates AF. 1889. 275; id. Entom. 1890. 213; Lesne Miss. Pav. 77; Kuntz. Trig. 63 et. seqq.; Andr. AM. (9). xiii. 1924. 469.
Laos.
Cochin-China.

O* *crenata* Chaud. Rév. 159; Kuntz. Trig. 63; Andr. AM. (9). xiv. 1924. 591.
Indo-China.
China.

L **fulgidicollis** Cast. Etud. Ent. 1834. 77; Chaud. Rév 162; Gestro. AG. 1882. 309; Tchitch. HER. xxiv. 1900. 192; Kuntz. Trig. 52 et seqq.

X* **funebris** Tchitch. HER. xxxiv. 1900. 153; Kuntz., Trig. 61 et seqq.
Java (in description)
Laos (on label).

G **igneicollis** Bates AG. xxxii. 1892. 353; Kuntz., Trig. 63.
Siam.

X* **india** Brullé HN. iv. 1834. 333; Chaud. Rév. 158; Bates AM. (5). xvii. 1886. 145.
Palon.

O* *viridicollis* ‡ Dej. (not Macl.) Sp. iii. 1828. 183; Tchitch. HER. xxxiv. 1900. 176 (note 26); Kuntz. Trig. 63 et seqq.
Tezpur., Mandelli (Darjiling)., Calcutta.
Malda., Murshidabad.
Dehra Dun.
Gorakhpur U. P.).
Lucknow.
"Hurdwar & Roorkee."
Nagpur.
Koyna Valley (Satara).
Madras., Madura., Dindigul.
Ceylon : Colombo, Anuradhapura.

O* **iodes** Bates AG. xxxii. 1892. 352 (note); Kuntz., Trig. 63.

L' **kuntzeni** Hubthl. DEZ. 1914. 437.
Assam. Noa Dehing Valley.

G* **lamprodera** Bates AG. xxxii. 1892. 353; Kuntz. Trig. 63.
Harmatti (Dafla Hills).

X* **leotaudi** Tchitch. HER. xxxiv. 1900. 158 and 185; Kuntz. Trig. 64 et seqq; Heller PJS. xxiii. 1923. 301.
Boreno.

B* **lewisi** Bates TES. 1873. 284; Har. DEZ. 1877. 339; Kuntz. Ent. Rundsch. 1911. 182; id. Trig. 63 et seqq.
Karen Hills.

O* **luzonica** Chaud. Rév. 161; Tchitch. HER. xxxiv. 1900. 181; Kuntz. Trig. 64 et seqq.; Heller PJS. xxiii. 1923. 301.

X* **Morosa** Tchitch. HER. xxxiv. 1900. 164; Lesne Miss. Pav. 78; Kuntz. Trig. 61 et seqq.

Q **niasana** Tchitch. HER. xxxii. 1898. 35; id. HER. xxxiv. 1900. 184.

O* **nitidicollis** Chaud. Rév. 160; Bates AG. xxxii. 1892. 354; Kuntz., Ent. Rundsch. 1911. 182; id. Trig. 52, 63 et seqq.

Q* **oberthüri** Tchitch. HER. xxviii. 1894. 444; Kuntz. Trig. 63.

Q **palavanica** Tchitch. HER. xxx. 1897. 263; id. HER. xxxiv. 1900. 157. and 184; Kuntz., Ent. Rundsch. 1911. 182; id. Trig. 64 et seqq.; Heller PJS. xxiii. 1923. 301.

X* **perraudieri** Bates AF. 1889. 275; Tchitch. HER. xxxiv. 1900. 155; Kuntz. Ent. Rundsch. 1911, 182; id. Trig. 61 et. seqq.

L **peteli** Cast. Etud. Ent. 1834. 78; Chaud. Rév. 159; Tchitch. HER. xxxiv. 1900. 183; Kuntz. Ent. Rundsch. 1911. 182; id. Trig. 65 et seqq.

Q **psyche** Tchitch. HER. xxx. 1897. 262; id. HER. xxxiv. 1900. 182; Kuntz. Ent. Rundsch. 1911. 182; id. Trig. 65. et seqq.

Q var. **rectangula** Tchitch. HER. xxx. 1897. 263.

Q **peulla** Tchitch. HER. xxxii. 1898. 37; id. HER. xxxiv. 1900. 186; Kuntz. Ent. Rundsch. 1911. 182; id. Trig. 65 et seqq.

O* **similis** Chaud. Rév. 158; Kuntz. Trig. 63 et seqq.

Q **venus** Tchitch. HER. xxx. 1897. 260; id. HER. xxxiv. 1900. 181; Kuntz. Ent. Rundsch. 1911. 182; id. Trig. 64 et seqq.

J* var. **preyeri** Kuntz. Ent. Rundsch. 1911. 182; id. Trig. 65; Andr. AM. (9) xix. 1927. 110.

Philippine Is.

China., Japan. Formosa.

Philippine Is.

Cambodia.

Nias. I.

Palon., Siam. Indo-China (all Provinces).

Sikkim : Pedong, Gopaldhara, Mungphu, Kurseong, Lebong.

Philippine Is.

Siam. Indo-China.

Java.

Borneo.

Borneo.

Java.

Dacca.

J* **verberifera** Schauf. HER. xxi. 1887. 106; Tchitch. HER. xxxiv. 1906. 185 (note 43); Kuntz. Ent. Rundsch. 1911. 182; id. Trig. 64 et seqq; Andr. AM. (9). xix. 1927. 110. — Borneo.

O *modesta* Heyne in Heyne Tasch. EK. 1895. 24. — Java. Celbes.

TRILOPHUS* Andrewes.

Trilophus Andr. AM. (9). xx. 1927. 263; id. FBI. 344 and 381.

Oxydrepanus ‡ Bates (not Putz.) AG. xxxii. 1892. 283; AM. (9). xvii. 1926. 378.

O* **interpunctatus** *(Dyschirius)* Putz. AB. x. 1867. 97; id. CRB. 1878. 174; Andr. AM. (9). xvii. 1926. 378; id. AM. (9). xx. 1927. 265; id. FBI. 383. f. 57.

G* *birmanicus (Oxydrepanus)* Bates AG. xxxii. 1892. 283.

G* *debilis (Dyschirius)* ‡ Bouch (not Schm. Goeb.) AF. 1902. 170. — Ferozepore., Almora. Kierpur (Purnea). Sardah., Katha. Meetan (Tenasserim). Annam., Sumatra.

O* var. **hispidulus** *(Dyschirius)* Putz. AB. x. 1867. 98; Andr. AM. (9). xvii. 1926. 378; id. AM. (9). xx. 1927. 265; id. FBI. 383 and 385.

G* var. **impunctatus** *(Dyschirius)* Putz. AB. xi. 1868. 10; Andr. AM. (9). xvii. 1926. 378; id. AM. (9). xx. 1927. 265; id. FBI. 383 and 385. — Nagpur. Siam. Sumatra.

S* var. **schmidti** *(Dyschirius)* Putz. CRB. 1877. 41; Andr. AM. (9). xvii. 1926. 378; id. AM. (9). xx. 1927. 265; id. FBI. 383 and 384. — Calcutta. Kierpur (Purnea). Ceylon : Negombo. Shwegu., Insien. Siam., Tonkin. Sumatra.

TROPIDOCERUS* Chaudoir.

Tropidocerus Chaud. BM. 1878. ii. 9; Tchitch. HER. xxxiv. 1900. 534.

O* **indicus** Chaud. BM. 1878. ii. 13. — "Between Sorbhog and Golakganj." Shwegu., Annam.

VELINDA* Andrewes.

Velinda Andr. RIM. xxii. 1921. 347.

A* **lirata** Andr. RIM. xxii. 1921. 347. — "North India."

Gothangaon (Bhandara).
Barkuda I. (Lake Chilka).

XENODUS* Andrewes.

Xenodus Andr. MB. 92.

A* **d'abreui** Andr. MB. 94. t. 3. ff. 1. and 1. — Chapra.
"Hurdwar and Roorkee."
Delhi., Rawalpindi.
Kasara., Ratnagiri.
Mormugao.
C. P. : Manikpur, Jubbulpore, Nagpur, Kunavaram., Madras.
Secunderabad.

ZELMA* Andrewes.

Zelma Andr. AM. (9). v. 1920. 451; id. FBI. 416.

A* **miranda** Andr. AM. (9). v. 1920. 453; id. FBI. 418. ff 62 and 62a. — Calcutta. Tharrawaddy.

ZUPHIUM* Latreille.

Zuphium Latr. Gen. Crust. et Ins. i. 1806. 198; Latr. et Dej. HN. et Ic. 121; Dej. Sp. i. 1825. 192; Cast., Notemonographique sur le genre Zuphium, Silb. Rev. i. 1833. 251; (Zophium) Schm. Goeb. FCB. 27; Lac. 85; Chaud. BM. 1862. ii. 310; K. and J. Daniel, Uebersicht der pal. Artn der Gattung Zuphium, Col. Stud. ii. München, 1898. 24; (Zoyphium) Bedel Fn. med 1913. 295.

P* **bimaculatum** Schm. Goeb. FCB. 28; Gestro. AG. 1875. 864; Bates AG. xxxii. 1892. 386; Andr. TES. 1023. 10. — Tharrawaddy. Martaban., Palon. Siam.

P* *vittigerum* Schm. Goeb. FCB. 28; Chuad. RMZ. 1872. 105; Andr. TES. 1923. 10. — Cambodia.

O* **celebense** Chaud. BM. 1862. ii. 312; Gestro. AG. 1875. 864. — Celebes.

A* **d'abreui** Andr. AM. (9). x. 1922. 168. — Nagpur., Belgaum. Madras.
Ceylon : Vavuniya, Horawupotana.

B* **erebeum** Andr. Spol. Zeyl. 1923. 243. — Ceylon.

O* **erythrocephalum** Chaud. BM. 1862. ii. 311. — Neemuch., Nagpur. Malabar. Ootacamund. Madras., Pondicherry.

G* **formosum** Bates AG. xxxii. 1892. 386. — Tharrawaddy, Palon and Tikekee (Pegu).

P* **inconspicuum** Schm. Goeb. FCB. 30; Bates AG. xxxii. 1892. 387; Andr. TES. 1923. 10. — Belgaum., Bhamo. Palon., Siam.

A* **indicum** Andr. AM. (9). 1922. 167. — Nagpur.

A* **modestum** Schm. Goeb. FCB. 29; Chaud. BM. 1862. ii. 312; Gestro AG. 1875. 864; Bates AG. xxxii. 1892. 387; Andr. TES. 1923. 10. — Calcutta., Nagpur. Yenna Valley (Satara). Tharrawaddy. Palon., Penang. Siam., Cambodia. Cochin-China.

J* **olens** *(Carabus)* Rossi Fn. Etrusc. i. 1790. 217. t. 5. f. 2.; F. Ent. Syst. i. (1). 1792. 139; *(Galerita)* id. Syst. Eleuth. i. 1801. 215; Oliv. Ent. 35. 94. t. 13. f. 156; Latr. et Dej. HN. et Ic. 121. t. 10. f. 3.; Dej. Sp. i. 1825. 192; Dej. et Boisd. Ic. et HN. i. 70. t. 8. f. 1; Cast. Mon. 253; Brullé HN. iv. 1834. 175. t. 6. f. 1; Schm. Goeb. FCB. 28; Chaud. BM. 1862. ii. 311; Bates AF. 1889. 280; id. AG. xxxii. 1892. 386; Bedel Fn. med. 1914. 298; Andr. TES. 1921. 155; id. TES. 1923. 9; id. AM. (9). xix, 1927. 99. — Calcutta., Pusa. Chapra., Balugaon. Ceylon., Bhamo. Tharrawaddy. Moulmein. Palon and Tikekee (Pegu). Tenasserim. Langkawi I. Siam., Cochin-China. Annam., Java. Turkestan., Persia. Mesopotamia. Erythrea., Syria. Asia Minor. North Africa. Southern Europe.

O* *longiusculum* Chaud. BM. 1842. 804; id. BM. 1862. ii. 312.

J* var **pubescens** Nietn. AM. (3). ii 1858. 182; Chaud. RMZ. 1872. 105; Andr. Spol. Zeyl. 1924. 138; id. AM. (9). xix. 1927. 107. — Calcutta. Ceylon : Murunkam.

O* *rufifrons* Chaud. BM. 1862. ii. 311.

P*	**piceum** Schm. Goeb. FCB. 29; Bates AG. xxxii. 1892. 387; Andr. TES. 1923. 10.	Toungoo. Palon.
G*	**praestans** Bates AG. xxxii. 1892. 387.	Tharrawaddy. Paungde., Palon.
O*	**siamense** Chaud. RMZ. 1872. 104; Gestro. AG. 1875. 864; Dupuis. AB. 1912. 285.	Siam. Formosa (Dupuis).

www.ingramcontent.com/pod-product-compliance
Ingram Content Group UK Ltd.
Pitfield, Milton Keynes, MK11 3LW, UK
UKHW020246290926
15470UKWH00002B/766

9 789386 071088